Waste
Treatment
in the Food
Processing
Industry

Waste Treatment in the Food Processing Industry

edited by
Lawrence K. Wang
Yung-Tse Hung
Howard H. Lo
Constantine Yapijakis

CRC Press
Taylor & Francis Group
Boca Raton London New York

CRC Press is an imprint of the
Taylor & Francis Group, an **informa** business

A TAYLOR & FRANCIS BOOK

This material was previously published in the *Handbook of Industrial and Hazardous Wastes Treatment, Second Edition* © Taylor and Francis Group, 2004.

CRC Press
Taylor & Francis Group
6000 Broken Sound Parkway NW, Suite 300
Boca Raton, FL 33487-2742

First issued in paperback 2019

© 2006 by Taylor & Francis Group, LLC
CRC Press is an imprint of Taylor & Francis Group, an Informa business

No claim to original U.S. Government works

ISBN-13: 978-0-8493-7236-0 (hbk)
ISBN-13: 978-0-367-39202-4 (pbk)
Library of Congress Card Number 2005049975

Library of Congress Cataloging-in-Publication Data

Waste treatment in the food processing industry / edited by Lawrence K. Wang ... [et al.].
    p. cm.
  Includes bibliographical references and index.
  ISBN 0-8493-7236-4
  1. Food industry and trade--Waste disposal. I. Wang, Lawrence K.

TD899.F585W37 2005
664'.0028'6--dc22

2005049975

**Visit the Taylor & Francis Web site at
http://www.taylorandfrancis.com**

**and the CRC Press Web site at
http://www.crcpress.com**

Preface

Environmental managers, engineers, and scientists who have had experience with food industry waste management problems have noted the need for a book that is comprehensive in its scope, directly applicable to daily waste management problems of the industry, and widely acceptable by practicing environmental professionals and educators.

Many standard industrial waste treatment texts adequately cover a few major technologies for conventional in-plant environmental control strategies in food industry, but no one book, or series of books, focuses on new developments in innovative and alternative technology, design criteria, effluent standards, managerial decision methodology, and regional and global environmental conservation.

This book emphasizes in-depth presentation of environmental pollution sources, waste characteristics, control technologies, management strategies, facility innovations, process alternatives, costs, case histories, effluent standards, and future trends for the food industry, and in-depth presentation of methodologies, technologies, alternatives, regional effects, and global effects of important pollution control practice that may be applied to the industry. This book covers new subjects as much as possible.

Important waste treatment topics covered in this book include: dairies, seafood processing plants, olive oil manufacturing factories, potato processing installations, soft drink production plants, bakeries and various other food processing facilities. Special efforts were made to invite experts to contribute chapters in their own areas of expertise. Since the areas of food industry waste treatment is broad, no one can claim to be an expert in all areas; collective contributions are better than a single author's presentation for a book of this nature.

This book is one of the derivative books of the *Handbook of Industrial and Hazardous Wastes Treatment*, and is to be used as a college textbook as well as a reference book for the food industry professional. It features the major food processing plants or installations that have significant effects on the environment. Professors, students, and researchers in environmental, civil, chemical, sanitary, mechanical, and public health engineering and science will find valuable educational materials here. The extensive bibliographies for each type of food waste treatment or practice should be invaluable to environmental managers or researchers who need to trace, follow, duplicate, or improve on a specific food waste treatment practice.

The intention of this book is to provide technical and economical information on the development of the most feasible total environmental control program that can benefit both food industry and local municipalities. Frequently, the most economically feasible methodology is combined industrial-municipal waste treatment.

We are indebted to Dr. Mu Hao Sung Wang at the New York State Department of Environmental Conservation, Albany, New York, who co-edited the first edition of the

Handbook of Industrial and Hazardous Wastes Treatment, and to Ms. Kathleen Hung Li at NEC Business Network Solutions, Irving, Texas, who is the Consulting Editor for this new book.

Lawrence K. Wang
Yung-Tse Hung
Howard H. Lo
Constantine Yapijakis

Contents

Contributors

Adel Awad   Tishreen University, Lattakia, Syria

Renbi Bai   National University of Singapore, Singapore

Charles J. Banks   University of Southampton, Southampton, England

Trevor J. Britz   University of Stellenbosch, Matieland, South Africa

J. Paul Chen   National University of Singapore, Singapore

Mohd Ali Hassan   University Putra Malaysia, Serdang, Malaysia

Yung-Tse Hung   Cleveland State University, Cleveland, Ohio, U.S.A.

Tsuyoshi Imai   Yamaguchi University, Yamaguchi, Japan

Howard H. Lo   Cleveland State University, Cleveland, Ohio, U.S.A.

Hana Salman   Tishreen University, Lattakia, Syria

Swee-Song Seng   National University of Singapore, Singapore

Yoshihito Shirai   Kyushu Institute of Technology, Kitakyushu, Japan

Kuan-Yeow Show   Nanyang Technological University, Singapore

Joo-Hwa Tay   Nanyang Technological University, Singapore

Masao Ukita   Yamaguchi University, Yamaguchi, Japan

Corné van Schalkwyk   University of Stellenbosch, Matieland, South Africa

Zhengjian Wang   University of Southampton, Southampton, England

Shahrakbah Yacob   University Putra Malaysia, Serdang, Malaysia

Lei Yang   National University of Singapore, Singapore

1

Treatment of Dairy Processing Wastewaters

Trevor J. Britz and Corné van Schalkwyk
University of Stellenbosch, Matieland, South Africa

Yung-Tse Hung
Cleveland State University, Cleveland, Ohio, U.S.A.

1.1 INTRODUCTION

The dairy industry is generally considered to be the largest source of food processing wastewater in many countries. As awareness of the importance of improved standards of wastewater treatment grows, process requirements have become increasingly stringent. Although the dairy industry is not commonly associated with severe environmental problems, it must continually consider its environmental impact — particularly as dairy pollutants are mainly of organic origin. For dairy companies with good effluent management systems in place [1], treatment is not a major problem, but when accidents happen, the resulting publicity can be embarrassing and very costly.

All steps in the dairy chain, including production, processing, packaging, transportation, storage, distribution, and marketing, impact the environment [2]. Owing to the highly diversified nature of this industry, various product processing, handling, and packaging operations create wastes of different quality and quantity, which, if not treated, could lead to increased disposal and severe pollution problems. In general, wastes from the dairy processing industry contain high concentrations of organic material such as proteins, carbohydrates, and lipids, high concentrations of suspended solids, high biological oxygen demand (BOD) and chemical oxygen demand (COD), high nitrogen concentrations, high suspended oil and/or grease contents, and large variations in pH, which necessitates "specialty" treatment so as to prevent or minimize environmental problems. The dairy waste streams are also characterized by wide fluctuations in flow rates, which are related to discontinuity in the production cycles of the different products. All these aspects work to increase the complexity of wastewater treatment.

The problem for most dairy plants is that waste treatment is perceived to be a necessary evil [3]; it ties up valuable capital, which could be better utilized for core business activity. Dairy wastewater disposal usually results in one of three problems: (a) high treatment levies being charged by local authorities for industrial wastewater; (b) pollution might be caused when untreated wastewater is either discharged into the environment or used directly as irrigation water; and (c) dairy plants that have already installed an aerobic biological system are faced with the problem of sludge disposal. To enable the dairy industry to contribute to water conservation, an efficient and cost-effective wastewater treatment technology is critical.

Presently, plant managers may choose from a wide variety of technologies to treat their wastes. More stringent environmental legislation as well as escalating costs for the purchase of fresh water and effluent treatment has increased the impetus to improve waste control. The level of treatment is normally dictated by environmental regulations applicable to the specific area. While most larger dairy factories have installed treatment plants or, if available, dispose of their wastewater into municipal sewers, cases of wastewater disposal into the sea or disposal by means of land irrigation do occur. In contrast, most smaller dairy factories dispose of their wastewater by irrigation onto lands or pastures. Surface and groundwater pollution is, therefore, a potential threat posed by these practices.

Because the dairy industry is a major user and generator of water, it is a candidate for wastewater reuse. Even if the purified wastewater is initially not reused, the dairy industry will still benefit from in-house wastewater treatment management, because reducing waste at the source can only help in reducing costs or improving the performance of any downstream treatment facility.

1.2 DAIRY PROCESSES AND COMPOSITION OF DAIRY PRODUCTS

Before the methods of treatment of dairy processing wastewater can be appreciated, it is important to be acquainted with the various production processes involved in dairy product manufacturing and the pollution potential of different dairy products (Table 1.1). A brief summary of the most common processes [8] is presented below.

1.2.1 Pasteurized Milk

The main steps include raw milk reception (the first step of any dairy manufacturing process), pasteurization, standardization, deaeration, homogenization and cooling, and filling of a variety of different containers. The product from this point should be stored and transported at 4°C.

1.2.2 Milk and Whey Powders

This is basically a two-step process whereby 87% of the water in pasteurized milk is removed by evaporation under vacuum and the remaining water is removed by spray drying. Whey powder can be produced in the same way. The condensate produced during evaporation may be collected and used for boiler feedwater.

1.2.3 Cheese

Because there are a large variety of different cheeses available, only the main processes common to all types will be discussed. The first process is curd manufacturing, where pasteurized milk is mixed with rennet and a suitable starter culture. After coagulum formation and heat and mechanical treatment, whey separates from the curd and is drained. The finished curd is then salted, pressed, and cured, after which the cheese is coated and wrapped. During this process two types of wastewaters may arise: whey, which can either be disposed of or used in the production of whey powder, and wastewater, which can result from a cheese rinse step used during the manufacturing of certain cheeses.

Table 1.1 Reported BOD and COD Values for Typical Dairy Products and Domestic Sewage

Product	BOD$_5$ (mg/L)	COD (mg/L)	Reference
Whole milk	114,000	183,000	4
	110,000	190,000	5
	120,000		6
	104,000		7
Skim milk	90,000	147,000	4
	85,000	120,000	5
	70,000		6
	67,000		7
Buttermilk	61,000	134,000	4
	75,000	110,000	5
	68,000		7
Cream	400,000	750,000	4
	400,000	860,000	5
	400,000		6
	399,000		7
Evaporated milk	271,000	378,000	4
	208,000		7
Whey	42,000	65,000	4
	45,000	80,000	5
	40,000		6
	34,000		7
Ice cream	292,000		7
Domestic sewage	300	500	4, 5

BOD, biochemical oxygen demand; COD, chemical oxygen demand.
Source: Refs. 4–7.

1.2.4 Butter

Cream is the main raw material for manufacturing butter. During the churning process it separates into butter and buttermilk. The drained buttermilk can be powdered, cooled, and packed for distribution, or discharged as wastewater.

1.2.5 Evaporated Milk

The milk is first standardized in terms of fat and dry solids content after which it is pasteurized, concentrated in an evaporator, and homogenized, then packaged, sterilized, and cooled for storage. In the production of sweetened condensed milk, sugar is added in the evaporation stage and the product is cooled.

1.2.6 Ice Cream

Raw materials such as water, cream, butter, milk, and whey powders are mixed, homogenized, pasteurized, and transferred to a vat for ageing, after which flavorings, colorings, and fruit are added prior to freezing. During primary freezing the mixture is partially frozen and air is incorporated to obtain the required texture. Containers are then filled and frozen.

1.2.7 Yogurt

Milk used for yogurt production is standardized in terms of fat content and fortified with milk solids. Sugar and stabilizers are added and the mixture is then heated to 60°C, homogenized, and heated again to about 95°C for 3–5 minutes [9]. It is then cooled to 30–45°C and inoculated with a starter culture. For set yogurts, the milk base is packed directly and the retail containers are incubated for the desired period, after which they are cooled and dispatched. For stirred yogurts, the milk base is incubated in bulk after which it is cooled and packaged, and then distributed.

1.2.8 Wastewater from Associated Processes

Most of the water consumed in a dairy processing plant is used in associated processes such as the cleaning and washing of floors, bottles, crates, and vehicles, and the cleaning-in-place (CIP) of factory equipment and tanks as well as the inside of tankers. Most CIP systems consist of three steps: a prerinse step to remove any loose raw material or product remains, a hot caustic wash to clean equipment surfaces, and a cold final rinse to remove any remaining traces of caustic.

1.3 CHARACTERISTICS AND SOURCES OF WASTEWATER

The volume, concentration, and composition of the effluents arising in a dairy plant are dependent on the type of product being processed, the production program, operating methods, design of the processing plant, the degree of water management being applied, and, subsequently, the amount of water being conserved. Dairy wastewater may be divided into three major categories:

1. Processing waters, which include water used in the cooling and heating processes. These effluents are normally free of pollutants and can with minimum treatment be reused or just discharged into the storm water system generally used for rain runoff water.
2. Cleaning wastewaters emanate mainly from the cleaning of equipment that has been in contact with milk or milk products, spillage of milk and milk products, whey, pressings and brines, CIP cleaning options, and waters resulting from equipment malfunctions and even operational errors. This wastewater stream may contain anything from milk, cheese, whey, cream, separator and clarifier dairy waters [10], to dilute yogurt, starter culture, and dilute fruit and stabilizing compounds [9].
3. Sanitary wastewater, which is normally piped directly to a sewage works.

Dairy cleaning waters may also contain a variety of sterilizing agents and various acid and alkaline detergents. Thus, the pH of the wastewaters can vary significantly depending on the cleaning strategy employed. The most commonly used CIP chemicals are caustic soda, nitric acid, phosphoric acid, and sodium hypochloride [10]; these all have a significant impact on wastewater pH. Other concerns related to CIP and sanitizing strategies include the biochemical oxygen demand (BOD) and chemical oxygen demand (COD) contributions (normally <10% of total BOD concentration in plant wastewater), phosphorus contribution resulting from the use of phosphoric acid and other phosphorus-containing detergents, high water volume usage for cleaning and sanitizing (as high as 30% of total water discharge), as well as general concerns regarding the impact of detergent biodegradability and toxicity on the specific waste treatment facility and the environment in general [11].

Dairy industry wastewaters are generally produced in an intermittent way; thus the flow and characteristics of effluents could differ between factories depending on the kind of products produced and the methods of operation [12]. This also influences the choice of the wastewater treatment option, as specific biological systems have difficulties dealing with wastewater of varying organic loads.

Published information on the chemical composition of dairy wastewater is scarce [10]. Some of the more recent information available is summarized in Tables 1.2 and 1.3. Milk has a BOD content 250 times greater than that of sewage [23]. It can, therefore, be expected that dairy wastewaters will have relatively high organic loads, with the main contributors being lactose, fats, and proteins (mainly casein), as well as high levels of nitrogen and phosphorus that are largely associated with milk proteins [12,17]. The COD and BOD for whey have, for instance, been established to be between 35,000–68,000 mg/L and 30,000–60,000 mg/L, respectively, with lactose being responsible for 90% of the COD and BOD contribution [24].

1.4 TREATMENT OPTIONS

The highly variable nature of dairy wastewaters in terms of volumes and flow rates (which is dependent on the factory size and operation shifts) and in terms of pH and suspended solid (SS) content (mainly the result of the choice of cleaning strategy employed) makes the choice of an effective wastewater treatment regime difficult. Because dairy wastewaters are highly biodegradable, they can be effectively treated with biological wastewater treatment systems, but can also pose a potential environmental hazard if not treated properly [23]. The three main options for the dairy industry are: (a) discharge to and subsequent treatment of factory wastewater at a nearby sewage treatment plant; (b) removal of semisolid and special wastes from the site by waste disposal contractors; or (c) the treatment of factory wastewater in an onsite wastewater treatment plant [25,26]. According to Robinson [25], the first two options are continuously impacted by increasing costs, while the control of allowable levels of SS, BOD, and COD in discharged wastewaters are also becoming more stringent. As a result, an increasing number of dairy industries must consider the third option of treating industrial waste onsite. It should be remembered, however, that the treatment chosen should meet the required demands and reduce costs associated with long-term industrial wastewater discharge.

1.4.1 Direct Discharge to a Sewage Treatment Works

Municipal sewage treatment facilities are capable of treating a certain quantity of organic substances and should be able to deal with certain peak loads. However, certain components found in dairy waste streams may present problems. One such substance is fat, which adheres to the walls of the main system and causes sedimentation problems in the sedimentation tanks. Some form of onsite pretreatment is, therefore, advisable to minimize the fat content of the industrial wastewater that can be mixed with the sanitary wastewater going to the sewage treatment facility [6].

Dairy industries are usually subjected to discharge regulations, but these regulations differ significantly depending on discharge practices and capacities of municipal sewage treatment facilities. Sewer charges are based on wastewater flow rate, BOD_5 mass, SS, and total P discharged per day [10]. Some municipal treatment facilities may demand treatment of high-strength industrial effluents to dilute the BOD load of the water so that it is comparable to that of domestic sewage [7].

Table 1.2 Chemical Characteristics of Different Dairy Plant Wastewaters

Industry	BOD₅ (mg/L)	COD (mg/L)	pH	FOG (g/L)	TS (mg/L)	TSS (mg/L)	Alkalinity (mg/L as CaCO₃)	Reference
Cheese								
14 Cheese/whey plants	565–5722	785–7619	6.2–11.3	–	1837–14,205	326–3560	225–1550	10
Cheese/whey plant	377–2214	189–6219	5.2	–	–	188–2330	–	13
Cheese factory	–	5340	5.22	–	4210	–	335	14
Cheese factory	–	2830	4.99	–	–	–	–	15
Cheese processing industry	–	63,300	3.38	2.6	53,200	12,500	–	16
Cheese/casein product plant	–	5380	6.5	0.32	–	–	–	15
Cheese/casein product plant	8000	–	4.5–6.0	0.4	–	–	–	17
Milk								
Milk processing plant	–	713–1410	7.1–8.1	–	900–1470	360–920	–	18
Milk/yogurt plant	–	4656	6.92	–	2750	–	546	14
Milk/cream bottling plant	1200–4000	2000–6000	8–11	3–5	–	350–1000	150–300	19, 20
Butter/milk powder								
Butter/milk powder plant	–	1908	5.8	–	1720	–	532	14
Butter/milk powder plant	1500	–	10–11	0.4	–	–	–	17
Butter/Comté cheese plant	1250	2520	5–7	–	–	–	–	21
Whey								
Whey wastewater	35,000	–	4.6	0.8	–	–	–	17
Raw cheese whey	–	68,814	–	–	3190	1300	–	22

BOD, biological oxygen demand; COD, chemical oxygen demand; TS, total solids; TSS, total suspended solids; FOG, fats, oil and grease.

Table 1.3 Concentrations of Selected Elements in Different Dairy Wastewaters

Industry	Total P (mg/L)	PO$_4$-P (mg/L)	TKN (mg/L)	NH$_4$-N (mg/L)	Na$^+$ (mg/L)	K$^+$ (mg/L)	Ca^{2+} (mg/L)	Mg^{2+} (mg/L)	Reference
Cheese									
14 Cheese/whey plants	29–181	6–35	14–140	1–34	263–1265	8.6–155.5	1.4–58.5	6.5–46.3	16
Cheese/whey plant	0.2–48.0	0.2–7.9	13–172	0.7–28.5	–	–	–	–	13
Cheese factory	45	–	102	–	550	140	30	35	15
Cheese/casein product plant	85	–	140	–	410	125	70	12	15
Cheese/casein product plant	100	–	200	–	380	160	95	14	17
Milk									
Milk/cream bottling plant	–	20–50	50–60	–	170–200	35–40	35–40	5–8	19, 20
Butter/milk powder									
Butter/milk powder plant	35	–	70	–	560	13	8	1	17
Butter/Comté cheese plant	50	–	66	–	–	–	–	–	21
Whey									
Whey wastewater	640	–	1400	–	430	1500	1250	100	17
Raw cheese whey	379	327	1462	64.3	–	–	–	–	22

In a recent survey conducted by Danalewich et al. [10] at 14 milk processing plants in Minnesota, Wisconsin, and South Dakota, it was reported that four facilities directed both their mixed sanitary and industrial wastewater directly to a municipal treatment system, while the rest employed some form of wastewater treatment. Five of the plants that treated their wastewater onsite did not separate their sanitary wastewater from their processing wastewater, which presents a major concern when it comes to the final disposal of the generated sludge after the wastewater treatment, since the sludge may contain pathogenic microorganisms [10]. It would thus be advisable for factories that employ onsite treatment to separate the sanitary and processing wastewaters, and dispose of the sanitary wastewater by piping directly to a sewage treatment facility.

1.4.2 Onsite Pretreatment Options

Physical Screening

The main purpose of screens in wastewater treatment is to remove large particles or debris that may cause damage to pumps and downstream clogging [27]. It is also recommended that the physical screening of dairy wastewater should be carried out as quickly as possible to prevent a further increase in the COD concentration as a result of the solid solubilization [28]. Wendorff [7] recommended the use of a wire screen and grit chamber with a screen aperture size of 9.5 mm, while Hemming [28] recommended the use of even finer spaced mechanically brushed or inclined screens of 40 mesh (about 0.39 mm) for solids reduction. According to Droste [27], certain precautionary measures should be taken to prevent the settling of coarse matter in the wastewater before it is screened. These requirements include the ratio of depth to width of the approach channel to the screen, which should be 1 : 2, as well as the velocity of the water, which should not be less than 0.6 m/sec. Screens can be cleaned either manually or mechanically and the screened material disposed of at a landfill site.

pH Control

As shown in Table 1.2, large variations exist in wastewater pH from different dairy factories. This may be directly attributed to the different cleaning strategies employed. Alkaline detergents generally used for the saponification of lipids and the effective removal of proteinacous substances would typically have a pH of 10–14, while a pH of 1.5–6.0 can be encountered with acidic cleaners used for the removal of mineral deposits and acid-based sanitizers [11,29]. The optimum pH range for biological treatment plants is between 6.5 and 8.5 [30,31]. Extreme pH values can be highly detrimental to any biological treatment facility, not only for the negative effect that it will have on the microbial community, but also due to the increased corrosion of pipes that will occur at pH values below 6.5 and above 10 [6]. Therefore, some form of pH adjustment as a pretreatment step is strongly advised before wastewater containing cleaning agents is discharged to the drain or further treated onsite. In most cases, flow balancing and pH adjustment are performed in the same balancing tank. According to the International Dairy Federation (IDF) [30], a near-neutral pH is usually obtained when water used in different production processes is combined. If pH correction needs to be carried out in the balancing tank, the most commonly used chemicals are H_2SO_4, HNO_3, NaOH, CO_2, or lime [30].

Flow and Composition Balancing

Because discharged dairy wastewaters can vary greatly with respect to volume, strength, temperature, pH, and nutrient levels, flow and composition balancing is a prime requirement for any subsequent biological process to operate efficiently [28]. pH adjustment and flow balancing can be achieved by keeping effluent in an equalization or balancing tank for at least 6–12 hours [7]. During this time, residual oxidants can react completely with solid particles, neutralizing cleaning solutions. The stabilized effluent can then be treated using a variety of different options.

According to the IDF [30], balance tanks should be adequately mixed to obtain proper blending of the contents and to prevent solids from settling. This is usually achieved by the use of mechanical aerators. Another critical factor is the size of the balance tank. This should be accurately determined so that it can effectively handle a dairy factory's daily flow pattern at peak season. It is also recommended that a balancing tank should be large enough to allow a few hours extra capacity to handle unforeseen peak loads and not discharge shock loads to public sewers or onsite biological treatment plants [30].

Fats, Oil, and Grease Removal

The presence of fats, oil, and grease (FOG) in dairy processing wastewater can cause all kinds of problems in biological wastewater treatment systems onsite and in public sewage treatment facilities. It is, therefore, essential to reduce, if not remove FOG completely, prior to further treatment. According to the IDF [32], factories processing whole milk, such as milk separation plants as well as cheese and butter plants, whey separation factories, and milk bottling plants, experience the most severe problems with FOG. The processing of skim milk seldom presents problems in this respect.

As previously mentioned, flow balancing is recommended for dairy processing plants. An important issue, however, is whether the FOG treatment unit should be positioned before or after the balancing tank [32]. If the balancing tank is placed before the FOG unit, large fat globules can accumulate in the tank as the discharged effluent cools down and suspended fats aggregate during the retention period. If the balancing tank is placed after the FOG removal unit, the unit should be large enough to accommodate the maximum anticipated flow from the factory. According to the IDF [32], it is generally accepted that flow balancing should precede FOG removal. General FOG removal systems include the following.

Gravity Traps. In this extremely effective, self-operating, and easily constructed system, wastewater flows through a series of cells, and the FOG mass, which usually floats on top, is removed by retention within the cells. Drawbacks include frequent monitoring and cleaning to prevent FOG buildup, and decreased removal efficiency at pH values above 8 [32].

Air Flotation and Dissolved Air Flotation. Mechanical removal of FOG with dissolved air flotation (DAF) involves aerating a fraction of recycled wastewater at a pressure of about 400–600 kPa in a pressure chamber, then introducing it into a flotation tank containing untreated dairy processing wastewater. The dissolved air is converted to minute air bubbles under the normal atmospheric pressure in the tank [6,32]. Heavy solids form sediment while the air bubbles attach to the fat particles and the remaining suspended matter as they are passed through the effluent [6,9,25]. The resulting scum is removed and will become odorous if stored in an open tank. It is an unstable waste material that should preferably not be mixed with sludge from biological and chemical treatment processes since it is very difficult to dewater. FOG waste should be removed and disposed of according to approved methods [32]. DAF components require regular maintenance and the running costs are usually fairly high.

Air flotation is a more economical variation of DAF. Air bubbles are introduced directly into the flotation tank containing the untreated wastewater, by means of a cavitation aerator coupled to a revolving impeller [32]. A variety of different patented air flotation systems are available on the market and have been reviewed by the IDF [32]. These include the "Hydrofloat," the "Robosep," vacuum flotation, electroflotation, and the "Zeda" systems.

The main drawback of the DAF [25], is that only SS and free FOG can be removed. Thus, to increase the separation efficiency of the process, dissolved material and emulsified FOG solutions must undergo a physico-chemical treatment during which free water is removed and waste molecules are coagulated to form larger, easily removable masses. This is achieved by recirculating wastewater prior to DAF treatment in the presence of different chemical solutions such as ferric chloride, aluminum sulfate, and polyelectrolytes that can act as coalescents and coagulants. pH correction might also be necessary prior to the flotation treatment, because a pH of around 6.5 is required for efficient FOG removal [32].

Enzymatic Hydrolysis of FOG. Cammarota et al. [33] and Leal et al. [34] utilized enzymatic preparations of fermented babassu cake containing lipases produced by a *Penicillium restrictum* strain for FOG hydrolysis in dairy processing wastewaters prior to anaerobic digestion. High COD removal efficiencies as well as effluents of better quality were reported for a laboratory-scale UASB reactor treating hydrolyzed dairy processing wastewater, and compared to the results of a UASB reactor treating the same wastewater without prior enzymatic hydrolysis treatment.

1.4.3 Treatment Methods

Biological Treatment

Biological degradation is one of the most promising options for the removal of organic material from dairy wastewaters. However, sludge formed, especially during the aerobic biodegradation processes, may lead to serious and costly disposal problems. This can be aggravated by the ability of sludge to adsorb specific organic compounds and even toxic heavy metals. However, biological systems have the advantage of microbial transformations of complex organics and possible adsorption of heavy metals by suitable microbes. Biological processes are still fairly unsophisticated and have great potential for combining various types of biological schemes for selective component removal.

Aerobic Biological Systems. Aerobic biological treatment methods depend on micro-organisms grown in an oxygen-rich environment to oxidize organics to carbon dioxide, water, and cellular material. Considerable information on laboratory- and field-scale aerobic treatments has shown aerobic treatment to be reliable and cost-effective in producing a high-quality effluent. Start-up usually requires an acclimation period to allow the development of a competitive microbial community. Ammonia-nitrogen can successfully be removed, in order to prevent disposal problems. Problems normally associated with aerobic processes include foaming and poor solid–liquid separation.

The conventional *activated sludge process* (ASP) is defined [35] as a continuous treatment that uses a consortium of microbes suspended in the wastewater in an aeration tank to absorb, adsorb, and biodegrade the organic pollutants ((Fig. 1.1). Part of the organic composition will be completely oxidized to harmless endproducts and other inorganic substances to provide energy to sustain the microbial growth and the formation of biomass (flocs). The flocs are kept in suspension either by air blown into the bottom of the tank (diffused air system) or by mechanical aeration. The dissolved oxygen level in the aeration tank is critical and should preferably be $1-2$ mg/L and the tank must always be designed in terms of the aeration period and cell residence time. The mixture flows from the aeration tank to a sedimentation tank where the activated sludge flocs form larger particles that settle as sludge. The biological aerobic metabolism mode is extremely efficient in terms of energy recovery, but results in large quantities of sludge being produced (0.6 kg dry sludge per kg of BOD_5 removed). Some of the sludge is returned to the aeration tank but the rest must be processed and disposed of in an environmentally acceptable

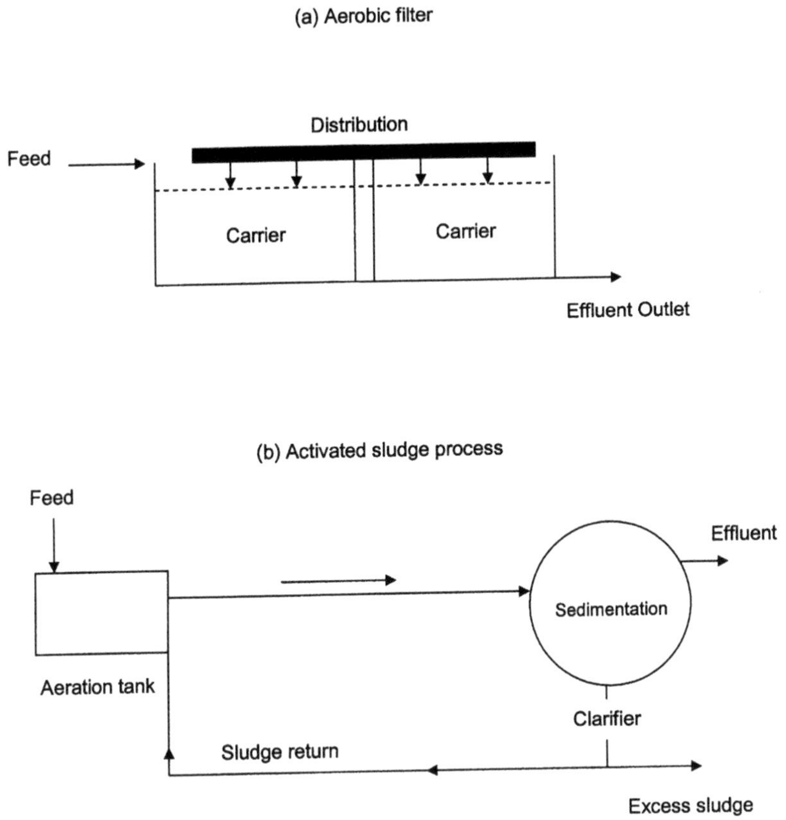
(a) Aerobic filter
Distribution
Feed
Carrier
Carrier
Effluent Outlet
(b) Activated sludge process
Feed
Effluent
Sedimentation
Aeration tank
Clarifier
Sludge return
Excess sludge

Figure 1.1 Simplified illustrations of aerobic wastewater treatment processes: (a) aerobic filter, (b) activated sludge process (from Refs. 31, and 35–37).

manner, which is a major operating expense. Many variations of the ASP exist, but in all cases, the oxygen supplied during aeration is the major energy-consuming operation. With ASPs, problems generally encountered are bulking [17], foam production, precipitation of iron and carbonates, excessive sludge production, and a decrease in efficiency during winter periods.

Many reports show that ASP has been used successfully to treat dairy industry wastes. Donkin and Russell [36] found that reliable COD removals of over 90% and 65% reductions in total nitrogen could be obtained with a milk powder/butter wastewater. Phosphorus removals were less reliable and appeared to be sensitive to environmental changes.

Aerobic filters such as conventional trickling or percolating filters (Fig. 1.1) are among the oldest biological treatment methods for producing high-quality final effluents [35]. The carrier media (20–100 mm diameter) may consist of pumice, rock, gravel, or plastic pieces, which is populated by a very diverse microbial consortium. Wastewater from a storage tank is normally dosed over the medium and then trickles downward through a 2-m medium bed. The slimy microbial mass growing on the carrier medium absorbs the organic constituents of the wastewater and decomposes them aerobically. Sludge deposits require removal from time to time. Aerobic conditions are facilitated by the downward flow and natural convection currents resulting from temperature differences between the air and the added wastewater. Forced ventilation may be employed to enhance the decomposition, but the air must be deodorized by passing through clarifying tanks. Conventional filters, with aerobic microbes growing on rock or gravel, are limited in depth to about 2 m, as deeper filters enhance anaerobic growth with subsequent odor problems. In contrast, filters with synthetic media can be fully aerobic up to about 8 m [37]. The final effluent flows to a sedimentation or clarifying tank to remove sludge and solids from the carrier medium.

It is generally recommended that organic loading for dairy wastewaters not exceed 0.28–0.30 kg BOD/m^3 and that recirculation be employed [38]. A 92% BOD removal of a dairy wastewater was reported by Kessler [4], but since the BOD of the final effluent was still too high, it was further treated in an oxidation pond.

An inherent problem is that trickling filters can be blocked by precipitated ferric hydroxide and carbonates, with concomitant reduction of microbial activity. In the case of overloading with dairy wastewater, the medium becomes blocked with heavy biological and fat films. Maris et al. [39] reported that biological filters are not appropriate for the treatment of high-strength wastewaters, as filter blinding by organic deposition on the filter medium is generally found.

The *rotating biological contactors* (RBC) design contains circular discs (Fig. 1.2) made of high-density plastic or other lightweight material [35]. The discs, rotating at 1–3 rpm, are placed on a horizontal shaft so that about 40–60% of the disc surface protrudes out of the tank; this allows oxygen to be transferred from the atmosphere to the exposed films. A biofilm develops on the disc surface, which facilitates the oxidation of the organic components of the wastewater. When the biofilm sludge becomes too thick, it is torn off and removed in a sedimentation tank. Operation efficiency is based on the g BOD per m^2 of disc surface per day [35]. Rusten and his coworkers [40] reported 85% COD removal efficiency with an organic loading rate (OLR) of 500 g COD/m^3 hour while treating dairy wastewater.

The RBC process offers several advantages over the activated sludge process for use in dairy wastewater treatment. The primary advantages are the low power input required, relative ease of operation and low maintenance. Furthermore, pumping, aeration, and wasting/recycle of solids are not required, leading to less operator attention. Operation for nitrogen removal is also relatively simple and routine maintenance involves only inspection and lubrication.

The *sequencing batch reactor* (SBR) is a single-tank fill-and-draw unit that utilizes the same tank (Fig. 1.2) to aerate, settle, withdraw effluent, and recycle solids [35]. After the tank is filled, the wastewater is mixed without aeration to allow metabolism of the fermentable compounds. This is followed by the aeration step, which enhances the oxidation and biomass formation. Sludge is then settled and the treated effluent is removed to complete the cycle. The SBR relies heavily on the site operator to adjust the duration of each phase to reflect fluctuations in the wastewater composition [41]. The SBR is seen as a good option with low-flow applications and allows for wider wastewater strength variations. Eroglu et al. [42] and Samkutty et al. [43] reported the SBR to be a cost-effective primary and secondary treatment option to handle dairy plant wastewater with COD removals of 91–97%. Torrijos et al. [21] also demonstrated the efficiency of the SBR system for the treatment of wastewater from small cheese-making dairies with treatment levels of >97% being obtained at a loading rate of 0.50 kg COD/m^3 day. In another study, Li and Zhang [44] successfully operated an SBR at a hydraulic retention time (HRT) of 24 hours to treat dairy waste with a COD of 10 g/L. Removal efficiencies of 80% in COD, 63% in total solids, 66% in volatile solids, 75% Kjeldahl nitrogen, and 38% in total nitrogen, were obtained.

In areas where land is available, *lagoons/ponds/reed beds* (Fig. 1.2) constitute one of the least expensive methods of biological degradation. With the exception of aerated ponds, no mechanical devices are used and flow normally occurs by gravity. As result of their simplicity and absence of a sludge recycle facility, lagoons are a favored method for effective wastewater treatment. However, the lack of a controlled environment slows the reaction times, resulting in

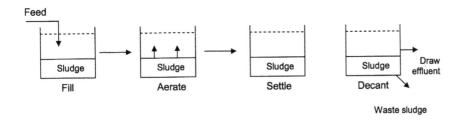

(a) Sequencing Batch Reactor
(same tank)

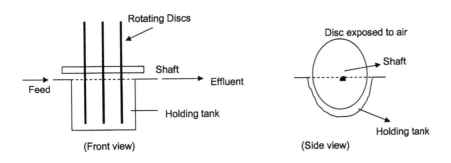

(b) Rotating Biological Contactor

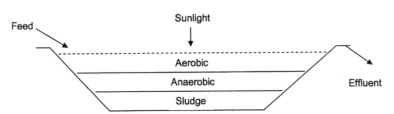

(c) Treatment pond

Figure 1.2 Simplified illustrations of aerobic wastewater treatment processes: (a) sequencing batch reactor, (b) rotating biological contactor, (c) treatment pond (from Refs. 35, 40, 42, 45, 47–49).

long retention times of up to 60 days. Operators of sites in warmer climates may find the use of lagoons a more suitable and economical wastewater treatment option. However, the potential does exist for surface and groundwater pollution, bad odors, and insects that may become a nuisance.

Aerated ponds are generally 0.5–4.0 m deep [45]. Evacuation on the site plus lining is a simple method of lagoon construction and requires relatively unskilled attention. Floating aerators may be used to allow oxygen and sunlight penetration. According to Bitton [46], aeration for 5 days at 20°C in a pond normally gives a BOD removal of 85% of milk wastes. Facultative ponds are also commonly used for high-strength dairy wastes [47]. Although ponds/lagoons are simple to operate, they are the most complex of all biologically engineered degradation systems [48]. In these systems, both aerobic and anaerobic metabolisms occur in addition to photosynthesis and sedimentation. Although most of the organic carbon is converted to microbial biomass, some is lost as CO_2 or CH_4. It is thus essential to remove sludge regularly to prevent buildup and clogging. The HRT in facultative ponds can vary between 5 and 50 days depending on climatic conditions.

Reed-bed or wetland systems have also found widespread application [49]. A design manual and operating guidelines were produced in 1990 [49,50]. Reed beds are designed to treat wastewaters by passing the latter through rhizomes of the common reed in a shallow bed of soil or gravel. The reeds introduce oxygen and as the wastewater percolates through it, aerobic microbial communities establish among the roots and degrade the contaminants. Nitrogen and phosphorus are thus removed directly by the reeds. However, reed beds are poor at removing ammonia, and with high concentrations of ammonia being toxic, this may be a limiting factor. The precipitation of large quantities of iron, manganese, and calcium within the reed beds will also affect rhizome growth and, in time, reduce the permeability of the bed. According to Robinson et al. [49], field studies in the UK have shown that reed beds have enormous potential and, in combination with aerobic systems, provide high effluent quality at reasonable cost.

Anaerobic Biological Systems. Anaerobic digestion (AD) is a biological process performed by an active microbial consortium in the absence of exogenous electron acceptors. Up to 95% of the organic load in a waste stream can be converted to biogas (methane and carbon dioxide) and the remainder is utilized for cell growth and maintenance [51,52]. Anaerobic systems are generally seen as more economical for the biological stabilization of dairy wastes [14], as they do not have the high-energy requirements associated with aeration in aerobic systems. Anaerobic digestion also yields methane, which can be utilized as a heat or power source. Furthermore, less sludge is generated, thereby reducing problems associated with sludge disposal. Nutrient requirements (N and P) are much lower than for aerobic systems [37], pathogenic organisms are usually destroyed, and the final sludge has a high soil conditioning value if the concentration of heavy metals is low. The possibility of treating high COD dairy wastes without previous dilution, as required by aerobic systems, reduces space requirements and the associated costs [53]. Bad odors are generally absent if the system is operated efficiently [51,54].

The disadvantages associated with anaerobic systems are the high capital cost, long start-up periods, strict control of operating conditions, greater sensitivity to variable loads and organic shocks, as well as toxic compounds [55]. The operational temperature must be maintained at about 33–37°C for efficient kinetics, because it is important to keep the pH at a value around 7, as a result of the sensitivity of the methanogenic population to low values [48]. As ammonia-nitrogen is not removed in an anaerobic system, it is consequently discharged with the digester effluent, creating an oxygen demand in the receiving water. Complementary treatment to achieve acceptable discharge standards is also required.

The *anaerobic lagoon* (anaerobic pond) (Fig. 1.3) is the simplest type of anaerobic digester. It consists of a pond, which is normally covered to exclude air and to prevent methane loss to the atmosphere. Lagoons are far easier to construct than vertical digester types, but the biggest drawback is the large surface area required.

In New Zealand, dairy wastewater [51] was treated at 35°C in a lagoon (26,000 m^3) covered with butyl rubber at an organic load of 40,000 kg COD per day, pH of 6.8–7.2, and HRT of 1–2 days. The organic loading rate (OLR) of 1.5 kg COD/m^3 day was on the low side. The pond's effluent was clarified and the settled biomass recycled through the substrate feed. The clarified effluent was then treated in an 18,000 m^3 aerated lagoon. The efficiency of the total system reached a 99% reduction in COD.

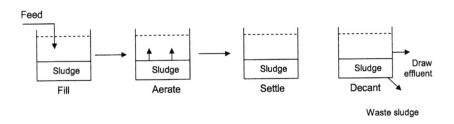

(a) Sequencing Batch Reactor
(same tank)

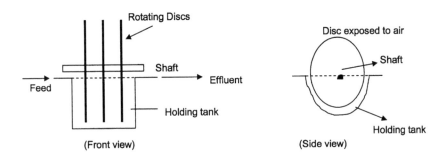

(b) Rotating Biological Contactor

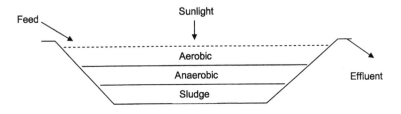

(c) Treatment pond

Figure 1.3 Simplified illustrations of anaerobic wastewater treatment processes: (a) anaerobic filter digester, (b) fluidized-bed digester, (c) UASB digester, (d) anaerobic lagoon/pond (from Refs. 31, 35, 51, 58, 70).

Completely stirred tank reactors (CSTR) [56] are, next to lagoons, the simplest type of anaerobic digester (Fig. 1.4). According to Sahm [57], the OLR rate ranges from 1–4 kg organic dry matter m^{-3} day^{-1} and the digesters usually have capacities between 500 and 700 m^3. These reactors are normally used for concentrated wastes, especially those where the polluting matter is present mainly as suspended solids and has COD values of higher than 30,000 mg/L. In the CSTR, there is no biomass retention; consequently, the HRT and sludge retention time (SRT) are not separated, necessitating long retention times that are dependent on the growth rate of the

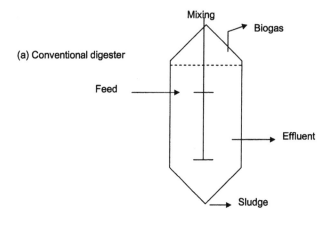

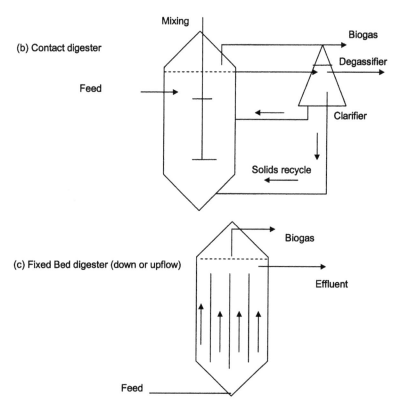

Figure 1.4 Simplified illustrations of anaerobic wastewater treatment processes: (a) conventional digester, (b) Contact digester, (c) fixed-bed digester (from Refs. 31, 57, 58, 60, 64, 66, 79).

slowest-growing bacteria involved in the digestion process. Ross [58] found that the HRT of the conventional digesters is equal to the SRT, which can range from 15–20 days.

This type of digester has in the past been used by Lebrato et al. [59] to treat cheese factory wastewater. While 90% COD removal was achieved, the digester could only be operated at a minimum HRT of 9.0 days, most probably due to biomass washout. The wastewater, consisting of 80% washing water and 20% whey, had a COD of 17,000 mg/L. While the CSTR is very useful for laboratory studies, it is hardly a practical option for full-scale treatment due to the HRT limitation.

The anaerobic *contact process* (Fig. 1.4) was developed in 1955 [60]. It is essentially an anaerobic activated sludge process that consists of a completely mixed anaerobic reactor followed by some form of biomass separator. The separated biomass is recycled to the reactor, thus reducing the retention time from the conventional 20–30 days to <1.0 days. Because the bacteria are retained and recycled, this type of plant can treat medium-strength wastewater (200–20,000 mg/L COD) very efficiently at high OLRs [57]. The organic loading rate can vary from 1 to 6 kg/m^3 day COD with COD removal efficiencies of 80–95%. The treatment temperature ranges from 30–40°C. A major difficulty encountered with this process is the poor settling properties of the anaerobic biomass from the digester effluent. Dissolved air flotation [61] and dissolved biogas flotation techniques [62] have been attempted as alternative sludge separation techniques. Even though the contact digester is considered to be obsolete there are still many small dairies all over the world that use the system [63].

The upflow *anaerobic filter* (Fig. 1.3) was developed by Young and McCarty in 1969 [64] and is similar to the aerobic trickling filter process. The reactor is filled with inert support material such as gravel, rocks, coke, or plastic media and thus there is no need for biomass separation and sludge recycling. The anaerobic filter reactor can be operated either as a downflow or an upflow filter reactor with OLR ranging from 1–15 kg/m^3 day COD and COD removal efficiencies of 75–95%. The treatment temperature ranges from 20 to 35°C with HRTs in the order of 0.2–3 days. The main drawback of the upflow anaerobic filter is the potential risk of clogging by undegraded suspended solids, mineral precipitates or the bacterial biomass. Furthermore, their use is restricted to wastewaters with COD between 1000 and 10,000 mg/L [58]. Bonastre and Paris [65] listed 51 anaerobic filter applications of which five were used for pilot plants and three for full-scale dairy wastewater treatment. These filters were operated at HRTs between 12 and 48 hours, while COD removal ranged between 60 and 98%. The OLR varied between 1.7 and 20.0 kg COD/m^3 day.

The *expanded bed* and/or *fluidized-bed digesters* (Fig. 1.3) are designed so that wastewaters pass upwards through a bed of suspended media, to which the bacteria attach [66]. The carrier medium is constantly kept in suspension by powerful recirculation of the liquid phase. The carrier media include plastic granules, sand particles, glass beads, clay particles, and activated charcoal fragments. Factors that contribute to the effectiveness of the fluidized-bed process include: (a) maximum contact between the liquid and the fine particles carrying the bacteria; (b) problems of channeling, plugging, and gas hold-up commonly encountered in packed-beds are avoided; and (c) the ability to control and optimize the biological film thickness [57]. OLRs of 1–20 kg/m^3 day COD can be achieved with COD removal efficiencies of 80–87% at treatment temperatures from 20 to 35°C.

Toldrá et al. [67] used the process to treat dairy wastewater with a COD of only 200–500 mg/L at an HRT of 8.0 hours with COD removal of 80%. Bearing in mind the wide variations found between different dairy effluents, it can be deduced that this particular dairy effluent is at the bottom end of the scale in terms of its COD concentration and organic load. The dairy wastewater was probably produced by a dairy with very good product-loss control and rather high water use [68].

The *upflow anaerobic sludge blanket* (UASB) reactor was developed for commercial purposes by Lettinga and coworkers at the Agricultural University in Wageningen, The Netherlands. It was first used to treat maize-starch wastewaters in South Africa [69], but its full potential was only realized after an impressive development program by Lettinga in the late 1970s [70,71]. The rather simple design of the UASB bioreactor (Fig. 1.3) is based on the superior settling properties of a granular sludge. The growth and development of granules is the key to the success of the UASB digester. It must be noted that the presence of granules in the UASB system ultimately serves to separate the HRT from the solids retention time (SRT). Thus, good granulation is essential to achieve a short HRT without inducing biomass washout. The wastewater is fed from below and leaves at the top via an internal baffle system for separation of the gas, sludge, and liquid phases. With this device, the granular sludge and biogas are separated. Under optimal conditions, a COD loading of 30 kg/m^3 day can be treated with a COD removal efficiency of 85–95%. The methane content of the biogas is between 80 and 90% (v/v). HRTs of as low as 4 hours are feasible, with excellent settling sludge and SRT of more than 100 days. The treatment temperature ranges from 7–40°C, with the optimum being at 35°C.

Goodwin et al. [72] treated a synthetic ice cream wastewater using the UASB process at HRTs of 18.4 hours and an organic carbon removal of 86% was achieved. The maximum OLR was 3.06 kg total organic carbon (TOC) per m^3 day. Cheese effluent has also been treated in the UASB digester at a cheese factory in Wisconsin, USA [73]. The UASB was operated at an HRT of 16.0 hours and an OLR of 49.5 kg COD/m^3 day with a plant wastewater COD of 33,000 mg/L and a COD removal of 86% was achieved. The UASB digester was, however, only a part of a complete full-scale treatment plant. The effluent from the UASB was recycled to a mixing tank, which also received the incoming effluent. Although the system is described as an UASB system, it could also pass as a separated or two-phase system, since some degree of pre-acidification is presumably attained in the mixing tank. Furthermore, the pH in the mixing tank was controlled by means of lime dosing when necessary. The effluent emerging from the mixing tank was treated in an aerobic system, serving as a final polishing step, to provide an overall COD removal of 99%.

One full-scale UASB treatment plant [51] in Finland at the Mikkeli Cooperative Dairy, produces Edam type cheese, butter, pasteurized and sterilized milk, and has a wastewater volume of 165 million liters per year. The digester has an operational volume of 650 m^3, which includes a balancing tank of 300 m^3 [74,75]. The COD value was reduced by 70–90% and 400 m^3 biogas is produced daily with a methane content of 70%, which is used to heat process water in the plant.

One of the most successful full-scale 2000 m^3 UASB described in the literature was in the UK at South Caernarvon Creameries to treat whey and other wastewaters [76]. The whey alone reached volumes of up to 110 kiloliters (kL) per day. In the system, which included a combined UASB and aerobic denitrification system, COD was reduced by 95% and sufficient biogas was produced to meet the total energy need of the whole plant. The final effluent passed to a sedimentation tank, which removed suspended matter. From there, it flowed to aerobic tanks where the BOD was reduced to 20.0 mg/L and the NH$_3$-nitrogen reduced to 10.0 mg/L. The effluent was finally disposed of into a nearby river. The whey disposal costs, which originally amounted to £30,000 per year, were reduced to zero; the biogas also replaced heavy fuel oil costs. On full output, the biogas had a value of up to £109,000 per year as an oil replacement and a value of about £60,000 as an electricity replacement. These values were, however, calculated in terms of the oil and electricity prices of 1984, but this illustrates the economic potential of the anaerobic digestion process.

The *fixed-bed digester* (Fig. 1.4) contains permanent porous carrier materials and by means of extracellular polysaccharides, bacteria can attach to the surface of the packing material and still remain in close contact with the passing wastewater. The wastewater is added either at the bottom or at the top to create upflow or downflow configurations.

A downflow fixed-film digester was used by Cánovas-Diaz and Howell [77] to treat deproteinized cheese whey with an average COD of 59,000 mg/L. At an OLR of 12.5 kg COD/m^3 day, the digester achieved a COD reduction of 90–95% at an HRT of 2.0–2.5 days. The deproteinized cheese whey had an average pH of 2.9, while the digester pH was consistently above pH 7.0 [78].

A laboratory-scale fixed-bed digester with an inert polyethylene bacterial carrier was also used by De Haast et al. [79] to treat cheese whey. The best results were obtained at an HRT of 3.5 days, with 85–87% COD removal. The OLR was 3.8 kg COD/m^3 day and biogas yield amounted to 0.42 m^3/kg COD_{added} per day. The biogas had a methane content of between 55 and 60%, and 63.7% of the calorific value of the substrate was conserved in the methane.

In a *membrane anaerobic reactor system* (MARS), the digester effluent is filtrated by means of a filtration membrane. The use of membranes enhances biomass retention and immediately separates the HRT from the SRT [68].

Li and Corrado (80) evaluated the MARS (completely mixed digester with operating volume of 37,850 L combined with a microfiltration membrane system) on cheese whey with a COD of up to 62,000 mg/L. The digester effluent was filtrated through the membrane and the permeate discharged, while the retentate, containing biomass and suspended solids, was returned to the digester. The COD removal was 99.5% at an HRT of 7.5 days. The most important conclusion the authors made was that the process control parameters obtained in the pilot plant could effectively be applied to their full-scale demonstration plant.

A similar membrane system, the anaerobic digestion ultrafiltration system (ADUF) has successfully been used in bench- and pilot-scale studies on dairy wastewaters [81]. The ADUF system does not use microfiltration, but rather an ultrafiltration membrane; therefore, far greater biomass retention efficiency is possible.

Separated phase digesters are designed to spatially separate the acid-forming bacteria and the acid-consuming bacteria. These digesters are useful for the treatment of wastes either with unbalanced carbon to nitrogen (C : N) ratios, such as wastes with high protein levels, or wastes such as dairy wastewaters that acidify quickly [51,68]. High OLRs and short HRTs are claimed to be the major advantages of the separated phase digester.

Burgess [82] described two cases where dairy wastewaters were treated using a separated phase full-scale process. One dairy had a wastewater with a COD of 50,000 mg/L and a pH of 4.5. Both digester phases were operated at 35°C, while the acidogenic reactor was operated at an HRT of 24 hours and the methanogenic reactor at an HRT of 3.3 days. In the acidification tank, 50% of the COD was converted to organic acids while only 12% of the COD was removed. The OLR for the acidification reactor was 50.0 kg COD/m^3 day, and for the methane reactor, 9.0 kg COD/m^3 day. An overall COD reduction of 72% was achieved. The biogas had a methane content of 62%, and from the data supplied, it was calculated that a methane yield ($Y_{CH4}/COD_{removed}$) of 0.327 m^3/kg $COD_{removed}$ was obtained.

Lo and Liao [83,84] also used separated phase digesters to treat cheese whey. The digesters were described as anaerobic rotating biological contact reactors (AnRBC), but can really be described as tubular fixed-film digesters orientated horizontally, with internally rotating baffles. In the methane reactor, these baffles were made from cedar wood, as the authors contend that the desired bacterial biofilms develop very quickly on wood. The acidogenic reactor was mixed by means of the recirculation of the biogas. However, it achieved a COD reduction of only 4%. More importantly, the total volatile fatty acids concentration was increased from 168 to 1892 mg/L. This was then used as substrate for the second phase where a COD reduction of up to 87% was achieved. The original COD of the whey was 6720 mg/L, which indicates that the whey was diluted approximately tenfold.

Many other examples of two-phase digesters are found in the literature. It was the opinion of Kisaalita et al. [85] that two-phase processes may be more successful in the treatment of lactose-containing wastes. The researchers studied the acidogenic fermentation of lactose, determined the kinetics of the process [86], and also found that the presence of whey protein had little influence on the kinetics of lactose acidogenesis [87]. Venkataraman et al. [88] also used a two-phase packed-bed anaerobic filter system to treat dairy wastewater. Their main goals were to determine the kinetic constants for biomass and biogas production rates and substrate utilization rates in this configuration.

Land Treatment

Dairy wastewater, along with a wide variety of other food processing wastewaters, has been successfully applied to land in the past [31]. Interest in the land application of wastes is also increasing as a direct result of the general move of regulatory authorities to restrict waste disposal into rivers, lakes, and the ocean, but also because of the high costs of incineration and landfilling [89]. Nutrients such as N and P that are contained in biodegradable processing wastewaters make these wastes attractive as organic fertilizers, especially since research has shown that inorganic fertilizers might not be enough to stem soil degradation and erosion in certain parts of the world [89,90]. Land application of these effluents may, however, be limited by the presence of toxic substances, high salt concentrations, or extreme pH values [89]. It might be, according to Wendorff [7], the most economical option for dairy industries located in rural areas.

Irrigation

The distribution of dairy wastewaters by irrigation can be achieved through spray nozzles over flat terrain, or through a ridge and furrow system [7]. The nature of the soil, topography of the land and the waste characteristics influence the specific choice of irrigation method. In general, loamy well-drained soils, with a minimum depth to groundwater of 1.5 m, are the most suitable for irrigation. Some form of crop cover is also desirable to maintain upper soil layer porosity [31]. Wastewater would typically percolate through the soil, during which time organic substances are degraded by the heterotrophic microbial population naturally present in the soil [7]. An application period followed by a rest period (in a 1 : 4 ratio) is generally recommended.

Eckenfelder [31] reviewed two specific dairy factory irrigation regimes. The first factory produced cream, butter, cheese, and powdered milk, and irrigated their processing wastewaters after pretreatment by activated sludge onto coarse and fine sediments covered with reed and canary grass in a 1 : 3 application/rest ratio. The second factory, a Cheddar cheese producer, employed only screening as a pretreatment method and irrigated onto Chenango gravel with the same crop cover as the first factory, in a 1 : 6 application/rest ratio.

Specific wastewater characteristics can have an adverse effect on a spray irrigation system that should also be considered. Suspended solids, for instance, may clog spray nozzles and render the soil surface impermeable, while wastewater with an extreme pH or high salinity might be detrimental to crop cover. Highly saline wastewater might further cause soil dispersion, and a subsequent decrease in drainage and aeration, as a result of ion exchange with sodium replacing magnesium and calcium in the soil [31]. The land application of dairy factory wastewater, which typically contains high concentrations of sodium ions, might thus be restricted [89]. And although milk proteins and lactose are readily degradable by anaerobic bacteria naturally present in the soil, FOG tends to be more resistant to degradation and will accumulate under anaerobic conditions [7].

According to Sparling et al. [15] there is little published information relating the effect that long-term irrigation of dairy factory effluent may have on soil properties. Based on the irrigation data Degens et al. [91] and Sparling et al. [15] investigated the effect that long-term dairy wastewater irrigation can have on the storage and distribution of nutrients such as C, N, and P, and the differences existing between key soil properties of a long-term irrigation site (22 years) and a short-term irrigation site (2 years). Degens et al. [91] reported that irrigation had no effect on total soil C in the 0–0.75 m layer, although redistribution of C from the top 0–0.1 m soil had occurred, either as a result of leaching caused by the irrigation of highly alkaline effluents, or as a result of increased earthworm activity. The latter were probably promoted by an increased microbial biomass in the soil, which were mostly lactose and glucose degraders. It was also reported that about 81% of the applied P were stored in the 0–0.25 m layer compared to only 8% of the total applied N. High nitrate concentrations were measured in the groundwater below the site, and reduced nitrogen loadings were recommended in order to limit nitrogen leaching to the environment [91]. In contrast to the results reported by Degens et al. (2000) for a long-term irrigated site, Sparling et al. [15] found no redistribution of topsoil C in short-term irrigated soils, which was probably the result of a lower effluent loading. Generally, it was found that hydraulic conductivity, microbial content, and N-cycling processes all increased substantially in long-term irrigated soils. Since increases in infiltration as well as biochemical processing were noted in all the irrigated soils, most of the changes in soil properties were considered to be beneficial. A decrease in N-loading was, however, also recommended [15].

1.4.4 Sludge Disposal

Different types of sludge arise from the treatment of dairy wastewaters. These include: (a) sludge produced during primary sedimentation of raw effluents (the amounts of which are usually low); (b) sludge produced during the precipitation of suspended solids after chemical treatment of raw wastewaters; (c) stabilized sludge resulting from the biological treatment processes, which can be either aerobic or anaerobic; and (d) sludge generated during tertiary treatment of wastewater for final suspended solid or nutrient removal after biological treatment [92]. Primary sedimentation of dairy wastewater for BOD reduction is not usually an efficient process, so in most cases the settleable solids reach the next stage in the treatment process directly. An important advantage of anaerobic processes is that the sludge generated is considerably less than the amount produced by aerobic processes, and it is easier to dewater. Final wastewater polishing after biological treatment usually involves chemical treatment of the wastewater with calcium, iron, or aluminum salts to remove dissolved nutrients such as nitrogen and phosphorus. The removal of dissolved phosphorus can have a considerable impact on the amount of sludge produced during this stage of treatment [92].

The application of dairy sludge as fertilizer has certain advantages when compared to municipal sludge. It is a valuable source of nitrogen and phosphorous, although some addition of potassium might be required to provide a good balance of nutrients. Sludge from different factories will also contain different levels of nutrients depending on the specific products manufactured. Dairy sludge seldom contains the same pathogenic bacterial load as domestic sludge, and also has considerably lower heavy metal concentrations. The recognition of dairy sludge as a fertilizer does, however, depend on local regulations. Some countries have limited the amount of sludge that can be applied as fertilizer to prevent nitrates from leaching into groundwater sources [92].

According to the IDF [92], dairy sludge disposal must be reliable, legally acceptable, economically viable, and easy to conduct. Dairy wastewater treatment facilities are usually small compared to sewage treatment works, which means that thermal processes such as drying and incineration can be cost-prohibitive for smaller operations. It is generally agreed that disposal of sludge by land spraying or as fertilizer is the least expensive method. If the transport and disposal of liquid sludge cannot be done within reasonable costs, other treatment options such as sludge thickening, dewatering, drying, or incineration must be considered. Gravity thickeners are most commonly used for sludge thickening, while the types of dewatering machines most commonly applied are rotary drum vacuum filters, filter presses, belt presses, and decanter centrifuges [92].

1.5 POLLUTION PREVENTION

Reduction of wastewater pollution levels may be achieved by more efficiently controlling water and product wastage in dairy processing plants. Comparisons of daily water consumption records vs. the amount of milk processed will give an early indication of hidden water losses that could result from defective subfloor and underground piping. An important principle is to prevent wastage of product rather than flush it away afterwards. Spilled solid material such as curd from the cheese production area, and spilled dry product from the milk powder production areas should be collected and treated as solid waste rather than flushing them down the drain [6].

Small changes could also be made to dairy manufacturing processes to reduce wastewater pollution loads, as reviewed by Tetrapak [6]. In the cheese production area, milk spillage can be restricted by not filling open cheese vats all the way to the rim. Whey could also be collected sparingly and used in commercial applications instead of discharging it as waste.

Manual scraping of all accessible areas after a butter production run and before cleaning starts would greatly reduce the amount of residual cream and butter that would enter the wastewater stream. In the milk powder production area, the condensate formed could be reused as cooling water (after circulation through the cooling tower), or as feedwater to the boiler. Returned product could be emptied into containers and used as animal feed [6]. Milk and product spillage can further be restricted by regular maintenance of fittings, valves, and seals, and by equipping fillers with drip and spill savers. Pollution levels could also be limited by allowing pipes, tanks, and transport tankers adequate time to drain before being rinsed with water [8].

1.6 CASE STUDIES

1.6.1 Case Study 1

A summary of a case study as reported by Rusten et al. [93] is presented for the upgrading of a cheese factory additionally producing casein granules.

Background

The authors described how a wastewater treatment process of a Norwegian cheese factory, producing casein granules as a byproduct, was upgraded to meet the wastewater treatment demands set by large increases in production and stricter environmental regulations. The design criteria were based on the assumption that the plant produced an average amount of 150 m^3/day of wastewater, which had an average organic load of 200 kg BOD/day with an average total phosphorous (TP) load of 3.5 kg TP/day and a pH range between 2 and 12.

Requirements

It was required that the treatment plant be able to remove more than 95% of the total BOD (>95% total COD). The specific amount of phosphorous that could be allowed in the discharged wastewater was still being negotiated with the authorities. The aim however, was to remove as much phosphorous as possible. The pH of the final effluent had to be between 6.5 and 8.0.

The Final Process

A flow diagram of the final process is summarized in Figure 1.5.

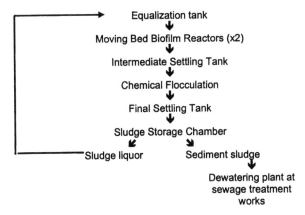

Figure 1.5   Flow diagram of the final process of Case Study 1.

Process Efficiency

After modifications, the average organic load was 347 kg COD/day with average removal efficiency of 98% for both the total COD and the total phosphorous content. Extreme pH values in the incoming wastewater were also efficiently neutralized in the equalization tank, resulting in a 7.0–8.0 pH range in the reactors.

1.6.2   Case Study 2

A summary of a case study reported by Monroy et al. [94] is presented.

Background

As with the first case study, the authors reported on how an existing wastewater treatment system of a cheese manufacturing industry in Mexico, which was operating below the consents, could be upgraded so that the treated wastewater could meet the discharge limits imposed by local environmental authorities. The factory produced an average wastewater volume of 500 m^3/day with an average composition (mg/L) of 4430 COD, 3000 BOD$_5$, 1110 TSS, and 754 FOG.

Requirements

Environmental regulations required the treated wastewater to have less than 100 mg/L BOD, 300 mg/L COD, 100 mg/L TSS, and 15 mg/L FOG. The pH of the discharged effluent had to be between 6.0 and 9.0. The old treatment system was not effective enough to reduce the BOD, COD, TSS, and FOG to acceptable levels, although the final pH of 7.5 was within the recommended range. The factory was looking for a more effective treatment system that could utilize preexisting installations, thereby reducing initial investment costs, and also have low operation costs.

The Final Process

A flow diagram of the final process is summarized in Figure 1.6.

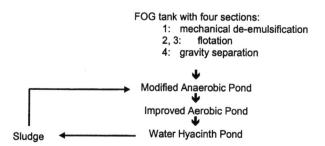

Figure 1.6 Flow diagram of the final process of Case Study 2.

Process Efficiency

Pollution levels in the raw wastewater were first reduced by initiating an "in-factory" wastewater management program, which resulted in greater pH stability and lower phosphorous levels (by recycling certain cleaning chemicals and substituting others) as well as reduced levels of salt (by concentrating and drying brine). The modified wastewater treatment process resulted in an overall removal efficiency of 98% BOD (final concentration = 105 mg/L), 96% COD (final concentration = 225 mg/L), 98% TSS (final concentration = 24 mg/L), and 99.8% FOG (final concentration = 1.7 mg/L). The modifications ultimately resulted in a total operating cost increase of 0.4% at the factory.

1.6.3 General Conclusions: Case Studies

All wastewater treatment systems are unique. Before a treatment strategy is chosen, careful consideration should be given to proper wastewater sampling and composition analysis as well as a process survey. This would help prevent an expensive and unnecessary or overdesigned treatment system [95]. A variety of different local and international environmental engineering firms are able to assist in conducting surveys. These firms can also be employed to install effective patented industrial-scale installations for dairy processing wastewater treatment.

1.7 CONCLUSIONS

As management of dairy wastes becomes an ever-increasing concern, treatment strategies will need to be based on state and local regulations. Because the dairy industry is a major water user and wastewater generator, it is a potential candidate for wastewater reuse. Purified wastewater can be utilized in boilers and cooling systems as well as for washing plants, and so on. Even if the purified wastewater is initially not reused, the dairy industry will still benefit directly from in-house wastewater treatment, since levies charged for wastewater reception will be significantly reduced. In the United Kingdom, 70% of the total savings that have already been achieved with anaerobic digestion are due to reduced discharge costs [96]. The industry will also benefit where effluents are currently used for irrigation of pastures, albeit in a more indirect way. All these facts underline the need for efficient dairy wastewater management.

Before selecting any treatment method, a complete process evaluation should be undertaken along with economic analysis. This should include the wastewater composition, concentrations, volumes generated, and treatment susceptibility, as well as the environmental impact of the solution to be adopted. All options are expensive, but an economic analysis may indicate that slightly higher maintenance costs may be less than increased operating costs. What is appropriate for one site may be unsuitable for another.

The most useful processes are those that can be operated with a minimum of supervision and are inexpensive to construct or even mobile enough to be moved from site to site. The changing quantity and quality of dairy wastewater must also be included in the design and operational procedures. From the literature it appears as if biological methods are the most cost-effective for the removal of organics, with aerobic methods being easier to control, but anaerobic methods having lower energy requirements and lower sludge production rates. Since no single process for treatment of dairy wastewater is by itself capable of complying with the minimum effluent discharge requirements, it is necessary to choose a combined process especially designed to treat a specific dairy wastewater.

REFERENCES

1. Russell, P. Effluent and waste water treatment. Milk Ind. Int. **1998**, *100 (10)*, 36–39.
2. Strydom, J.P.; Mostert, J.F.; Britz, T.J. Effluent production and disposal in the South African dairy industry–a postal survey. Water SA **1993**, *19 (3)*, 253–258.
3. Robinson, T. The real value of dairy waste. Dairy Ind. Int. **1997**, *62 (3)*, 21–23.
4. Kessler, HG. (Ed.) *Food Engineering and Dairy Technology*; Verlag: Freisburg, Germany, 1981.
5. Odlum, C.A. Reducing the BOD level from a dairy processing plant. *Proc. 23rd Int. Dairy Cong.*, Montreal, Canada, October 1990.
6. Tetrapak. *TetraPak Dairy Processing Handbook*; TetraPak Printers: London, UK, 1995.
7. Wendorff, W.L. Treatment of dairy wastes. In *Applied Dairy Microbiology*, 2nd ed.; Marth, E.H., Steele, J.L., Eds.; Marcel Dekker Inc: New York, 2001; 681–704.
8. Steffen, Robertson, Kirsten Inc. *Water and Waste-water Management in the Dairy Industry*, WRC Project No. 145 TT38/89. Water Research Commission: Pretoria, South Africa, 1989.
9. Tamime, A.Y.; Robinson, R.K. (Eds.) *Yoghurt Science and Technology*; Woodhead Publishing Ltd: Cambridge, England, 1999.
10. Danalewich, J.R.; Papagiannis, T.G.; Belyea, R.L.; Tumbleson, M.E.; Raskin, L. Characterization of dairy waste streams, current treatment practices, and potential for biological nutrient removal. Water Res. **1998**, *32 (12)*, 3555–3568.
11. Bakka, R.L. Wastewater issues associated with cleaning and sanitizing chemicals. Dairy Food Environ. Sanit. **1992**, *12 (5)*, 274–276.
12. Vidal, G.; Carvalho, A.; Méndez, R.; Lema, J.M. Influence of the content in fats and proteins on the anaerobic biodegradability of dairy wastewaters. Biores. Technol. **2000**, *74, 231–239*.
13. Andreottola, G.; Foladori, P.; Ragazzi, M.; Villa, R. Dairy wastewater treatment in a moving bed biofilm reactor. Wat. Sci. Technol. **2002**, *45 (12)*, 321–328.
14. Strydom, J.P.; Britz, T.J.; Mostert, J.F. Two-phase anaerobic digestion of three different dairy effluents using a hybrid bioreactor. Water SA **1997**, *23*, 151–156.
15. Sparling, G.P.; Schipper, L.A.; Russell, J.M. Changes in soil properties after application of dairy factory effluent to New Zealand volcanic ash and pumice soils. Aust. J. Soil. Res. **2001**, *39*, 505–518.
16. Hwang, S.; Hansen, C.L. Characterization of and bioproduction of short-chain organic acids from mixed dairy-processing wastewater. Trans. ASAE **1998**, *41 (3)*, 795–802.
17. Donkin, J. Bulking in aerobic biological systems treating dairy processing wastewaters. Int. J. Dairy Tech. **1997**, *50*, 67–72.
18. Samkutty, P.J.; Gough, R.H. Filtration treatment of dairy processing wastewater. J. Environ. Sci. Health. **2002**, *A37 (2)*, 195–199.
19. Ince, O. Performance of a two-phase anaerobic digestion system when treating dairy wastewater. Wat. Res. **1998**, *32 (9)*, 2707–2713.
20. Ince, O. Potential energy production from anaerobic digestion of dairy wastewater. J. Environ. Sci. Health. **1998**, *A33 (6)*, 1219–1228.

21. Torrijos, M.; Vuitton, V.; Moletta, R. The SBR process: an efficient and economic solution for the treatment of wastewater at small cheese making dairies in the Jura Mountains. Wat. Sci. Technol. **2001**, *43*, 373–380.

22. Malaspina, F.; Cellamare, C.M.; Stante, L.; Tilche, A. Anaerobic treatment of cheese whey with a downflow-upflow hybrid reactor. Biores. Technol. **1996**, *55*, 131–139.

23. Burton, C. FOG clearance. Dairy Ind. Int. **1997**, *62* (*12*), 41–42.

24. Berruga, M.I.; Jaspe, A.; San-Jose, C. Selection of yeast strains for lactose hydrolysis in dairy effluents. Int. Biodeter. Biodeg. **1997**, *40* (*2–4*), 119–123.

25. Robinson, T. How to be affluent with effluent. The Milk Ind. **1994**, *96* (*4*), 20–21.

26. Gough, R.H.; McGrew, P. Preliminary treatment of dairy plant waste water. J. Environ. Sci. Health. **1993**, *A28* (*1*), 11–19.

27. Droste, R.L. (Ed.) *Theory and Practice of Water and Wastewater Treatment*; John Wiley & Sons Inc: New York, USA, 1997.

28. Hemming, M.L. The treatment of dairy wastes. In *Food Industry Wastes: Disposal and Recovery*; Herzka, A., Booth, R.G., Eds.; Applied Science Publishers Ltd: Essex, 1981.

29. Graz, C.J.M.; McComb, D.G. Dairy CIP–A South African review. Dairy, Food Environ. Sanit. **1999**, *19* (*7*), 470–476.

30. IDF. Balance tanks for dairy effluent treatment plants. Bull. Inter. Dairy Fed. **1984**, Doc. No. 174.

31. Eckenfelder, W.W. (Ed.) *Industrial Water Pollution Control*; McGraw-Hill Inc: New York, USA, 1989.

32. IDF. Removal of fats, oils and grease in the pretreatment of dairy wastewaters. Bull. Inter. Dairy Fed. **1997**, Doc. No. 327.

33. Cammarota, M.C.; Teixeira, G.A.; Freire, D.M.G. Enzymatic pre-hydrolysis and anaerobic degradation of wastewaters with high fat contents. Biotech. Lett. **2001**, *23*, 1591–1595.

34. Leal, M.C.M.R.; Cammarota, M.C.; Freire, D.M.G.; Sant'Anna Jr, G.L. Hydrolytic enzymes as coadjuvants in the anaerobic treatment of dairy wastewaters. Brazilian J. Chem. Eng. **2002**, *19* (*2*), 175–180.

35. Smith, P.G.; Scott J.S. (Eds.) *Dictionary of Water and Waste Management*; IWA Publishing. Butterworth Heinemann: Oxford, UK, 2002.

36. Donkin, J.; Russell, J.M. Treatment of a milk powder/butter wastewater using the AAO activated sludge configuration. Water Sci. Tech. **1997**, *36*, 79–86.

37. Thirumurthi, D. Biodegradation of sanitary landfill leachate. In *Biological Degradation of Wastes*, A.M. Martin, Ed. Elsevier Appl. Sci.; London, UK, 1991; 208.

38. Herzka, A.; Booth, R.G. (Eds.) *Food Industry Wastes: Disposal and Recovery*; Applied Science Publishers Ltd: Essex, UK, 1981.

39. Maris, P.J.; Harrington, D.W.; Biol, A.I.; Chismon, G.L. Leachate treatment with particular reference to aerated lagoons. Water Poll. Cont. **1984**, *83*, 521–531.

40. Rusten, B.; Odegaard, H.; Lundar, A. Treatment of dairy wastewater in a novel moving-bed biofilm reactor. Water Sci. Tech. **1992**, *26* (*3/4*), 703–709.

41. Gough, R.H.; Samkutty, P.J.; McGrew, P.; Arauz, A.; Adkinson, A. Prediction of effluent biochemical oxygen demand in a dairy plant SBR waste water system. J. Environ. Sci. Health. **2000**, *A35*, 169–175.

42. Eroglu, V.; Ozturk, I.; Demir, I.; Akca, A. Sequencing batch and hybrid reactor treatment of dairy wastes. In *Proc 46th Purdue Ind. Wast. Conf.*, West Lafayette, IN, 1992; 413–420.

43. Samkutty, P.J.; Gough, R.H.; McGrew, P. Biological treatment of dairy plant wastewater. J. Environ. Sci. Health **1996**, *A31*, 2143–2153.

44. Li, X.; Zhang, R. Aerobic treatment of dairy wastewater with sequencing batch reactor systems. Bioproc. Biosys. Eng. **2002**, *25*, 103–109.

45. Tanaka, T. Use of aerated lagoons for dairy effluent treatment. *Sym. Dairy Effl. Treat*, Kollenbolle, Denmark, May 1973.

46. Bitton, G. (Ed.) *Wastewater Microbiology*; Wiley Press: New York, 1994.

47. Sterritt, R.M.; Lester, J.N. (Eds.) *Microbiology for Environmental and Public Health Engineers*; E & FN Spon., London, UK, 1988.

48. Thirumurthi, D. Biodegradation in waste stabilization ponds (facultative lagoons). In *Biological Degradation of Wastes*; Martin, A.M., Ed. Elsevier Applied Sci.; New York, 1991; 231–235.

49. Robinson, H.D.; Barr, M.J.; Formby, B.W.; Formby, B.W.; Moag, A. The treatment of landfill leachates using reed bed systems. *IWEM Annual Training Day*, October 1992.

50. European Water Pollution Control Association (EWPCA). *European Design and Operations Guidelines for Reed-Bed Treatment Systems*, Report to EC/EWPCA Treatment Group, P.F. Cooper, Ed.; August 1990.

51. IDF. Anaerobic treatment of dairy effluents – The present stage of development. Bull. Inter. Dairy Fed., Doc. **1990**, *252*.

52. Weber, H.; Kulbe, K.D.; Chmiel, H.; Trösch, W. Microbial acetate conversion to methane: kinetics, yields and pathways in a two-step digestion process. Appl. Microb. Biotech. **1984**, *19*, 224–228.

53. Lema, J.M.; Mendez, R.; and Blazquez, R. Characteristics of landfill leachates and alternatives for their treatment: a review. Water Air Soil Pollut. **1988**, *40*, 223–227.

54. Strydom, J.P.; Mostert, J.F.; Britz, T.J. Anaerobic treatment of a synthetic dairy effluent using a hybrid digester. Water SA **1995**, *21* (2), 125–130.

55. Britz, T.J.; Van Der Merwe, M.; Riedel, K.-H.J. Influence of phenol additions on the efficiency of an anaerobic hybrid digester treating landfill leachate. Biotech. Lett. **1992**, *14*, 323–327.

56. Feilden, N.E.H. The theory and practice of anaerobic reactor design. Proc. Biochem. **1983**, *18*, 34–37.

57. Sahm, H. Anaerobic wastewater treatment. Adv. Biochem. Eng. Biotech. **1984**, *29*, 83–115.

58. Ross, W.R. *Anaerobic Digestion of Industrial Effluents With Emphasis on Solids-Liquid Separation and Biomass Retention*, Ph.D. Thesis, University of the Orange Free State Press, South Africa, 1991.

59. Lebrato, J.; Perez-Rodriguez, J.L.; Maqueda, C.; Morillo, E. Cheese factory wastewater treatment by anaerobic semicontinuous digestion. Res. Cons. Recyc. **1990**, *3*, 193–199.

60. Schroepfer, G.J.; Fuller, W.J.; Johnson, A.S.; Ziemke, N.R.; Anderson, J.J. The anaerobic contact process as applied to packinghouse wastes. Sew. Ind. Was. **1955**, *27*, 460–486.

61. Speece, R.E. Advances in anaerobic biotechnology for industrial waste water treatment. *Proc. 2nd Int. Conf. Anaerobic Treat. Ind. Wast. Wat.*, Chicago, II, USA, 1986; 6–17.

62. Ross, W.R.; De Villiers, H.A.; Le Roux, J.; Barnard, J.P. Sludge separation techniques in the anaerobic digestion of wine distillery waste. *Proc. 5th Int. Symp. Anaerobic Digestion*, Bologna, Italy, May 1988, 571–574.

63. Ross, W.R. Anaerobic treatment of industrial effluents in South Africa. Water SA **1989**, *15*, 231–246.

64. Young, J.C.; McCarty, P.L. The anaerobic filter for waste treatment. J. Wat. Poll. Cont. Fed. **1969**, *41*, 160–173.

65. Bonastre, N.; Paris, J.M. Survey of laboratory, pilot and industrial anaerobic filter installations. Proc. Biochem. **1989**, *24*, 15–20.

66. Switzenbaum, M.S.; Jewell, W.J. Anaerobic attached-film expanded-bed reactor treatment. J. Wat. Poll. Cont. Fed. **1980**, *52*, 1953–1965.

67. Toldrá, F.; Flors, A.; Lequerica, J.L.; Vall S.S. Fluidized bed anaerobic biodegradation of food industry wastewaters. Biol. Wast. **1987**, *21*, 55–61.

68. Strydom, J.P.; Mostert, J.F.; Britz, T.J. Anaerobic digestion of dairy factory effluents. WRC Report No 455/1/01. ISBN 1868457249; Water Research Commission: Pretoria, South Africa, 2001.

69. Hemens, J.; Meiring, P.G.; Stander, G.J. Full-scale anaerobic digestion of effluents from the production of maize-starch. Wat. Wast. Treat. **1962**, *9* (*1*), 16–35.

70. Lettinga, G.; Van Velsen, A.F.M.; Hobma, S.W.; De Zeeuw, W.; Klapwijk, A. Use of the upflow sludge blanket (USB) reactor concept for biological wastewater treatment especially for anaerobic treatment. Biotech. Bioeng. **1980**, *22*, 699–734.

71. Lettinga, G.; Hulshoff-Pol, L.W. UASB-process design for various types of wastewaters. Water Sci. Tech. **1991**, *24*, 87–107.

72. Goodwin, J.A.S.; Wase, D.A.J.; Forster, C.F. Anaerobic digestion ice-cream wastewaters using the UASB process. Biol. Wast. **1990**, *32*, 125–144.

73. De Man, G.; De Bekker, P.H.A.M.J. New technology in dairy wastewater treatment. Dairy Ind. Int. **1986**, *51* (*5*), 21–25.

74. Carballo-Caabeira, J. Depuracion de augas residuales de centrales lecheras. Rev. Española de Lech. **1990**, *13* (*12*), 13–16.

75. Ikonen, M.; Latola, P.; Pankakoski, M.; Pelkonen, J. Anaerobic treatment of waste water in a Finnish dairy. Nord. Mejeriind. **1985**,*12* (*8*), 81–82.

76. Anon. South Caernarvon Creameries converts whey into energy. Dairy Ind. Int. **1984**, *49* (*10*), 16–17.

77. Cánovas-Diaz, M.; Howell, J.A. Down-flow anaerobic filter stability studies with different reactor working volumes. Proc. Biochem. **1987**, *22*, 181–184.

78. Cánovas-Diaz, M.; Howell, J.A. Stratified ecology techniques in the start-up of an anaerobic down-flow fixed film percolating reactor. Biotech. Bioeng. **1987**, *10*, 289–296.

79. De Haast, J.; Britz, T.J.; Novello, J.C. Effect of different neutralizing treatments on the efficiency of an anaerobic digester fed with deproteinated cheese whey. J. Dairy Res. **1986**, *53*, 467–476.

80. Li, A.Y.; Corrado J.J. Scale up of the membrane anaerobic reactor system. *Proc. 40th Annu. Purdue Ind. Wast. Conf.*, West Lafayette, IN, 1985; 399–404.

81. Ross, W.R.; Barnard, J.P.; De Villiers, H.A. The current status of ADUF technology in South Africa. In *Proc. 2nd Anaerobic Digestion Symp*, University of the Orange Free State Press: Bloemfontein, South Africa, 1989; 65–69.

82. Burgess, S. Anaerobic treatment of Irish creamery effluents. Proc. Biochem. **1985**, *20*, 6–7.

83. Lo, K.V.; Liao, P.H. Digestion of cheese whey with anaerobic rotating biological contact reactor. Biomass **1986**, *10*, 243–252.

84. Lo, K.V.; Liao, P.H. Laboratory scale studies on the mesophilic anaerobic digestion of cheese whey in different digester configurations. J. Agric. Eng. Res. **1988**, *39*, 99–105.

85. Kisaalita, W.S.; Pinder, K.L.; Lo, K.V. Acidogenic fermentation of lactose. Biotech. Bioeng. **1987**, *30*, 88–95.

86. Kissalita, W.S.; Lo, K.V.; Pinder, K.L. Kinetics of whey-lactose acidogenesis. Biotech. Bioeng. **1989**, *33*, 623–630.

87. Kisaalita, W.S.; Lo, K.V.; Pinder, K.L. Influence of whey protein on continuous acidogenic degradation of lactose. Biotech. Bioeng. **1990**, *36*, 642–646.

88. Venkataraman, J.; Kaul, S.N.; Satyanarayan, S. Determination of kinetic constants for a two-stage anaerobic up-flow packed bed reactor for dairy wastewater. Biores. Technol. **1992**, *40*, 253–261.

89. Cameron, K.C.; Di, H.J.; McLaren, R.G. Is soil an appropriate dumping ground for our wastes? Aust. J. Soil Res. **1997**, *35*, 995–1035.

90. Obi, M.E.; Ebo, P.O. The effects of organic and inorganic amendments on soil physical properties and maize production in a severely degraded sandy soil in Southern Nigeria. Biores. Technol. **1995**, *51*, 117–123.

91. Degens, B.P.; Schipper, L.A.; Claydon, J.J.; Russell, J.M.; Yeates, G.W. Irrigation of an allophonic soil with dairy factory effluent for 22 years: responses of nutrient storage and soil biota. Aust. J. Soil. Res. **2000**, *38*, 25–35.

92. IDF. Sludge from dairy effluent treatment plants – 1998 survey. International Dairy Federation Draft paper: IDF-group B 18/19, 1999.

93. Rusten, B.; Siljudalen, J.G.; Strand, H. Upgrading of a biological-chemical treatment plant for cheese factory wastewater. Wat. Sci. Tech. **1996**, *43* (*11*), 41–49.

94. Monroy H.O.; Vázquez M.F.; Derramadero, J.C.; Guyot, J.P. Anaerobic-aerobic treatment of cheese wastewater with national technology in Mexico: the case of "El Sauz". Wat. Sci. Tech. **1995**, *32* (*12*), 149–156.

95. Ardundel, J. (Ed.) *Sewage and Industrial Effluent Treatment*; Blackwell Science Ltd: Oxford, England, 1995.

96. Senior, E. Wealth from Waste. In *Proc. 1st Anaerobic Digestion Symp*; University of the Orange Free State Press: Bloemfontein, South Africa, 1986; pp. 19–30.

2

Seafood Processing Wastewater Treatment

Joo-Hwa Tay and Kuan-Yeow Show
Nanyang Technological University, Singapore

Yung-Tse Hung
Cleveland State University, Cleveland, Ohio, U.S.A.

2.1 INTRODUCTION

The seafood industry consists primarily of many small processing plants, with a number of larger plants located near industry and population centers. Numerous types of seafood are processed, such as mollusks (oysters, clams, scallops), crustaceans (crabs and lobsters), saltwater fishes, and freshwater fishes. As in most processing industries, seafood-processing operations produce wastewater containing substantial contaminants in soluble, colloidal, and particulate forms. The degree of the contamination depends on the particular operation; it may be small (e.g., washing operations), mild (e.g., fish filleting), or heavy (e.g., blood water drained from fish storage tanks).

Wastewater from seafood-processing operations can be very high in biochemical oxygen demand (BOD), fat, oil and grease (FOG), and nitrogen content. Literature data for seafood processing operations showed a BOD production of 1–72.5 kg of BOD per tonne of product [1]. White fish filleting processes typically produce 12.5–37.5 kg of BOD for every tonne of product. BOD is derived mainly from the butchering process and general cleaning, and nitrogen originates predominantly from blood in the wastewater stream [1].

It is difficult to generalize the magnitude of the problem created by these wastewater streams, as the impact depends on the strength of the effluent, the rate of discharge, and the assimilatory capacity of the receiving water body. Nevertheless, key pollution parameters must be taken into account when determining the characteristics of a wastewater and evaluating the efficiency of a wastewater treatment system. Section 2.2 discusses the parameters involved in the characterization of the seafood processing wastewater.

Pretreatment and primary treatment for seafood processing wastewater are presented in Section 2.3. These are the simplest operations to reduce contaminant load and remove oil and grease from an effluent of seafood processing wastewater. Common pretreatments for seafood-processing wastewater include screening, settling, equalization, and dissolved air flotation.

Section 2.4 focuses on biological treatments for seafood processing wastewater, namely aerobic and anaerobic treatments. The most common operations of biological treatments are also described in this section.

Section 2.5 discusses the physico-chemical treatments for seafood processing waste-water. These operations include coagulation, flocculation, and disinfection. Direct disposal of seafood processing wastewaters is discussed in Section 2.6. Potential problems in land application are highlighted. General seafood processing plant schemes are presented in Section 2.7. Economic considerations are always the most important factors that influence the final decision for selecting processes for wastewater treatment. The economic issues related to wastewater treatment process are discussed in Section 2.8.

2.2 SEAFOOD-PROCESSING WASTEWATER CHARACTERIZATION

Seafood-processing wastewater characteristics that raise concern include pollutant parameters, sources of process waste, and types of wastes. In general, the wastewater of seafood-processing wastewater can be characterized by its physicochemical parameters, organics, nitrogen, and phosphorus contents. Important pollutant parameters of the wastewater are five-day biochemical oxygen demand (BOD_5), chemical oxygen demand (COD), total suspended solids (TSS), fats, oil and grease (FOG), and water usage [2]. As in most industrial wastewaters, the contaminants present in seafood-processing wastewaters are an undefined mixture of substances, mostly organic in nature. It is useless or practically impossible to have a detailed analysis for each component present; therefore, an overall measurement of the degree of contamination is satisfactory.

2.2.1 Physicochemical Parameters

pH

pH serves as one of the important parameters because it may reveal contamination of a wastewater or indicate the need for pH adjustment for biological treatment of the wastewater. Effluent pH from seafood processing plants is usually close to neutral. For example, a study found that the average pH of effluents from blue crab processing industries was 7.63, with a standard deviation of 0.54; for non-Alaska bottom fish, it was about 6.89 with a standard deviation of 0.69 [2]. The pH levels generally reflect the decomposition of proteinaceous matter and emission of ammonia compounds.

Solids Content

Solids content in a wastewater can be divided into dissolved solids and suspended solids. However, suspended solids are the primary concern since they are objectionable on several grounds. Settleable solids may cause reduction of the wastewater duct capacity; when the solids settle in the receiving water body, they may affect the bottom-dwelling flora and the food chain. When they float, they may affect the aquatic life by reducing the amount of light that enters the water.

Soluble solids are generally not inspected even though they are significant in effluents with a low degree of contamination. They depend not only on the degree of contamination but also on the quality of the supply water used for the treatment. In one analysis of fish filleting wastewater, it was found that 65% of the total solids present in the effluent were already in the supply water [3].

Odor

In seafood-processing industries, odor is caused by the decomposition of the organic matter, which emits volatile amines, diamines, and sometimes ammonia. In wastewater that has become septic, the characteristic odor of hydrogen sulfide may also develop. Odor is a very important issue in relation to public perception and acceptance of any wastewater treatment plant. Although relatively harmless, it may affect general public life by inducing stress and sickness.

Temperature

To avoid affecting the quality of aquatic life, the temperature of the receiving water body must be controlled. The ambient temperature of the receiving water body must not be increased by more than 2 or 3°C, or else it may reduce the dissolved oxygen level. Except for wastewaters from cooking and sterilization processes in canning factories, fisheries do not discharge wastewaters above ambient temperatures. Therefore, wastewaters from canning operations should be cooled if the receiving water body is not large enough to restrict the change in temperature to 3°C [4].

2.2.2 Organic Content

The major types of wastes found in seafood-processing wastewaters are blood, offal products, viscera, fins, fish heads, shells, skins, and meat "fines." These wastes contribute significantly to the suspended solids concentration of the waste stream. However, most of the solids can be removed from the wastewater and collected for animal food applications. A summary of the raw wastewater characteristics for the canned and preserved seafood processing industry is presented in Table 2.1.

Wastewaters from the production of fish meal, solubles, and oil from herring, menhaden, and alewives can be divided into two categories: high-volume, low-strength wastes and low-volume, high-strength wastes [5].

High-volume, low-strength wastes consist of the water used for unloading, fluming, transporting, and handling the fish plus the washdown water. In one study, the fluming flow was estimated to be 834 L/tonne of fish with a suspended solids loading of 5000 mg/L. The solids consisted of blood, flesh, oil, and fat [2]. The above figures vary widely. Other estimates listed herring pump water flows of 16 L/sec with total solids concentrations of 30,000 mg/L and oil concentrations of 4000 mg/L. The boat's bilge water was estimated to be 1669 L/ton of fish with a suspended solids level of 10,000 mg/L [2].

Stickwaters comprise the strongest wastewater flows. The average BOD_5 value for stickwater has been listed as ranging from 56,000 to 112,000 mg/L, with average solids concentrations, mainly proteinaceous, ranging up to 6%. The fish-processing industry has found the recovery of fish solubles from stickwater to be at least marginally profitable. In most instances, stickwater is now evaporated to produce condensed fish solubles. Volumes have been estimated to be about 500 L/ton of fish processed [2].

The degree of pollution of a wastewater depends on several parameters. The most important factors are the types of operation being carried out and the type of seafood being processed. Carawan [2] reported on an EPA survey with BOD_5, COD, TSS, and fat, oil and grease (FOG) parameters. Bottom fish was found to have a BOD_5 of 200–1000 mg/L, COD of 400–2000 mg/L, TSS of 100–800 mg/L, and FOG of 40–300 mg/L. Fish meal plants were reported to have a BOD_5 of 100–24,000 mg/L, COD of 150–42,000 mg/L, TSS of 70–20,000 mg/L, and FOG of 20–5000 mg/L. The higher numbers were representative of bailwater only. Tuna plants were reported to have a BOD_5 of 700 mg/L, COD of 1600 mg/L,

Table 2.1 Raw Wastewater Characteristics of the Canned and Preserved Seafood-Processing Industries

Effluent	Flow (L/day)	BOD$_5$ (mg/L)	COD (mg/L)	TSS (mg/L)	FOG (mg/L)
Farm-raised catfish	79.5K–170K	340	700	400	200
Conventional blue crab	2650	4400	6300	420	220
Mechanized blue crab	75.7K–276K	600	1000	330	150
West coast shrimp	340K–606K	2000	3300	900	700
Southern nonbreaded shrimp	680K–908K	1000	2300	800	250
Breaded shrimp	568K–757K	720	1200	800	–
Tuna processing	246K–13.6M	700	1600	500	250
Fish meal	348K–378.5K[a]	100–24M[a]	150–42K[a]	70–20K[a]	20K–5K[a]
All salmon	220K–1892.5K	253–2600	300–5500	120–1400	20–550
Bottom and finfish (all)	22.71K–1514K	200–1000	400–2000	100–800	40–300
All herring	110K	1200–6000	3000–10,000	500–5000	600–5000
Hand shucked clams	325.5K–643.5K	800–2500	1000–4000	600–6000	16–50
Mechanical clams	1135.5K–11.4M	500–1200	700–1500	200–400	20–25
All oysters	53K–1211K	250–800	500–2000	200–2000	10–30
All scallops	3.785K–435K	200K–10M	300–11,000	27–4000	15–25
Abalone	37.85K–53K	430–580	800–1000	200–300	22–30

BOD$_5$, five day biochemical oxygen demand; COD, chemical oxygen demand; TSS, total suspended solids; FOG, fat, oil, and grease.
[a] Higher range is for bailwater only; K = 1000; M = 1,000,000.
Source: Ref. 2.

TSS of 500 mg/L, and FOG of 250 mg/L. Seafood-processing wastewater was noted to sometimes contain high concentrations of chlorides from processing water and brine solutions, and organic nitrogen of up to 300 mg/L from processing water.

Several methods are used to estimate the organic content of the wastewater. The two most common methods are biochemical oxygen demand (BOD) and chemical oxygen demand (COD).

Biochemical Oxygen Demand

Biochemical oxygen demand (BOD) estimates the degree of contamination by measuring the oxygen required for oxidation of organic matter by aerobic metabolism of the microbial flora. In seafood-processing wastewaters, this oxygen demand originates mainly from two sources. One is the carbonaceous compounds that are used as substrate by the aerobic microorganisms; the other source is the nitrogen-containing compounds that are normally present in seafood-processing wastewaters, such as proteins, peptides, and volatile amines. Standard BOD tests are conducted at 5-day incubation for determination of BOD$_5$ concentrations.

Wastewaters from seafood-processing operations can be very high in BOD_5. Literature data for seafood processing operations show a BOD_5 production of one to 72.5 kg of BOD_5 per ton of product [1]. White fish filleting processes typically produce 12.5–37.5 kg BOD_5 for every ton of product. The BOD is generated primarily from the butchering process and from general cleaning, while nitrogen originates predominantly from blood in the wastewater stream [1].

Chemical Oxygen Demand

Another alternative for measuring the organic content of wastewater is the chemical oxygen demand (COD), an important pollutant parameter for the seafood industry. This method is more convenient than BOD_5 since it needs only about 3 hours for determination compared with 5 days for BOD_5 determination. The COD analysis, by the dichromate method, is more commonly used to control and continuously monitor wastewater treatment systems. Because the number of compounds that can be chemically oxidized is greater than those that can be degraded biologically, the COD of an effluent is usually higher than the BOD_5. Hence, it is common practice to correlate BOD_5 vs. COD and then use the analysis of COD as a rapid means of estimating the BOD_5 of a wastewater.

Depending on the types of seafood processing, the COD of the wastewater can range from 150 to about 42,000 mg/L. One study examined a tuna-canning and byproduct rendering plant for five days and observed that the average daily COD ranged from 1300–3250 mg/L [2].

Total Organic Carbon

Another alternative for estimating the organic content is the total organic carbon (TOC) method, which is based on the combustion of organic matter to carbon dioxide and water in a TOC analyzer. After separation of water, the combustion gases are passed through an infrared analyzer and the response is recorded. The TOC analyzer is gaining acceptance in some specific applications as the test can be completed within a few minutes, provided that a correlation with the BOD_5 or COD contents has been established. An added advantage of the TOC test is that the analyzer can be mounted in the plant for online process control. Owing to the relatively high cost of the apparatus, this method is not widely used.

Fats, Oil, and Grease

Fats, oil, and grease (FOG) is another important parameter of seafood-processing wastewater. The presence of FOG in an effluent is mainly due to the processing operations such as canning, and the seafood being processed. The FOG should be removed from wastewater because it usually floats on the water's surface and affects the oxygen transfer to the water; it is also objectionable from an aesthetic point of view. The FOG may also cling to wastewater ducts and reduce their capacity in the long term. The FOG of a seafood-processing wastewater varies from zero to about 17,000 mg/L, depending on the seafood being processed and the operation being carried out.

2.2.3 Nitrogen and Phosphorus

Nitrogen and phosphorus are nutrients that are of environmental concern. They may cause proliferation of algae and affect the aquatic life in a water body if they are present in excess. However, their concentration in the seafood-processing wastewater is minimal in most cases. It is recommended that a ratio of N to P of 5 : 1 be achieved for proper growth of the biomass in the biological treatment [6,7].

Sometime the concentration of nitrogen may also be high in seafood-processing wastewaters. One study shows that high nitrogen levels are likely due to the high protein content (15–20% of wet weight) of fish and marine invertebrates [8]. Phosphorus also partly originates from the seafood, but can also be introduced with processing and cleaning agents.

2.2.4 Sampling

Of equal importance is the problem of obtaining a truly representative sample of the stream effluent. The samples may be required not only for the 24-hour effluent loads, but also to determine the peak load concentrations, the duration of peak loads, and the occurrence of variation throughout the day. The location of sampling is usually made at or near the point of discharge to the receiving water body, but in the analysis prior to the design of a wastewater treatment, facility samples will be needed from each operation in the seafood-processing facility. In addition, samples should be taken more frequently when there is a large variation in flow rate, although wide variations may also occur at constant flow rate.

The particular sampling procedure may vary, depending on the parameter being monitored. Samples should be analyzed as soon as possible after sampling because preservatives often interfere with the test. In seafood-processing wastewaters, there is no single method of sample preservation that yields satisfactory results for all cases, and all of them may be inadequate with effluents containing suspended matter. Because samples contain an amount of settleable solids in almost all cases, care should be taken in blending the samples just prior to analysis. A case in which the use of preservatives is not recommended is that of BOD_5 storage at low temperatures (4°C), which may be used with caution for very short periods, and chilled samples should be warmed to 20°C before analysis. For COD determination, the samples should be collected in clean glass bottles, and can be preserved by acidification to a pH of 2 with concentrated sulfuric acid. Similar preservation can also be done for organic nitrogen determination. For FOG determination, a separate sample should be collected in a wide-mouth glass bottle that is well rinsed to remove any trace of detergent. For solids determination, an inspection should be done to ensure that no suspended matter adheres to the walls and that the solids are refrigerated at 4°C to prevent decomposition of biological solids. For the analysis of phosphorus, samples should be preserved by adding 40 mg/L of mercuric chloride and stored in well-rinsed glass bottles at −10°C [4].

2.3 PRIMARY TREATMENT

In the treatment of seafood-processing wastewater, one should be cognizant of the important constituents in the waste stream. This wastewater contains considerable amounts of insoluble suspended matter, which can be removed from the waste stream by chemical and physical means. For optimum waste removal, primary treatment is recommended prior to a biological treatment process or land application. A major consideration in the design of a treatment system is that the solids should be removed as quickly as possible. It has been found that the longer the detention time between waste generation and solids removal, the greater the soluble BOD_5 and COD with corresponding reduction in byproduct recovery. For seafood-processing wastewater, the primary treatment processes are screening, sedimentation, flow equalization, and dissolved air flotation. These unit operations will generally remove up to 85% of the total suspended solids, and 65% of the BOD_5 and COD present in the wastewater.

2.3.1 Screening

The removal of relatively large solids (0.7 mm or larger) can be achieved by screening. This is one of the most popular treatment systems used by food-processing plants, because it can reduce the amount of solids being discharged quickly. Usually, the simplest configuration is that of flow-through static screens, which have openings of about 1 mm. Sometimes a scrapping mechanism may be required to minimize the clogging problem in this process.

Generally, tangential screening and rotary drum screening are the two types of screening methods used for seafood-processing wastewaters. Tangential screens are static but less prone to clogging due to their flow characteristics (Fig. 2.1), because the wastewater flow tends to avoid clogging. The solids removal rates may vary from 40 to 75% [4]. Rotary drum screens are mechanically more complex. They consist of a drum that rotates along its axis, and the effluent enters through an opening at one end. Screened wastewater flows outside the drum and the retained solids are washed out from the screen into a collector in the upper part of the drum by a spray of the wastewater.

Fish solids dissolve in water with time; therefore, immediate screening of the waste streams is highly recommended. Likewise, high-intensity agitation of waste streams should be minimized before screening or even settling, because they may cause breakdown of solids rendering them more difficult to separate. In small-scale fish-processing plants, screening is often used with simple settling tanks.

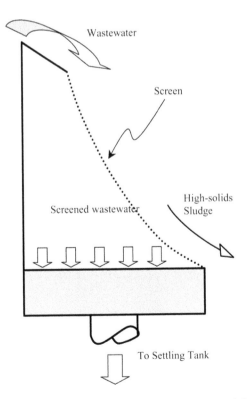

Figure 2.1 Diagram of an inclined or tangential screen.

2.3.2 Sedimentation

Sedimentation separates solids from water using gravity settling of the heavier solid particles [9]. In the simplest form of sedimentation, particles that are heavier than water settle to the bottom of a tank or basin. Sedimentation basins are used extensively in the wastewater treatment industry and are commonly found in many flow-through aquatic animal production facilities. This operation is conducted not only as part of the primary treatment, but also in the secondary treatment for separation of solids generated in biological treatments, such as activated sludge or trickling filters. Depending on the properties of solids present in the wastewater, sedimentation can proceed as discrete settling, flocculent settling, or zone settling. Each case has different characteristics, which will be outlined.

Discrete settling occurs when the wastewater is relatively dilute and the particles do not interact. A schematic diagram of discrete settling is shown in Figure 2.2.

Calculations can be made on the settling velocity of individual particles. In a sedimentation tank, settling occurs when the horizontal velocity of a particle entering the basin is less than the vertical velocity in the tank. The length of the sedimentation basin and the detention time can be calculated so that particles with a particular settling velocity (V_c) will settle to the bottom of the basin [9]. The relationship of the settling velocity to the detention time and basin depth is:

$$V_c = \frac{\text{depth}}{\text{detention time}} \qquad (2.1)$$

For flocculent suspension, the formation of larger particles due to coalescence depends on several factors, such as the nature of the particles and the rate of coalescence. A theoretical analysis is not feasible due to the interaction of particles, which depends, among other factors, on the overflow rate, the concentration of particles, and the depth of the tank.

Zone settling occurs when the particles do not settle independently. In this case, an effluent is initially uniform in solids concentration and settles in zones. The clarified effluent and compaction zones will increase in size while the other intermediate zones will eventually disappear.

The primary advantages of using sedimentation basins to remove suspended solids from effluents from seafood-processing plants are: the relative low cost of designing, constructing, and operating sedimentation basins; the low technology requirements for the operators; and the demonstrated effectiveness of their use in treating similar effluents. Therefore, proper design,

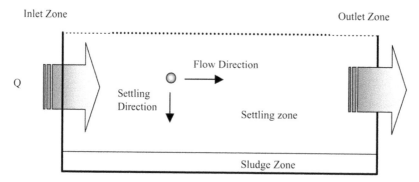

Figure 2.2 Schematics of discrete settling.

construction, and operation of the sedimentation basin are essential for the efficient removal of solids. Solids must be removed at proper intervals to ensure the designed removal efficiencies of the sedimentation basin.

Rectangular settling tanks (Fig. 2.3) are generally used when several tanks are required and there is space constraint, because they occupy less space than several circular tanks. Usually there is a series of chain-driven scrapers used for removal of solids. The sludge is collected in a hopper at the end of the tank, where it may be removed by screw conveyors or pumped out.

Circular tanks are reported to be more effective than rectangular ones. The effluent in a circular tank circulates radially, with the water introduced at the periphery or from the center. The configuration is shown in Figure 2.4. Solids are generally removed from near the center, and the sludge is forced to the outlet by two or four arms provided with scrapers, which span the radius of the tank. For both types of flows, a means of distributing the flow in all directions is provided. An even distribution of inlet and outlet flows is important to avoid short-circuiting in the tank, which would reduce the separation efficiency.

Generally, selection of a circular tank size is based on the surface-loading rate of the tank. It is defined as the average daily overflow divided by the surface area of the tank and is expressed as volume of wastewater per unit time and unit area of settler (m^3/m^2 day), as shown in Eq. (2.2). This loading rate depends on the characteristics of the effluent and the solids content. The retention time in the settlers is generally one to two hours, but the capacity of the tanks must be determined by taking into account the peak flow rates so that acceptable separation is obtained in these cases. Formation of scum is almost unavoidable in seafood-processing wastes, so some settling tanks are provided with a mechanism for scum removal.

Selection of the surface loading rate depends on the type of suspensions to be removed. The design overflow rates must be low enough to ensure satisfactory performance at peak rates of flow, which may vary from two to three times the average flow.

$$V_o = \frac{Q}{A} \tag{2.2}$$

where V_o = overflow rate (surface-loading rate) (m^3/m^2 day), Q = average daily flow (m^3/day), and A = total surface area of basin (m^2).

The area A is calculated by using inside tank dimensions, disregarding the central stilling well or inboard well troughs. The quantity of overflow from a primary clarifier Q is equal to the wastewater influent, and since the volume of the tank is established, the detention period in the tank is governed by water depth. The side water depth of the tank is

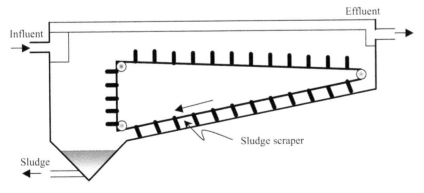

Figure 2.3 Diagram of a rectangular clarifier.

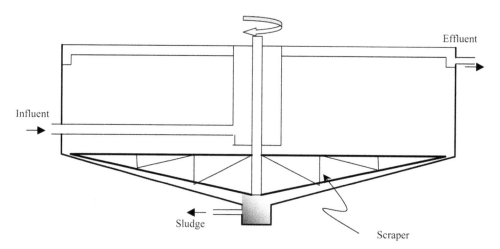

Figure 2.4 Diagram of radial flow sedimentation tank.

generally between 2.5 and 5 m. Detention time is computed by dividing the tank volume by influent flow uniform rate equivalent to the design average daily flow. A detention time of between 1.5 and 2.5 hours is normally provided based on the average rate of wastewater flow. Effluent weir loading is equal to the average daily quantity of overflow divided by the total weir length expressed in m^3/m day.

$$T = \frac{24V}{Q} \tag{2.3}$$

where T = detention time (hour), Q = average daily flow (m^3/day), and V = basin volume (m^3).

Temperature effects are normally not an important consideration in the design. However, in cold climates, the increase in water viscosity at lower temperatures retards particles settling and reduces clarifier performance.

In cases of small or elementary settling basins, the sludge can be removed using an arrangement of perforated piping placed at the bottom of the settling tank [10]. The pipes must be regularly spaced, as shown in Figure 2.5, to be of a diameter wide enough to be cleaned easily in case of clogging. The flow velocities should also be high enough to prevent sedimentation. Flow in individual pipes may be regulated by valves. This configuration is best used after screening and is also found in biological treatment tanks for sludge removal.

Inclined tube separators are an alternative to the above configurations for settling [11]. These separators consist of tilted tubes, which are usually inclined at 45–60°. When a settling particle reaches the wall of the tube or the lower plate, it coalesces with another particle and forms a larger mass, which causes a higher settling rate. A typical configuration for inclined media separators is shown in Figure 2.6.

2.3.3 Flow Equalization

A flow equalization step follows the screening and sedimentation processes and precedes the dissolved air flotation (DAF) unit. Flow equalization is important in reducing hydraulic loading in the waste stream. Equalization facilities consist of a holding tank and pumping equipment designed to reduce the fluctuations of the waste streams. The equalizing tank will store excessive

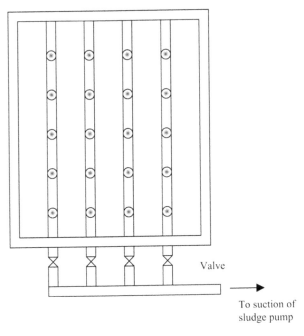

Figure 2.5   Pipe arrangement for sludge removal from settling tanks.

hydraulic flow surges and stabilize the flow rate to a uniform discharge rate over a 24-hour day. The tank is characterized by a varying flow into the tank and a constant flow out.

2.3.4   Separation of Oil and Grease

Seafood-processing wastewaters contain variable amounts of oil and grease, which depend on the process used, the species processed, and the operational procedure. Gravitational separation may be used to remove oil and grease, provided that the oil particles are large enough to float towards the surface and are not emulsified; otherwise, the emulsion must be first broken by pH adjustment. Heat may also be used for breaking the emulsion but it may not be economical unless there is excess steam available. The configurations of gravity separators of oil–water are similar to the inclined tubes separators discussed in the previous section.

2.3.5   Flotation

Flotation is one of the most effective removal systems for suspensions that contain oil and grease. The most common procedure is that of dissolved air flotation (DAF), which is a waste-treatment process in which oil, grease, and other suspended matter are removed from a waste stream. This treatment process has been in use for many years and has been most successful in removing oil from waste streams. Essentially, DAF is a process that uses minute air bubbles to remove the suspended matter from the wastewater stream. The air bubbles attach themselves to a discrete particle, thus effecting a reduction in the specific gravity of the aggregate particle to less than that of water. Reduction of the specific gravity for the aggregate particle causes separation from the carrying liquid in an upward direction. Attachment of the air bubble to the particle induces a vertical rate of rise. The mechanism of operation involves a clarification vessel where

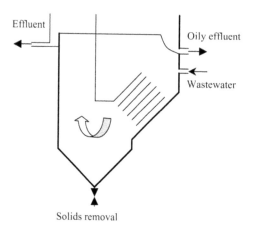

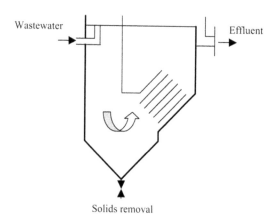

Figure 2.6 Typical configurations for inclined media separators.

the particles are floated to the surface and removed by a skimming device to a collection trough for removal from the system. The raw wastewater is brought in contact with a recycled, clarified effluent that has been pressurized through air injection in a pressure tank. The combined flow stream enters the clarification vessel and the release of pressure causes tiny air bubbles to form and ascend to the surface of the water, carrying the suspended particles with their vertical rise. A schematic diagram of the DAF system is shown in Figure 2.7.

Key factors in the successful operation of DAF units are the maintenance of proper pH (usually between 4.5 and 6, with 5 being most common to minimize protein solubility and break up emulsions), proper flow rates, and the continuous presence of trained operators.

In one case, oil removal was reported to be 90% [12]. In tuna processing wastewaters, the DAF removed 80% of oil and grease and 74.8% of suspended solids in one case, and a second case showed removal efficiencies of 64.3% for oil and grease and 48.2% of suspended solids. The main difference between these last two effluents was the usually lower solids content of the second [13]. However, although DAF systems are considered very effective, they are probably not suitable for small-scale, seafood-processing facilities due to the relatively high cost. It was reported that the estimated operating cost for a DAF system was about US$250,000 in 1977 [14].

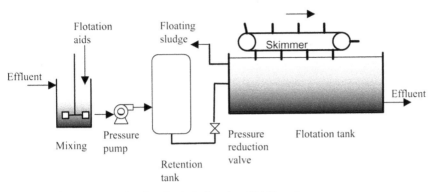

Figure 2.7   Diagram of a dissolved air flotation (DAF) system.

2.4 BIOLOGICAL TREATMENT

To complete the treatment of the seafood-processing wastewaters, the waste stream must be further processed by biological treatment. Biological treatment involves the use of microorganisms to remove dissolved nutrients from a discharge [15]. Organic and nitrogenous compounds in the discharge can serve as nutrients for rapid microbial growth under aerobic, anaerobic, or facultative conditions. The three conditions differ in the way they use oxygen. Aerobic microorganisms require oxygen for their metabolism, whereas anaerobic microorganisms grow in absence of oxygen; the facultative microorganism can proliferate either in absence or presence of oxygen although using different metabolic processes. Most of the microorganisms present in wastewater treatment systems use the organic content of the wastewater as an energy source to grow, and are thus classified as heterotrophes from a nutritional point of view. The population active in a biological wastewater treatment is mixed, complex, and interrelated. In a single aerobic system, members of the genera *Pseudomonas*, *Nocardia*, *Flavobacterium*, *Achromobacter*, and *Zooglea* may be present, together with filamentous organisms. In a well-functioning system, protozoas and rotifers are usually present and are useful in consuming dispersed bacteria or nonsettling particles.

Biological treatment systems can convert approximately one-third of the colloidal and dissolved organic matter into stable endproducts and convert the remaining two-thirds into microbial cells that can be removed through gravity separation. The organic load present is incorporated in part as biomass by the microbial populations, and almost all the rest is liberated gas. Carbon dioxide (CO_2) is produced in aerobic treatments, whereas anaerobic treatments produce both carbon dioxide and methane (CH_4). In seafood-processing wastewaters, the nonbiodegradable portion is very low.

The biological treatment processes used for wastewater treatment are broadly classified as aerobic and anaerobic treatments. Aerobic and facultative microorganisms predominate in aerobic treatments, while only anaerobic microorganisms are used for the anaerobic treatments.

If microorganisms are suspended in the wastewater during biological operation, this is known as a "suspended growth process," whereas the microorganisms that are attached to a surface over which they grow are said to undergo an "attached growth process."

Biological treatment systems are most effective when operating continuously 24 hours/ day and 365 days/year. Systems that are not operated continuously have reduced efficiency because of changes in nutrient loads to the microbial biomass. Biological treatment systems also generate a consolidated waste stream consisting of excess microbial biomass, which must be properly disposed. Operation and maintenance costs vary with the process used.

The principles and main characteristics of the most common processes used in seafood-processing wastewater treatment are explained in this section.

2.4.1   Aerobic Process

In seafood processing wastewaters, the need for adding nutrients (the most common being nitrogen and phosphorus) seldom occurs, but an adequate provision of oxygen is essential for successful operation. The most common aerobic processes are activated sludge systems, lagoons, trickling filters and rotating disc contactors. The reactions occurring during the aerobic process can be summarized as follows:

$$\text{Organic} + O_2 \longrightarrow \text{cells} + CO_2 + H_2O$$

Apart from economic considerations, several factors influence the choice of a particular aerobic treatment system. The major considerations are: the area availability; the ability to operate intermittently is critical for several seafood industries that do not operate in a continuous fashion or work only seasonally; the skill needed for operation of a particular treatment cannot be neglected; and finally the operating and capital costs are also sometimes decisive. Table 2.2 summarizes these factors when applied to aerobic treatment processes.

The considerations for rotating biological contactors (RBC) systems are similar to those of trickling filters.

Activated Sludge Systems

In an activated sludge treatment system, an acclimatized, mixed, biological growth of microorganisms (sludge) interacts with organic materials in the wastewater in the presence of excess dissolved oxygen and nutrients (nitrogen and phosphorus). The microorganisms convert the soluble organic compounds to carbon dioxide and cellular materials. Oxygen is obtained from applied air, which also maintains adequate mixing. The effluent is settled to separate

Table 2.2   Factors Affecting the Choice of Aerobic Processes

	(A) Operating characteristics		
System	Resistance to shock loads of organics or toxics	Sensitivity to intermittent operations	Degree of skill needed
Lagoons	Maximum	Minimum	Minimum
Trickling filters	Moderate	Moderate	Moderate
Activated	Minimum	Maximum	Maximum
	(B) Cost considerations		
System	Land needed	Initial costs	Operating costs
Lagoons	Maximum	Minimum	Minimum
Trickling filters	Moderate	Moderate	Moderate
Activated	Minimum	Maximum	Maximum

Source: Ref. 10.

biological solids and a portion of the sludge is recycled; the excess is wasted for further treatment such as dewatering. These systems originated in England in the early 1900s. The layout of a typical activated sludge system is shown in Figure 2.8.

Most of the activated sludge systems utilized in the seafood-processing industry are of the extended aeration types: that is, they combine long aeration times with low applied organic loadings. The detention times are 1 to 2 days. The suspended solids concentrations are maintained at moderate levels to facilitate treatment of the low-strength wastes, which usually have a BOD_5 of less than 800 mg/L.

It is usually necessary to provide primary treatment and flow equalization prior to the activated sludge process, to ensure optimum operation. A BOD_5 and suspended solids removals in the range of 95–98% can be achieved. However, pilot- or laboratory-scale studies are required to determine organic loadings, oxygen requirements, sludge yields, sludge settling rates, and so on, for these high-strength wastes.

In contrast to other food-processing wastewaters, seafood wastes appear to require higher oxygen availability to stabilize them. Whereas dairy, fruit, and vegetable wastes require approximately 1.3 kg of oxygen per kg of BOD_5, seafood wastes may demand as much as 3 kg of oxygen per kg of BOD_5 applied to the extended aeration system [2].

The most common types of activated sludge process are the conventional and the continuous flow stiffed tanks, as shown in Figure 2.8, in which the contents are fully mixed. In the conventional process, the wastewater is circulated along the aeration tank, with the flow being arranged by baffles in plug flow mode. This arrangement demands a maximum amount of oxygen and organic load concentration at the inlet. A typical conventional activated sludge process is shown in Figure 2.9. Unlike the conventional activated sludge process, the inflow streams in the completely mixed process are usually introduced at several points to facilitate the homogeneity of the mixing such that the properties are constant throughout the reactor if the mixing is completed. This configuration is inherently more stable in terms of perturbations because mixing causes dilution of the incoming stream into the tank. In seafood-processing wastewaters the perturbations that may appear are peaks of concentration of organic load or flow peaks. Flow peaks can be damped in the primary treatment tanks. The conventional configurations would require less reactor volume if smooth plug flow could be assured, which usually does not occur.

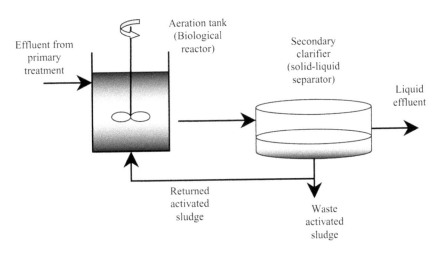

Figure 2.8   Diagram of a simple activated sludge system.

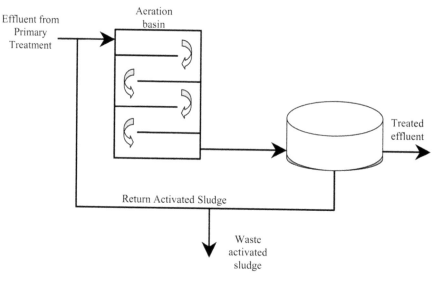

Figure 2.9 Diagram of a conventional activated sludge process.

In activated sludge systems, the cells are separated from the liquid and partially returned to the system; the relatively high concentration of cells then degrades the organic load in a relatively short time. Therefore, there are two different resident times that characterize the systems: one is the hydraulic residence time (θ_H) given by the ratio of reactor volume (V_R) to flow of wastewater (Q_R):

$$\theta_H = \frac{V_R}{Q_R} \tag{2.4}$$

The other is the cell residence time (θ_C), which is given by the ratio of cells present in the reactor to the mass of cells wasted per day. Typical θ_H values are in the order of 3–6 hours, while θ_C fluctuates between 3 and 15 days.

To ensure the optimum operation of the activated sludge process, it is generally necessary to provide primary treatment and flow equalization prior to the activated sludge process. Pilot- or laboratory-scale studies are required to determine organic loadings, oxygen requirements, sludge yields, and sludge settling rates for these high-strength wastes. There are several pieces of information required to design an activated sludge system through the bench-scale or pilot-scale studies:

- BOD_5 removal rate;
- oxygen requirements for the degradation of organic material and the degradation of dead cellular material (endogenous respiration);
- sludge yield, determined from the conservation of soluble organics to cellular material and the influx of inorganic solids in the raw waste;
- solid/liquid separation rate: the final clarifier would be designed to achieve rapid sedimentation of solids, which could be recycled or further treated. A maximum surface settling rate of 16.5 m^3/m^2 day has been suggested for seafood-processing wastes [2].

Typically, 85–95% of organic load removals can be achieved in activated sludge systems. Although used by some large seafood-processing industries that operate on a year-round basis, activated sludge may not be economically justified for small, seasonal seafood processors because of the requirement of a fairly constant supply of wastewater to maintain the microorganisms.

Aerated Lagoons

Aerated lagoons are used where sufficient land is not available for seasonal retention, or land application and economics do not justify an activated sludge system. Efficient biological treatment can be achieved by the use of the aerated lagoon system. It was reported to have removal efficiency of 90–95% of BOD_5 in seafood-processing wastewater treatment [2]. The major difference with respect to activated sludge systems is that the aerated lagoons are basins, normally excavated in earth and operated without solids recycling into the system. The ponds are between 2.4 and 4.6 m deep, with 2–10 days retention and achieve 55–90% reduction in BOD_5. Two types of aerated lagoons are commonly used in seafood-processing wastewater treatment: completely mixed lagoons and facultative lagoons. In the completely mixed lagoon, the concentrations of solids and dissolved oxygen are uniformly maintained and neither the incoming solids nor the biomass of microorganisms settle, whereas in the facultative lagoons, the power input is reduced, causing accumulation of solids in the bottom that undergo anaerobic decomposition, while the upper portions are maintained in an aerobic state (Fig. 2.10).

The major operational difference between these lagoons is the power input, which is in the order of 2.5–6 W/m^3 for aerobic lagoons, while the requirement for facultative lagoons is of the order 0.8–1 W/m^3. Reduction in biological activity can occur when the lagoons are exposed to low temperatures and eventually ice formation. This problem can be partially alleviated by increasing the depth of the basin.

If excavated basins are used for settling, care should be taken to provide a residence time long enough for the solids to settle, and provision should also be made for the accumulation of sludge. There is a very high possibility of offensive odor development due to the decomposition of the settled sludge, and algae might develop in the upper layers causing an increased content of suspended solids in the effluent. Odors can be minimized by using minimum depths of up to 2 m, whereas algae production can be reduced with a hydraulic retention time of fewer than 2 days.

Solids will also accumulate all along the aeration basins in the facultative lagoons and even at corners, or between aeration units in the completely mixed lagoon. These accumulated

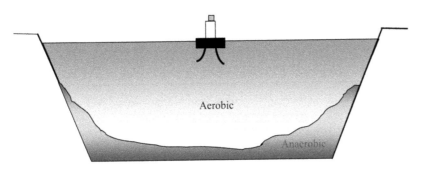

Figure 2.10   Diagram of facultative aerated lagoon.

solids will, on the whole, decompose at the bottom, but since there is always a nonbiodegradable fraction, a permanent deposit will build up. Therefore, periodic removal of these accumulated solids is necessary.

Stabilization/Polishing Ponds

A stabilization/polishing ponds system is commonly used to improve the effluent treated in the aerated lagoon. This system depends on the action of aerobic bacteria on the soluble organics contained in the waste stream. The organic carbon is converted to carbon dioxide and bacterial cells. Algal growth is stimulated by incident sunlight that penetrates to a depth of $1-1.5$ m. Photosynthesis produces excess oxygen, which is available for aerobic bacteria; additional oxygen is provided by mass transfer at the air–water interface.

Aerobic stabilization ponds are $0.18-0.9$ m deep to optimize algal activity and are usually saturated with dissolved oxygen throughout the depth during daylight hours. The ponds are designed to provide a detention time of $2-20$ days, with surface loadings of $5.5-22$ g BOD_5/day/m^2 [2]. To eliminate the possibility of shortcircuiting and to permit sedimentation of dead algal and bacterial cells, the ponds usually consist of multiple cell units operated in series. The ponds are constructed with inlet and outlet structures located in positions to minimize shortcircuiting due to wind-induced currents; the dimensions and geometry are designed to maximize mixing. These systems have been reported achieving $80-95\%$ removal of BOD_5 and approximately 80% removal of suspended solids, with most of the effluent solids discharged as algal cells [2].

During winter, the degree of treatment decreases markedly as the temperature decreases and ice cover eliminates algal growth. In regions where ice cover occurs, the lagoons may be equipped with variable depth overflow structures so that processing wastewater flows can be stored during the winter. An alternative method is to provide long retention storage ponds; the wastes can then be treated aerobically during the summer prior to discharge.

Aerobic stabilization ponds are utilized where land is readily available. In regions where soils are permeable, it is often necessary to use plastic, asphaltic, or clay liners to prevent contamination of adjacent groundwater.

Trickling Filters

The trickling filter is one of the most common attached cell (biofilm) processes. Unlike the activated sludge and aerated lagoons processes, which have biomass in suspension, most of the biomass in trickling filters are attached to certain support media over which they grow (Fig. 2.11).

Typical microorganisms present in trickling filters are *Zoogloea*, *Pseudomonas*, *Alcaligenes*, *Flavobacterium*, *Streptomyces*, *Nocardia*, fungi, and protozoa. The crux of the process is that the organic contents of the effluents are degraded by these attached growth populations, which absorb the organic contents from the surrounding water film. Oxygen from the air diffuses through this liquid film and enters the biomass. As the organic matter grows, the biomass layer thickens and some of its inner portions become deprived of oxygen or nutrients and separate from the support media, over which a new layer will start to grow. The separation of biomass occurs in relatively large flocs that settle relatively quickly in the supporting material. Media that can be used are rocks (low-rate filter) or plastic structures (high-rate filter). Denitrification can occur in low-rate filters, while nitrification occurs under high-rate filtration conditions; therefore, effluent recycle may be necessary in high-rate filters.

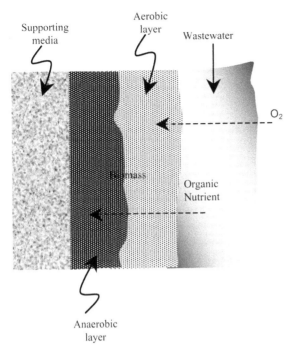

Figure 2.11 Cross-section of an attached growth biomass film.

In order to achieve optimum operation, several design criteria for trickling filters must be followed:

- roughing filters may be loaded at a rate of 4.8 kg $BOD_5/day/m^3$ filter media and achieve BOD_5 reductions of 40–50%;
- high-rate filters achieve BOD_5 reductions of 40–70% at organic loadings of 0.4–4.8 kg/BOD_5/day/m^3; and
- standard rate filters are loaded at 0.08–0.4 kg/BOD_5/day/m^3 and achieve BOD_5 removals greater than 70% [2].

The trickling filter consists of a circular tank filled with the packing media in depths varying from 1–2.5 m, or 10 m if synthetic packing is used. The bottom of the tank must be constructed rigidly enough to support the packing and designed to collect the treated wastewater, which is either sprayed by regularly spaced nozzles or by rotating distribution arms. The liquid percolates through the packing and the organic load is absorbed and degraded by the biomass while the liquid drains to the bottom to be collected.

With regard to the packing over which the biomass grows, the void fraction and the specific surface area are important features; the first is necessary to ensure a good circulation of air and the second is to accommodate as much biomass as possible to degrade the organic load of the wastewaters. Although more costly initially, synthetic packings have a larger void space, larger specific area, and are lighter than other packing media. Usually, the air circulates naturally, but forced ventilation is used with some high-strength wastewaters. The latter may be used with or without recirculation of the liquid after the settling tank. The need for recirculation is dictated by the strength of the wastewater and the rate of oxygen transfer to the biomass. Typically, recirculation is used when the BOD_5 of the seafood-processing wastewater to be treated exceeds 500 mg/L. The BOD_5 removal efficiency varies with the organic load imposed but usually fluctuates between 45 and 70% for a single-stage filter. Removal efficiencies of up to 90% can be achieved in two stages [4]. A typical unit of a trickling filter is shown in Figure 2.12.

Rotating Biological Contactors (RBC)

Increasingly stringent requirements for the removal of organic and inorganic substances from wastewater have necessitated the development of innovative, cost-effective wastewater treatment alternatives in recent years. The aerobic rotating biological contactor (RBC) is one of the biological processes for the treatment of organic wastewater. It is another type of attached growth process that combines advantages of biological fixed-film (short hydraulic retention time, high biomass concentration, low energy cost, easy operation, and insensitivity to toxic substance shock loads), and partial stir. Therefore, the aerobic RBC reactor is widely employed to treat both domestic and industrial wastewater [16–18]. A schematic diagram of the rotating biological contactor (RBC) unit is shown in Figure 2.13; it consists of closely spaced discs mounted on a common horizontal shaft, partially submerged in a semicircular tank receiving wastewater. When water containing organic waste and nutrients flows though the reactor, microorganisms consume the substrata and grow attached to the discs' surfaces to about 1–4 mm in thickness; excess is torn off the discs by shearing forces and is separated from the liquid in the secondary settling tank. A small portion of the biomass remains suspended in the liquid within the basin and is also responsible in minor part for the organic load removal.

Aeration of the culture is accomplished by two mechanisms. First, when a point on the discs rises above the liquid surface, a thin film of liquid remains attached to it and oxygen is transferred to the film as it passes through air; some amount of air is entrained by the bulk of liquid due to turbulence caused by rotation of discs. Rotation speeds of more than 3 rpm are seldom used because this increases electric power consumption while not sufficiently increasing oxygen transfer. The ratio of surface area of discs to liquid volume is typically 5 L/m^2. For high-strength wastewaters, more than one unit in series (staging) is used.

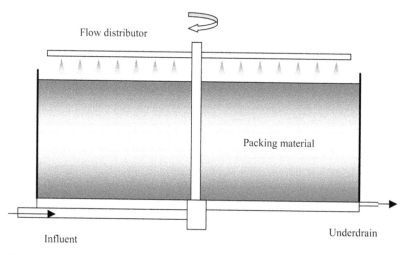

Figure 2.12   Sketch of a trickling filter unit.

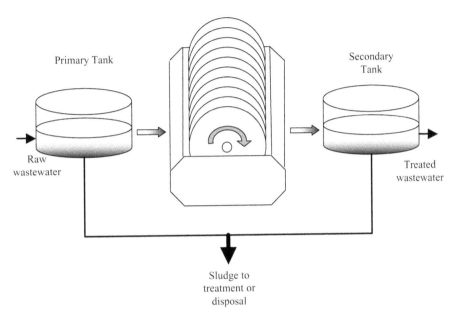

Figure 2.13 Diagram of a rotating biological contactor (RBC) unit.

2.4.2 Anaerobic Treatment

Anaerobic biological treatment has been applied to high BOD or COD waste solutions in a variety of ways. Treatment proceeds with degradation of the organic matter, in suspension or in a solution of continuous flow of gaseous products, mainly methane and carbon dioxide, which constitute most of the reaction products and biomass. Its efficient performance makes it a valuable mechanism for achieving compliance with regulations for contamination of recreational and seafood-producing wastes. Anaerobic treatment is the result of several reactions: the organic load present in the wastewater is first converted to soluble organic material, which in turn is consumed by acid-producing bacteria to produce volatile fatty acids, plus carbon dioxide and hydrogen. The methane-producing bacteria consume these products to produce methane and carbon dioxide. Typical microorganisms used in this methanogenic process are *Metanobacterium*, *Methanobacillus*, *Metanococcus*, and *Methanosarcina*. These processes are reported to be better applied to high-strength wastewaters, for example, blood water or stickwater. The scheme of reactions during anaerobic treatment is summarized in Figure 2.14.

Digestion Systems

Anaerobic digestion facilities have been used for the management of animal slurries for many years, they can treat most easily biodegradable waste products, including everything of organic or vegetable origin. Recent developments in anaerobic digestion technology have allowed the expansion of feedstocks to include municipal solid wastes, biosolids, and organic industrial waste (e.g., seafood-processing wastes). Lawn and garden, or "green" residues, may also be included, but care should be taken to avoid woody materials with high lignin content that requires a much longer decomposition time [19]. The digestion system seems to work best with a feedstock mixture of 15–25% solids. This may necessitate the addition of some liquid,

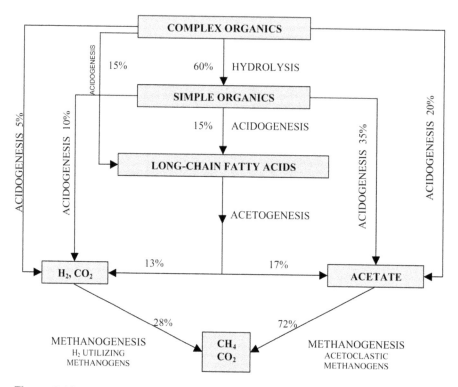

Figure 2.14 Scheme of reactions produced during anaerobic treatment.

providing an opportunity for the treatment of wastewater with high concentrations of organic contaminants. A typical anaerobic system diagram is shown in Figure 2.15.

The flow of anaerobic digestion resembles that of an activated sludge process except that it occurs in the absence of oxygen. Therefore, it is essential to have a good sealing of the digestion tanks since oxygen kills some of the anaerobic bacteria present and presence of air may easily disrupt the process. From the anaerobic digester the effluent proceeds to a degasifier and to a settler from which the wastewater is discharged and the solids are recycled. The need for recycling is attributed to the fact that anaerobic digestion proceeds at a much slower rate than aerobic processes, thereby requiring more time and more biomass to achieve high removal efficiencies. The amount of time required for anaerobic digestion depends upon its composition and the temperature maintained in the digester, because anaerobic processes are also sensitive to temperature. Mesophilic digestion occurs at approximately 35°C, and requires 12–30 days for processing. Thermophilic processes make use of higher temperatures (55°C) to speed up the reaction time to 6–14 days. Mixing the contents is not always necessary, but is generally preferred, as it leads to more efficient digestion by providing uniform conditions in the vessel and speeds up the biological reactions.

Anaerobic processes have been applied in seafood-processing wastewaters, obtaining high removal efficiencies (75–80%) with loads of 3 or 4 kg of COD/m^3 day [20,21].

In total, 60–70% of the gas produced by a balanced and well-functioning system consists of methane, with the rest being mostly carbon dioxide and minor amounts of nitrogen and hydrogen. This biogas is an ideal source of fuel, resulting in low-cost electricity and providing steam for use in the stirring and heating of digestion tanks.

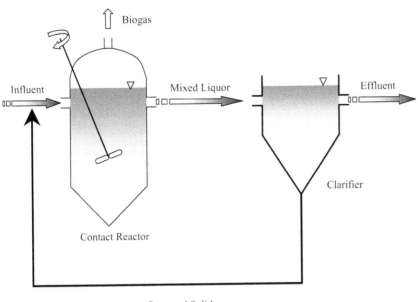

Figure 2.15 Diagram of an anaerobic digestion process.

Imhoff Tanks

The Imhoff tank is a relatively simple anaerobic system that was used to treat wastewater before heated digesters were developed. It is still used for plants of small capacity. The system consists of a two-chamber rectangular tank, usually built partially underground (Fig. 2.16).

Wastewater enters into the upper compartment, which acts as a settling basin while the settled solids are stabilized anaerobically at the lower part. Shortcircuiting of the wastewater can

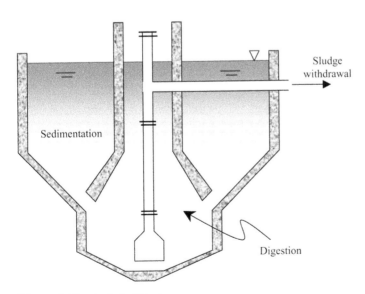

Figure 2.16 An Imhoff tank.

be prevented by using a baffle at the entrance with more than one port for discharge. The lower compartment is generally unheated. The stabilized sludge is removed from the bottom, generally twice a year, to provide ample time for the sludge to stabilize, although the removal frequency is sometimes dictated by the convenience of sludge disposal. In some cases, these tanks are designed with inlets and outlets at both ends, and the wastewater flow is reversed periodically so that the sludge at the bottom accumulates evenly. Although they are simple installations, Imhoff tanks are not without inconveniences; foaming, odor, and scum can form. These typically result when the temperature falls below 15°C and causes a process imbalance in which the bacteria that produce volatile acids predominate and methane production is reduced. This is why in some cases immersed heaters are used during cold weather. Scum forms because the gases that originate during anaerobic digestion are entrapped by the solids, causing the latter to float. This is usually overcome by increasing the depth in the lower chamber. At lower depths, bubbles form at a higher pressure, expand more when rising, and are more likely to escape from the solids. Odor problem is minimal when the two stages of the process of acid formation and gas formation are balanced.

2.5 PHYSICOCHEMICAL TREATMENTS

2.5.1 Coagulation/Flocculation

Coagulation or flocculation tanks are used to improve the treatability of wastewater and to remove grease and scum from wastewater [9]. In coagulation operations, a chemical substance is added to an organic colloidal suspension to destabilize it by reducing forces that keep them apart, that is, to reduce the surface charges responsible for particle repulsions. This reduction in charges is essential for flocculation, which has the purpose of clustering fine matter to facilitate its removal. Particles of larger size are then settled and clarified effluent is obtained. Figure 2.17 illustrates the coagulation/flocculation and settling of a seafood-processing wastewater.

In seafood processing wastewaters, the colloids present are of an organic nature and are stabilized by layers of ions that result in particles with the same surface charge, thereby increasing their mutual repulsion and stabilization of the colloidal suspension. This kind of wastewater may contain appreciable amounts of proteins and microorganisms, which become charged due to the ionization of carboxyl and amino groups or their constituent amino acids.

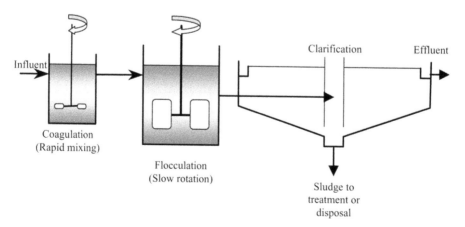

Figure 2.17 Chemical coagulation process.

The oil and grease particles, normally neutral in charge, become charged due to preferential absorption of anions, which are mainly hydroxyl ions.

Several steps are involved in the coagulation process. First, coagulant is added to the effluent, and mixing proceeds rapidly and with high intensity. The purpose is to obtain intimate mixing of the coagulant with the wastewater, thereby increasing the effectiveness of destabilization of particles and initiating coagulation. A second stage follows in which flocculation occurs for a period of up to 30 minutes. In the latter case, the suspension is stirred slowly to increase the possibility of contact between coagulating particles and to facilitate the development of large flocs. These flocs are then transferred to a clarification basin in which they settle and are removed from the bottom while the clarified effluent overflows.

Several substances may be used as coagulants. The pH of several wastewaters of the proteinaceous nature can be adjusted by adding acid or alkali. The addition of acid is more common, resulting in coagulation of the proteins by denaturing them, changing their structural conformation due to the change in their surface charge distribution. Thermal denaturation of proteins can also be used, but due to its high energy demand, it is only advisable if excess steam is available. In fact, the "cooking" of the blood–water in fishmeal plants is basically a thermal coagulation process.

Another commonly used coagulant is polyelectrolyte, which may be further categorized as cationic and anionic coagulants. Cationic polyelectrolytes act as a coagulant by lowering the charge of the wastewater particles, because wastewater particles are negatively charged. Anionic or neutral polyelectrolyte are used as bridges between the already formed particles that interact during the flocculation process, resulting in an increase of floc size.

Since the recovered sludges from coagulation/flocculation processes may sometimes be added to animal feeds, it is advisable to ensure that the coagulant or flocculant used is not toxic.

In seafood-processing wastewaters there are several reports on the use (at both pilot plant and working scale) of inorganic coagulants such as aluminum sulfate, ferric chloride, ferric sulfate, or organic coagulants [22–25].

On the other hand, fish scales are reported to be used effectively as an organic wastewater coagulant [26]. These are dried and ground before being added as coagulant in powder form. Another marine byproduct that can be used as coagulant is a natural polymer derived from chitin, a main constituent of the exoskeletons of crustaceans, which is also known as chitosan.

2.5.2 Electrocoagulation

Electrocoagulation (EC) has also been investigated as a possible means to reduce soluble BOD. It has been demonstrated to reduce organic levels in various food- and fish-processing waste streams [27]. During testing, an electric charge was passed through a spent solution in order to destabilize and coagulate contaminants for easy separation. Initial test results were quickly clarified with a small EC test cell – contaminants coagulated and floated to the top. Analytical test results showed some reduction in BOD_5, but not as much as originally anticipated when the pilot test was conducted. Additional testing was carried out on site on a series of grab samples; however, these runs did not appear to be as effective as originally anticipated. The pH was varied in an attempt to optimize the process, but BOD_5 reductions of only 21–33% were observed. Also, since metal electrodes (aluminum) were used in the process, the presence of metal in the spent solution and separated solids posed a concern for byproduct recovery. Initial capital outlays and anticipated operating costs were not unreasonable (US$140,000 and US$40,000, respectively), but satisfactory BOD_5 reductions could not be achieved easily. It was determined that long retention times would be needed in order to make EC work effectively.

2.5.3 Disinfection

Disinfection of seafood-processing wastewater is a process by which disease-causing organisms are destroyed or rendered inactive. Most disinfection systems work in one of the following four ways: (i) damage to the cell wall, (ii) alteration of cell permeability, (iii) alteration of the colloidal nature of protoplasm, or (iv) inhibition of enzyme activity [9,15].

Disinfection is often accomplished using bactericidal agents. The most common agents are chlorine, ozone (O_3), and ultraviolet (UV) radiation, which are discussed in the following sections.

Chlorination

Chlorination is a process commonly used in both industrial and domestic wastewaters for various reasons. In fisheries effluents, however, its primary purpose is to destroy bacteria or algae, or to inhibit their growth. Usually the effluents are chlorinated just before their final discharge to the receiving water bodies. For this process either chlorine gas or hypochlorite solutions may be used, the latter being easier to handle. In waste solutions, chlorine forms hypochlorous acid, which in turn forms hypochlorite.

$$Cl_2 + H_2O \longrightarrow HOCl + H^+ + Cl^-$$
$$HOCl \longrightarrow H^+ + OCl^-$$

A problem that may occur during chlorination of fisheries effluents is the formation of chloramines. These wastewaters may contain appreciable amounts of ammonia and volatile amines, which react with chlorine to give chloramines, resulting in an increased demand for chlorine to achieve a desired degree of disinfection. The proportions of these products depend on the pH and concentration of ammonia and the organic amines present. Chlorination also runs the risk of developing trihalomethanes, which are known carcinogens. Subsequently, the contact chamber must be cleaned regularly.

The degree of disinfection is attributed to the residual chlorine present in water. A typical plot of the breakpoint chlorination curve with detailed explanation is shown in Figure 2.18.

Initially, the presence of reducing agents reduce an amount of chlorine to chloride and makes the residual chlorine negligible (segment A–B). Further addition of chlorine may result in the formation of chloramines. These appear as residual chlorine but in the form of combined chlorine residual (segment B–C). Once all the ammonia and organic amines have reacted with the added chlorine, additional amounts of chlorine result in the destruction of the chloramines by oxidation, with a decrease in the chlorine residual as a consequence (segment C–D). Once this oxidation is completed, further addition of chlorine results in the appearance of free available chlorine. Point D on the curve is also known as "breakpoint chlorination." The goal in obtaining some free chlorine residual is to achieve disinfection purpose.

Chlorination units consist of a chlorination vessel in which the wastewater and the chlorine are brought into contact. In order to provide sufficient mixing, chlorine systems must have a chlorine contact time of 15–30 minutes, after which it must be dechlorinated prior to discharge. A schematic diagram of the systems is presented in Figure 2.19.

The channels in this contact basin are usually narrow in order to increase the water velocity and, hence, reduce accumulation of solids by settling. However, the space between the channels should allow for easy cleaning. The levels of available chlorine after the breakpoint should comply with the local regulations, which usually vary between 0.2 and 1 mg/L. This value strongly depends on the location of wastewater to be discharged, because residual chlorine in treated wastewater effluents was identified, in some cases, as the main toxicant suppressing

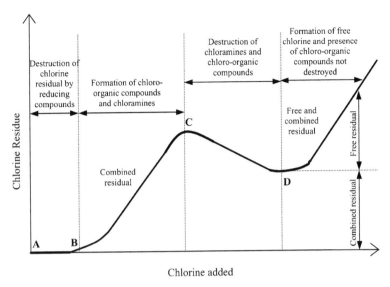

Figure 2.18 Breakpoint chlorinating curve (from Ref. 9).

the diversity, size, and quantity of fish in receiving streams [28]. Additionally, the chlorine dosage needed to achieve the residual effect required varies with the wastewater considered: 2–8 mg/L is common for an effluent from an activated sludge plant, and can be about 40 mg/L in the case of septic wastewater [6,7].

Ozonation

Ozone (O_3) is a strong oxidizing agent that has been used for disinfection due to its bactericidal properties and its potential for removal of viruses. It is produced by discharging air or oxygen across a narrow gap with application of a high voltage. An ozonation system is presented in Figure 2.20.

Ozonation has been used to treat a variety of wastewater streams and appears to be most effective when treating more dilute types of wastes [29]. It is a desirable application as a

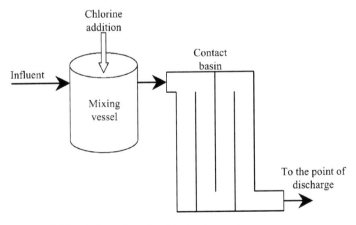

Figure 2.19 Schematics of a chlorination system.

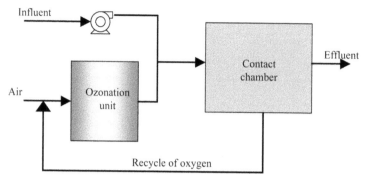

Figure 2.20 Simplified diagram of an ozonation system.

polishing step for some seafood-processing wastewaters, such as from squid-processing operations, which is fairly concentrated [30].

Ozone reverts to oxygen when it has been added and reacted, thus increasing somewhat the dissolved oxygen level of the effluent to be discharged, which is beneficial to the receiving water stream. Contact tanks are usually closed to recirculate the oxygen-enriched air to the ozonation unit. Advantages of ozonation over chlorination are that it does not produce dissolved solids and is affected neither by ammonia compounds present nor by the pH value of the effluent. On the other hand, ozonation has been used to oxidize ammonia and nitrites presented in fish culture facilities [31].

Ozonation also has limitations. Because ozone's volatility does not allow it to be transported, this system requires ozone to be generated onsite, which requires expensive equipment. Although much less used than chlorination in fisheries wastewaters, ozonation systems have been installed in particular in discharges to sensitive water bodies [4,32,33].

Ultraviolet (UV) Radiation

Disinfection can also be accomplished by using ultraviolet (UV) radiation as a disinfection agent. UV radiation disinfects by penetrating the cell wall of pathogens with UV light and completely destroying the cell and/or rendering it unable to reproduce.

However, a UV radiation system might have only limited value to seafood-processing wastewater without adequate TSS removal, because the effectiveness decreases when solids in the discharge block the light. This system also requires expensive equipment with high maintenance [34]. Nevertheless, UV radiation and other nontraditional disinfection processes are gaining acceptance due to stricter regulations on the amount of residual chlorine levels in discharged wastewaters.

2.6   LAND DISPOSAL OF WASTEWATER

Land application of wastewater is a low capital and operating cost method for treating seafood-processing wastes, provided that sufficient land with suitable characteristics is available. The ultimate disposal of wastewater applied to land is by one of the following methods:

- percolation to groundwater;
- overland runoff to surface streams;
- evaporation and evapo-transpiration to the atmosphere.

Generally, several methods are used for land application, including irrigation, surface ponding, groundwater recharge by injection wells, and subsurface percolation. Although each of these methods may be used in particular circumstances for specific seafood-processing waste streams, the irrigation method is most frequently used. Irrigation processes may be further divided into four subcategories according to the rates of application and ultimate disposal of liquid. These are overland flow, normal irrigation, high-rate irrigation, and infiltration — percolation.

Two types of land application techniques seem to be most efficient, namely infiltration and overland flow. As these land application techniques are used, the processor must be cognizant of potential harmful effects of the pollutants on the vegetation, soil, surface and groundwaters. On the other hand, in selecting a land application technique one must be aware of several factors such as wastewater quality, climate, soil, geography, topography, land availability, and return flow quality.

The treatability of seafood-processing wastewater by land application has been shown to be excellent for both infiltration and overland flow systems [2]. With respect to organic carbon removal, both systems have achieved pollutant removal efficiencies of approximately 98 and 84%, respectively. The advantage of higher efficiency obtained with the infiltration system is offset somewhat by the more expensive and complicated distribution system involved. Moreover, the overland flow system is less likely to pollute potable water supplies.

Nitrogen removal is found to be slightly more effective with infiltration land application when compared to overland flow application. However, the infiltration type of application has been shown to be quite effective for phosphorus and grease removal, and thus offers a definite advantage over the overland flow if phosphorus and grease removal are the prime factors. [One factor that may negate this advantage is that soil conditions are not favorable for phosphorus and grease removal and chemical treatment is required.]

Irrigation is a treatment process that consists of a number of segments:

- aerobic bacterial degradation of the deposited suspended materials and evaporation of water and concentration of soluble salts;
- filtration of small particles through the soil cover, and biological degradation of entrapped organics in the soil by aerobic and anaerobic bacteria;
- adsorption of organics on soil particles and uptake of nitrogen and phosphorus by plants and soil microorganisms;
- uptake of liquid wastes and transpiration by plants;
- percolation of water to groundwater.

The importance of these processes depends on the rate of application of waste, the characteristics of the waste, the characteristics of soil and substrata, and the type of cover crop grown on the land.

2.6.1 Loading Rates

Application rates should be determined by pilot plant testing for each particular location. The rate depends on whether irrigation techniques are to be used for roughing treatment or as an ultimate disposal method.

This method has both hydraulic and organic loading constraints for the ultimate disposal of effluent. If the maximum recommended hydraulic loading is exceeded, the surface runoff would increase. Should the specified organic loading be exceeded, anaerobic conditions could develop with resulting decrease in BOD_5 removal and the development of odor problem. The average applied loadings of organic suspended solids is approximately 8 g/m^2; however, loadings up to

22 g/m^2 have also been applied successfully [2]. A resting period between applications is important to ensure survival of the aerobic bacteria. The spray field is usually laid out in sections such that resting periods of 4–10 days can be achieved.

2.6.2   Potential Problems in Land Application with Seafood-Processing Wastewater

Two potential problems may be encountered with land application of seafood-processing wastewaters: the presence of disease-producing bacteria and unfavorable sodium absorption ratios of the soil. A key to minimizing the risk of spreading disease-producing bacteria can be accomplished by using low-pressure wastewater distribution systems to reduce the aerosol drift of the water spray. With respect to unfavorable sodium absorption ratios associated with the soil type, the seafood processor should be aware that clay-containing soils will cause the most serious sodium absorption problem. Sandy soils do not appear to be affected by unfavorable sodium absorption ratios and seem to be the best suited for accepting the high sodium chloride content found in most meat packing plant wastewaters.

As seafood-processing plant wastewaters are applied to land, certain types of grasses have been found to be compatible with these wastewaters. These are Bermuda NK-32, Kentucky-31 Tall Fescue, Jose Wheatgrass, and Blue Panicum [2]. In addition, it was reported that the southwestern coast of the United States, with its arid climate, mild winters, and vast available land areas, presents ideal conditions for land application treatment systems.

In some cases, the use of land application systems by today's seafood processors is feasible. However, in many cases, land disposal of seafood-processing wastes must be ruled out as a treatment alternative. Coastal topographic and soil characteristics, along with high costs of coastal property are the two major factors limiting the use of land application systems for treating seafood-processing wastes.

2.7   GENERAL SEAFOOD-PROCESSING PLANT SCHEMES

Seafood processing involves the capture and preparation of fish, shellfish, marine plants and animals, as well as byproducts such as fish meal and fish oil. The processes used in the seafood industry generally include harvesting, storing, receiving, eviscerating, precooking, picking or cleaning, preserving, and packaging [2]. Figure 2.21 shows a general process flow diagram for seafood processing. It is a summary of the major processes common to most seafood processing operations; however, the actual process will vary depending on the product and the species being processed.

There are several sources that produce wastewater, including:

- fish storage and transport;
- fish cleaning;
- fish freezing and thawing;
- preparation of brines;
- equipment sprays;
- offal transport;
- cooling water;
- steam generation;
- equipment and floor cleaning.

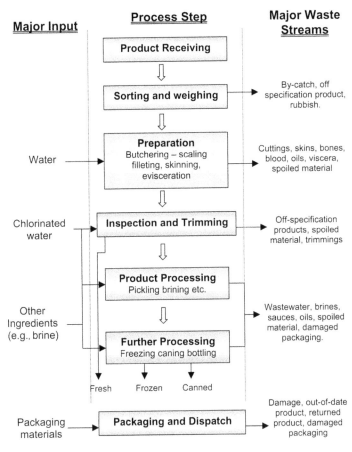

Figure 2.21 General process flow diagram for seafood processing operations.

Organic material in the wastewater is produced in the majority of these processes. However, most of it originates from the butchering process, which generally produces organic material such as blood and gut materials. The volume and quality of wastewater in each area is highly dependent on the products or species being processed and the production processes used.

Most seafood processors have a high baseline water use for cleaning plant and equipment. Therefore, water use per unit product decreases rapidly as production volume increases. Reducing wastewater volumes tends to have a significant impact on reducing organic loads as these strategies are typically associated with reduced product contact and better segregation of high-strength streams.

Water consumption in seafood-processing operations has traditionally been high to achieve effective sanitation. Industry literature indicates that water use varies widely throughout the sector, from 5–30 L/kg of product. Several factors affect water use, including the type of product processed, the scale of the operation, the process used, and the level of water minimization in place [1]. General cleaning contributes significantly to total water demand so smaller scale sites tend to have significantly higher water use per unit of production. Thawing operations can also account for up to 50% of the wastewater generated. A figure for water use of around 5–10 L/kg is typical of large operations with dedicated, automated, or semi-automated equipment that have implemented water minimization practises.

2.8 ECONOMIC CONSIDERATIONS OF SEAFOOD-PROCESSING WASTEWATER TREATMENT

Economic considerations are always the most important parameters that influence the final decision as to which process should be chosen for wastewater treatment. In order to estimate cost, data from the wastewater characterization should be available together with the design parameters for alternative processes and the associated costs. Costs related to these alternative processes and information on the quality of effluent should also be obtained prior to cost estimation in compliance with local regulations.

During the design phase of a wastewater treatment plant, different process alternatives and operating strategies could be evaluated by several methods. This cost evaluation can be achieved by calculating a cost index using commercially available software packages [36,37]. Nevertheless, actual cost indices are often restrictive, since only investment or specific operating costs are considered. Moreover, time-varying wastewater characteristics are not directly taken into account but rather through the application of large safety factors. Finally, the implementation of adequate control strategies such as a real-time control is rarely investigated despite the potential benefits [38,39]. In order to avoid these problems, a concept of MoSS-CC (Model-based Simulation System for Cost Calculation) was introduced by Gillot et al. [40], which is a modeling and simulation tool aimed at integrating the calculation of investment and fixed and variable operating costs of a wastewater treatment plant. This tool helps produce a holistic economic evaluation of a wastewater treatment plant over its life cycles.

2.8.1 Preliminary Costs of a Wastewater Treatment Plant

Several methods may be used to assess the preliminary costs of a wastewater treatment plant to facilitate a choice between different alternatives in the early phase of a process design. One method is cost functions [41–45]. Examples of different investment and operating cost functions are presented in Tables 2.3, 2.4, and 2.5. These cost functions were developed for the MoSS-CC modeling tool.

Another method was developed by EPA to estimate the construction costs for the most common unitary processes of wastewater treatment, as presented in Table 2.6. This was developed for municipal sewage treatment and may not be entirely applicable for small wastewater treatment plants, but it is useful for preliminary estimation and comparison among alternatives [4].

2.8.2 Cost of Operation and Maintenance

Several main factors influence the costs of operation and maintenance, including energy costs, labor costs, material costs, chemical costs, and cost of transportation of sludges for final disposal and discharge of treated wastewater. The relative importance of these items varies significantly depending on the location, the quality of the effluent discharged, and on the specific characteristics of the wastewater being treated [4].

The total operating cost of a wastewater treatment plant may be related to global plant parameters (e.g., average flow rate, population equivalent), generally through power laws [46–48]. However, such relationships apply to the average performance of plants and often suffer from a high uncertainty, unless very similar plant configurations are considered [40].

Table 2.3 Examples of Investment Cost Functions

Unit	Item	Cost function	Parameter	Parameter range	Reference	Cost unit
Influent pumping station	Concrete	$2334Q^{0.637}$	Q = flow rate (m³/hour)	250–4000	45	Euro of 1998
	Screws	$2123Q^{0.540}$				
	Screening	$3090Q^{0.349}$				
Any unit	Excavation	$2.9(\pi/4D^2H)$	D = diameter (m)	Not defined	44	Can\$ of 1995
	Compaction	$24.1 \times 0.4(\pi/4D^2)$	H = height (m)			
	Concrete base	$713.9 \times 0.5(\pi/4D^2)$				
	Concrete wall	$933.6 \times 0.5\pi DH$				
Oxidation ditch	Concrete	$10304V^{0.477}$	V = volume (m³)	1100–7700	45	Euro of 1998
	Electromech.[a]	$8590OC^{0.433}$	OC = oxygen capacity (kgO₂/hour)			
Settler	Concrete	$2630A^{0.678}$	A = area (m²)	175–1250	45	Euro of 1998
	Electromech.[a]	$6338A^{0.325}$				
	Concrete	$150(A/400)^{0.56}$	A	60–400	41, 42	Can\$*1000 of 1990
		$150(A/400)^{1.45}$		400–800		
	Electromech.[a]	$60(A/220)^{0.62}$		60–7000		
Sludge pump	Electromech.[a]	$9870IQ^{0.53}$	Q, I = Engin. Index[b]	Not defined	52	US\$ of 1971
	Electromech.[a]	$5038Q^{0.304}$	Q	35–2340	45	Euro of 1998

[a] Electromech. = electromechanical equipment; [b] Engineering News Record Index = index used to update costs in United States.
Source: Ref. 40

Table 2.4 Examples of Fixed Operating Cost Functions

| Cost item | Cost function | | | |
	Formula	Symbols	Units	Reference
Normal O&M	$L = U_c PE$	L = labor	man-hour/year	53
		U_c = unit cost	man-hour/year/PE	
		PE = population equivalent	—	
Clarifier mechanism	$P = \theta A^b$	P = power	kW	44
		θ, b = constant	—	
		A = area	m^2	
Mixers	$P = P_s V$	P = power	kW	53
		P_s = specific power	kW/m^3	
		V = volume	m^3	
Small equipment (supplies, spare parts…)	$C = U_c PE$	C = cost	Euro/year	5
		U_c = unit cost	Euro/year/PE	
		PE = population equivalent	—	
Analyses	$C = U_c PE$	C = cost	Euro/year	
		U_c = unit cost	Euro/year/PE	
		PE = population equivalent	—	

Source: Ref. 40

In terms of cost functions evaluations, some possible models in generic form for the fixed and variable operation costs are illustrated in Tables 2.4 and 2.5, respectively.

Capital Costs

These comprise mainly the unit construction costs, the land costs, the cost of the treatment units, and the cost of engineering, administration, and contingencies. The location should be carefully evaluated in each case because it affects the capital costs more than the operating costs [4]. When comparing different alternatives, special attention should be paid to the time and space scales chosen [38], since it may influence the choice of the implemented cost functions [49]. At best, an overall plant evaluation over the life span of the plant should be conducted [40].

Estimation of Total Costs

The total cost of a plant is normally determined by using the present worth method [50]. All annual operating costs for each process are converted into their corresponding present value and added to the investment cost of each process to yield the net present value (NPV). The net present value of a plant over a period of n years can be determined as:

$$\text{NPV} = \sum_{k=1}^{N} IC_k + \left[\frac{1 - (1 + i)^{-n}}{i} \right] \sum_{k=1}^{N} OC_k \tag{2.5}$$

Table 2.5 Example of Variable Operating Cost Functions

Cost item	Cost function		Units	Reference
	Formula	Symbols		
Pumping power	$P = Qwh/\eta$	Q = flow rate	m^3/s	54
		P = power	kW	
		w = specific liquid weight	N/m^3	
		h = dynamic head	m^3/s	
		η = pump efficiency	—	
Aeration power (fine bubble aeration)	$q_{air} = f(K_La_f)$ $P = f(q_{air})$	q_{air} = air flow rate	$Nm^3/hour$	53, 55
		P = power	kW	
		K_La_f = oxygen transfer coefficient in field conditions	1/hour	
Sludge thickening dewatering and disposal	$C = U_cTSS$	C = cost	Euro/year	5
		U_c = unit cost	Euro/t TSS	
		TSS = excess sludge	t	
Chemicals consumption	$C = U_cC_n$	C = cost	Euro/year	40
		U_c = unit cost	Euro/kg	
		C_n = consumption	kg	
Effluent taxes (organic matter and nutrient)	$L = U_c^*$ $(k_{org}{\cdot}N_{org} + k_{nut}{\cdot}N_{nut})$	U_c = unit cost	Euro/unit	38
		$N_{org} = f(Q, BOD, TSS, COD)$		
		$N_{nut} = f(Q, N, P)$		

Source: Ref. 40.

where IC_k represents the investment cost of a unit k, and OC_k the operating cost, i is the interest rate, and N is the number of units. The results could also be expressed as equivalent annual worth (AW):

$$AW = \frac{i(1+i)^n}{(1+i)^n - 1} \sum_{k=1}^{N} IC_k + \sum_{k=1}^{N} OC_k \tag{2.6}$$

For small wastewater treatments plants, an initial estimate of the total cost can be obtained from the cost of a similar plant with a different capacity, a relationship derived from costs relationships in chemical industries. The cost of plants of different sizes is related to the ratio of their capacity raised to the 0.6 power [4]:

$$\text{Capital}_2 = \text{Capital}_1 \times \left(\frac{\text{Capacity}_2}{\text{Capacity}_1}\right)^{0.6} \tag{2.7}$$

where $\text{Capital}_{1,2}$ = capital costs of plants 1 and 2, and $\text{Capacity}_{1,2}$ = capacity of plants 1 and 2.
The operation and maintenance costs can be estimated by a similar formula:

$$OM_2 = OM_1 \times \left(\frac{\text{Capacity}_2}{\text{Capacity}_1}\right)^{0.85} \tag{2.8}$$

where $OM_{1,2}$ = operation and maintenance costs of plants 1 and 2, $\text{Capacity}_{1,2}$ = capacity of plants 1 and 2.

Table 2.6 Construction Costs for Selected Unitary Operations of Wastewater
Treatment

Liquid stream	Correlation
Preliminary treatment	$C = 5.79 \times 10^4 \times Q^{1.17}$
Flow equalization	$C = 1.09 \times 10^5 \times Q^{0.49}$
Primary sedimentation	$C = 1.09 \times 10^5 \times Q^{1.04}$
Activated sludge	$C = 2.27 \times 10^5 \times Q^{0.17}$
Rotating biological contactor	$C = 3.19 \times 10^5 \times Q^{0.92}$
Chemical addition	$C = 2.36 \times 10^4 \times Q^{1.68}$
Stabilization pond	$C = 9.05 \times 10^5 \times Q^{1.27}$
Aerated lagoon	$C = 3.35 \times 10^5 \times Q^{1.13}$
Chlorination	$C = 5.27 \times 10^4 \times Q^{0.97}$
Solids stream	Correlation
Sludge handling	$C = 4.26 \times 10^4 \times Q^{1.36}$
Aerobic digestion	$C = 1.47 \times 10^5 \times Q^{1.14}$
Anaerobic digestion	$C = 1.12 \times 10^5 \times Q^{1.12}$
Incineration	$C = 8.77 \times 10^4 \times Q^{1.33}$

C represents the cost in USD and Q represents the flow rate of the wastewater to be treated.
Source: EPA, 1978.

An alternative procedure for developing cost models for wastewater treatment systems
includes the preparation of kinetic models for the possible treatment alternatives, in terms of area
and flow rates at various treatment efficiencies, followed by the computation of mechanical and
electrical equipment, as well as the operation and maintenance costs as a function of the flow
rates [51].

ACKNOWLEDGMENTS

The assistance provided by Mr. Lam Weh Yee is gratefully acknowledged.

REFERENCES

1. Environment Canada. *Canadian Biodiversity Strategy: Canadian response to the Convention on Biological Diversity*, Report of the Federal Provincial Territorial Biodiversity Working Group; Environmant Canada: Ottawa, 1994.
2. Carawan, R.E.; Chambers, J.V.; Zall, R.R. *Seafood Water and Wastewater Management*, 1979. The North Carolina, Agricultural Extension Service. U.S.A.
3. Gonzalez, J.F.; Civit, E.M.; Lupin, H.M. Composition of fish filleting wastewater. Environ. Technol. Lett. **1983**, *7*, 269–272.
4. Gonzalez, J.F. *Wastewater Treatment in the Fishery Industry*, FAO fisheries Technical Paper, 1996; 355.
5. Alexandre, O.; Grand d'Esnon, A. Le cout des services d'assinissement ruraux. Evaluation des couts d'investissement et d'exploitation. TSM, 7/8, 1998; 19–31. (In French.)
6. Metcalf and Eddy, Inc. *Wastewater Engineering: Treatment, Disposal, Reuse*. McGraw-Hill Book Co.: New York, 1979.
7. Eckenfelder, W.W. *Principles of Water Quality Management*; CBI Publishing Co.: Boston, 1980.

8. Sikorski, Z. *Seafood Resources: Nutritional Composition and Preservation*; CRC Press, Inc.: Boca Raton, FL, 1990.

9. Metcalf and Eddy, Inc. *Wastewater Engineering: Treatment and Disposal*, 3rd ed.; revised by Tchobanoglous, G., Burton, F.; McGraw-Hill, Inc.: New York, 1991.

10. Rich, L.G. *Low Maintenance, Mechanically Simple Wastewater Treatment Systems*; McGraw-Hill Book Co.: New York, 1980.

11. Hansen, S.P.; Culp, G.L. Applying shallow depth sedimentation theory. J. Am. Water Works Assoc. **1967**, *59*; 1134–1148.

12. Illet, K.J. Dissolved air flotation and hydrocyclones for wastewater treatment and by-product recovery in the food process industries. Water Services **1980**, *84*; 26–27.

13. Ertz, D.B.; Atwell, J.S.; Forsht, E.H. Dissolved air flotation treatment of seafood processing wastes – an assessment. In *Proceedings of the Eighth National Symposium on Food Processing Wastes*, EPA-600/Z-77-184, August 1977; p. 98.

14. Anon. *Environmental Assessment and Management of the Fish Processing Industry*, Sectoral studies series No. 28; UNIDO: Vienna, Austria, 1986.

15. Henry, J.G.; Heinke, G.W. *Environmental Science and Engineering*, 2nd Ed.; Prentice-Hall, Inc.: Upper Saddle River, NJ, 1996; 445–447.

16. Tokus, R.Y. Biodegradation and removal of phenols in rotating biological contactors. Water Sci. Technol. **1989**, *21*, 1751.

17. Gujer, W.; Boller, M. A mathematical model for rotating biological contactors. Water Sci. Technol. **1990**, *22*, 53–73.

18. Ahn, K.H.; Chang, J.S. Performance evaluation of compact RBC-settling tank system. Water Sci. Technol. **1991**, *23*, 1467–1476.

19. WRF (World Resource Foundation). Preserving Resources Through Integrated Sustainable Management of Waste; WRF, 1997.

20. Balslev-Olesen, P.; Lyngaard, A.; Neckelsen, C. Pilot-scale experiments on anaerobic treatment of wastewater from a fish processing plant. Water Sci. Technol. **1990**, *22*, 463–474.

21. Mendez, R.; Omil, F.; Soto, M.; Lema, J.M. Pilot plant studies on the anaerobic treatment of different wastewaters from a fish canning factory. Water Sci. Technol. **1992**, *25*, 37–44.

22. Johnson, R.A.; Gallager, S.M. Use of coagulants to treat seafood processing wastewaters. J. Water Pollut. Control Feder. **1984**, *56*, 970–976.

23. Nishide, E. Coagulation of fishery wastewater with inorganic coagulants. Bull. College of Agriculture and Veterinary Medicine: Nihon University, Japan, **1976**, *33*, 468–475.

24. Nishide, E. Coagulation of fishery wastewater with inorganic coagulants. Bull. College of Agriculture and Veterinary Medicine: Nihon University, Japan, **1977**, *34*, 291–294.

25. Ziminska, H. Protein recovery from fish wastewaters. In *Proceedings of the Fifth International Symposium on Agricultural Wastes*, American Society of Agriculture Engineering: St. Joseph, MI, 1985; 379.

26. Hood, L.F.; Zall, R.R. Recovery, utilization and treatment of seafood processing wastes. In *Advances in Fish Science and Technology*. Conell, J.J., Ed.; Fishing News Books, Ltd.: Surrey, England, 1980.

27. Beck, E.C.; Giannini, A.P.; Ramirez, E.R. Electrocoagulation clarifiers food wastewater. Food Technol. **1974**, *28* (2), 18–22.

28. Paller, M.H.; Lewis, W.M.; Heidinger, R.C.; Wawronowicz, J.L. Effects of ammonia and chlorine on fish in streams receiving secondary discharges. J. Water Pollut. Control Feder. **1983**, *55*, 1087–1097.

29. Ismond, A. End of pipe treatment options. Presented at *Wastewater Technology Conference and Exhibition*, Vancouver, BC, 1994.

30. Park, E.; Enander, R.; Barnett, S.M.; Lee, C. Pollution prevention and biochemical oxygen demand reduction in a squid processing facility. J. of Cleaner Production **2000**, 9 (*200*) 341–349.

31. Monroe, D.W.; Key, W.P. The feasibility of ozone for purification of hatchery waters. Ozone Sci. Engng. **1980**, *2*, 203–224.

32. Rosenthal, H.; Kruner, G. Efficiency of an improved ozonation unit applied to fish culture situations. Ozone, Sci. Eng. **1985**, *7*, 179–190.

33. Stover, E.L.; Jover, R.N. High level ozone disinfection of wastewater for shellfish discharges. Ozone Sci. Eng. **1980**, *1*, 335–346.

34. Whiteman, C.T.; Mehan, G.T.; Grubbs, G.H. et al. *Development Document for Proposed Effluent Limitations Guidelines and Standard for the Concentrated Aquatic Production Industry Point Source Category*, USEPA 2002; Chapter 7.

35. UNEP. 1998.

36. McGhee, T.J.; Mojgani, P.; Viicidomina, F. Use of EPA's CAPDET program for evaluation of wastewater treatment alternatives. J. Water Pollut. Control Fed. **1983**, *55* (*1*), 35–43.

37. Spearing, B.W. Sewage treatment optimization model – STOM – the sewage works in a personal computer. Proc. Instn. Civ. Engrs. Part 1 **1987**, *82*, 1145–1164.

38. Vanrolleghem, P.A.; Jeppsson, U.; Cartensen, J.; Carlsson, B.; Olsson, G. Integration of wastewater treatment plant design and operation — a systematic approach using cost functions. Water Sci. Technol. **1996**, 34 (*3–4*), 159–171.

39. Ekster, A. Automatic waste control. Water Environ. Technol. **1998**, *10* (*8*), 63–64.

40. Gillot, S.; Vermeire, P.; Grootaerd, H.; Derycke, D.; Simoens, F.; Vanrolleghem, P.A. *Integration of Wastewater Treatment Plant Investment and Operating Costs for Scenario Analysis Using Simulation*. In: Proceedings 13th Forum Applied Biotechnology. Med. Fac. Landbouww. Univ. Gent, Belgium, 64/5a, (1999), 13–20.

41. Wright, D.G.; Woods, D.R. Evaluation of capital cost data. Part 7: Liquid waste disposal with emphasis on physical treatment. Can. J. Chem. Eng. **1993**, *71*, 575–590.

42. Wright, D.G.; Woods, D.R. Evaluation of capital cost data. Part 8: Liquid waste disposal with emphasis on biological treatment. Can. J. Chem. Eng. **1993**, *72*, 342–351.

43. Agences de l'eau, Ministere de l'Environment. Approche technico-economique des couts d'investissement des stations d'epuration. Cahier Technique, 1995; 48 p. (In French.)

44. Fels, M.; Pinter, J.; Lycon, D.S. Optimized design of wastewater treatment systems: Application to the mechanical pulp and paper industry: I. Design and cost relationships. Can. J. Chem. Eng. **1997**, *75*, 437–451.

45. Vermeire, P. *Economishe optimalisatie van waterzuiveringsstations. Ontwikkeling van investeringskostenfunties voor Vlaanderen* (in Dutch). Engineers Thesis. Faculty of Agricultural and Applied Biological Sciences. Univ. Gent, Belgium, 1999, pp. 101.

46. Smeers, Y.; Tyteca, D. A geometric programming model for optimal design of wastewater treatment plants. Opn. Res. **1984**, *32* (2), 314–342.

47. Balmer, P.; Mattson, B. Wastewater treatment plant operation costs. Water Sci. Technol. **1994**, *30 (4)*, 7–15.

48. Water Environment Research Federation (WERF). Benchmarking wastewater operations — collection, treatment, and biosolids management – Final report. Project 96-CTS-5, 1997.

49. Rivas, A.; Ayesa, E. Optimum design of activated sludge plants using the simulator DAISY 2.0. In *Measurement and Modelling in Environmental Pollution*; San Jose, R., Brebbia, C.A., Ed.; Computational Mechanics Publications: Southampton, Boston, 1997.

50. White, J.A.; Agee, M.H.; Case, K.E. *Principles in Engineering Economic Analysis*; John Wiley & Sons, 1989.

51. Uluatam, S.S. Cost models for small wastewater treatment plants. Int. J. Environ. Studies **1991**, *37*, 171–181.

52. Tyteca, D. Mathematical models for cost effective biological wastewater treatment. In *Mathematical Models in Biological Wastewater Treatment*; Jorgensen and Gromiec: Amsterdam, 1985.

53. Jacquet, P. Een globale kostenfuctie voor tuning en evaluatie van op respirometrie gabaseerde controle algoritmen voor actiefsliprocessen. Engineers Thesis. Faculty of Agricultural and Applied Biological Sciences, University Gent: Belgium, 1999; 122. (In Dutch.)

54. ASCE. *ASCE Standard Measurement of Oxygen Transfer in Clean Water*; American Society of Civil Engineers, 1992.

55. Gillot, S.; De Clercq, B.; Defour, D.; Simoens, F.; Gernaey, K.; Vanrolleghem, P.A. Optimization of wastewater treatment plant design and operation using simulation and cost analysis. 72nd Annual Conference WEFTEC 1999, New Orleans, USA, 9–13 October, 1999.

56. Environmental Protection Agency (EPA). *Construction Costs for Municipal Wastewater Treatment Plants: 1973–1977*. Technical Report MCD-37, USEPA, Washington, D.C., USA, 1978

3
Treatment of Meat Wastes

Charles J. Banks and Zhengjian Wang
University of Southampton, Southampton, England

3.1 THE MEAT INDUSTRY

The meat industry is one of the largest producers of organic waste in the food processing sector and forms the interface between livestock production and a hygienically safe product for use in both human and animal food preparation. This chapter looks at this interface, drawing its boundaries at the point of delivery of livestock to the slaughterhouse and the point at which packaged meat is shipped to its point of use. The chapter deals with "meat" in accordance with the understanding of the term by the United States Environmental Protection Agency (USEPA) [1] as all animal products from cattle, calves, hogs, sheep and lambs, and from any meat that is not listed under the definition of poultry. USEPA uses the term "meat" as synonymous with the term "red meat." The definition also includes consumer products (e.g., cooked, seasoned, or smoked products, such as luncheon meat or hams). These specialty products, however, are outside the scope of the current text. The size of the meat industry worldwide, as defined above, can thus be judged by meat production (Table 3.1), which globally is around 140 million tons (143 million tonnes) for major species, with about one-third of production shared between the United States and the European Union. The single largest meat producer is China, which accounts for 36% of world production.

The first stages in meat processing occur in the slaughterhouse (abattoir) where a number of common operations take place, irrespective of the species. These include holding of animals for slaughter, stunning, killing, bleeding, hide or hair removal, evisceration, offal removal, carcass washing, trimming, and carcass dressing. Further secondary operations may also occur on the same premises and include cutting, deboning, grinding, and processing into consumer products.

There is no minimum or maximum size for a slaughterhouse, although the tendency in Europe is towards larger scale operations because EU regulations on the design and operation of abattoirs [2] have forced many smaller operators to cease work. In the United States there are approximately 1400 slaughterhouses employing 142,000 people, yet 3% of these provide 43% of the industry employment and 46% of the value of shipments [1]. In Europe slaughterhouses tend to process a mixed kill of animals; whereas in the United States larger operations specialize in processing one type of animal and, if a single facility does slaughter different types of meat animals, separate lines or even separate buildings are used [3].

Table 3.1 Meat Production Figures ($\times 1000$) and Percentage of Global Production by the United States and European Union (EU)

	Global tons/year (tonnes/year)	USA tons/year (tonnes/year)	%	EU tons/year (tonnes/year)	%
Beef[a]	49,427 (50,220)	12,138 (12,333)	24.6	7136 (7250)	14.4
Lamb[b]	6872 (6982)	111 (113)	1.6	1080 (1097)	15.7
Pork[a]	84,115 (85,465)	8831 (8973)	10.5	17,519 (17,800)	20.8
Total	140,414 (142,667)	21,081 (21,419)	15.0	25,734 (26,147)	18.3

Figures derived from a wide range of statistics provided by the U.S. Department of Agriculture Foreign Agricultural Service.
[a] Provisional figures for 2002.
[b] Figures for 1997.

3.2 PROCESSING FACILITIES AND WASTES GENERATED

As a direct result of its operation, a slaughterhouse generates waste comprised of the animal parts that have no perceived value to the slaughterhouse operator. It also generates wastewater as a result of washing carcasses, processing offal, and from cleaning equipment and the fabric of the building. The operations taking place within a slaughterhouse and the types of waste and products generated are summarized in Figure 3.1. Policies on the use of blood, gut contents, and meat and bone meal vary between different countries. Products that may be acceptable as a saleable product or for use in agriculture as a soil addition in one country may not be acceptable in another. Additionally, wastes and wastewaters are also generated from the stockyards, any rendering process, cooling facilities for refrigeration, compressors and pumps, vehicle wash facilities, wash rooms, canteen, and possibly laundry facilities.

3.2.1 Waste Characteristics and Quantities Generated

In general the characteristics of the solid wastes generated reflect the type of animal being killed, but the composition within a particular type of operation is similar regardless of the size of the plant. The reason for this is that the nature of the waste is determined by the animal itself and the quantity is simply a multiplication of the live weight of material processed. For example, the slaughter of a commercial steer would yield the products and byproducts shown in Table 3.2.

As can be seen the noncommercial sale material represents a little over 50% of the live weight of the animal, with about 25% requiring rendering or special disposal. The other 25% has a negative value and, because of its high water content, is not ideally suited to the rendering process. For this reason alternative treatment and disposal options have been sought for nonedible offal, gut fill, and blood, either separately or combined together, and in some cases combined with wastewater solids. The quantity of waste from sheep is again about 50% of the live weight, while pigs have only about 25% waste associated with slaughter.

Other solid waste requiring treatment or disposal arises mainly in the animal receiving and holding area, where regulations may demand that bedding is provided. In the European Union the volume of waste generated by farm animals kept indoors has been estimated by multiplying the number of animals by a coefficient depending on types of animals, function, sex, and age. Examples of coefficients that can be used for such calculations are given in Table 3.3 [5]. These

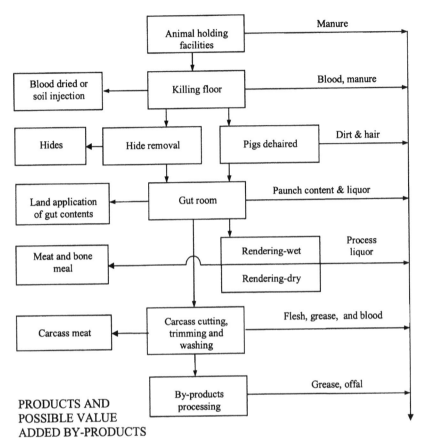

PRODUCTS AND
POSSIBLE VALUE
ADDED BY-PRODUCTS

Figure 3.1 Flow diagram indicating the products and sources of wastes from a slaughterhouse.

figures are for normal farm conditions and may vary for temporary holding accommodation depending on feeding and watering regimes.

For the purposes of waste treatment, volume is not as useful as knowing the pollution load. Denmead [6] estimated that 8.8 lb (4 kg) dry organic solids/cattle and 1.65 lb (0.75 kg) dry organic solids/sheep or lamb would be produced during an overnight stock of animals in the holding pens of a slaughterhouse.

Table 3.2 Raw Materials Segregated from a Commercial Steer (990 lb or 450 kg Live Weight)

Edible meat	Edible offals	Hide	High-grade fat	Bone and meat trim	Nonedible offal and gut fill	Blood	BSE suspect material
350 lb	35 lb	70 lb	100 lb	110 lb	245 lb	35 lb	45 lb
160 kg	15 kg	32 kg	45 kg	50 kg	112 kg	16 kg	20 kg
	Commercial sale		Byproducts for rendering		Waste		Special disposal

Source: Ref. 4.

Table 3.3 Waste Generated for Cattle and Pigs of Different
Ages and Sexes (*Source:* Ref. 5)

Animal category	Quantity (L/day)
Cattle	
Less than 1 year	11.4
Between 1 and 2 years	20
More than 2 years	40
Pigs	
Less than 44 lb (20 kg)	2.1
Fattening pigs more than 44 lb (20 kg)	4.3
Breeding pigs	8.6
Covered sows	14.3

Once on the slaughter line, the quantity of waste generated depends on the number of animals slaughtered and the type of animal. Considering the total annual tonnage of animals going to slaughter there is surprisingly little information in the scientific literature on the quantities of individual waste fractions destined for disposal. The average weight of wet solid material produced by cutting and emptying of the stomachs of ruminants was estimated by Fernando [7] as 60 lb (27 kg) for cattle, 6 lb (2.7 kg) for sheep and 3.7 lb (1.7 kg) for lambs. Pollack [8] gave a much higher estimate for the stomach contents of cattle at 154 lb (70 kg) per head, and 2.2 lb (1 kg) per animal for pigs. There is a more consistent estimate of the quantity of blood produced: Brolls and Broughton [9] reported average weight of wet blood produced is around 32 lb per 1000 lb of beef animal (14.5 kg per 454 kg); Grady and Lim [10] likewise reported 32.5 lb of blood produced per 1000 lb (14.7 kg per 453 kg) of live weight; and Banks [4] indicated 35 lb of blood produced per 990 lb (16 kg per 450 kg) of live weight.

Wastewater Flow

Water is used in the slaughterhouse for carcass washing after hide removal from cattle, calves, and sheep and after hair removal from hogs. It is also used to clean the inside of the carcass after evisceration, and for cleaning and sanitizing equipment and facilities both during and after the killing operation. Associated facilities such as stockyards, animal pens, the steam plant, refrigeration equipment, compressed air, boiler rooms, and vacuum equipment will also produce some wastewater, as will sanitary and service facilities for staff employed on site: these may include toilets, shower rooms, cafeteria kitchens, and laboratory facilities. The proportions of water used for each purpose can be variable, but as a useful guide the typical percentages of water used in a slaughterhouse killing hogs is shown in Figure 3.2 [11].

Johnson [12] classified meat plant wastewater into four major categories, defined as manure-laden; manure-free, high grease; manure-free, low grease; and clear water (Table 3.4).

The quantity of wastewater will depend very much on the slaughterhouse design, operational practise, and the cleaning methods employed. Wastewater generation rates are usually expressed as a volume per unit of product or per animal slaughtered and there is a reasonable degree of consistency between some of the values reported from reliable sources for different animal types (Table 3.5). These values relate to slaughterhouses in the United States

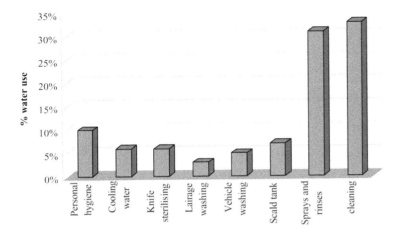

Figure 3.2 Percentage water use between different operations in a typical slaughterhouse killing hogs (from Ref. 11).

and Europe, but the magnitude of variation across the world is probably better reflected in the values given by the World Bank [13], which quotes figures between 2.5 and 40 m^3/ton or tonne for cattle and 1.5–10 m^3/ton or tonne for hogs.

The rate of water use and wastewater generation varies with both the time of day and the day of the week. To comply with federal requirements for complete cleaning and sanitation of equipment after each processing shift [1], typical practice in the United States is that a daily processing shift, usually lasting 8–10 hours, is followed by a 6–8 hours cleanup shift. Although the timing of the processing and cleanup stages may vary, the pattern is consistent across most

Table 3.4 Examples of Wastewater Types and Arisings from Slaughtering and Processing

Wastewater category	Examples
Manure-laden	Holding pens, gut room washwaters, scald tanks, dehairing and hair washing, hide preparation, bleed area cleanup, laundry, casing preparation, catch basins
Manure-free, high grease water	Drainage and washwater from slaughter floor area (except bleeding and dehairing), carcass washers, rendering operations
Manure-free, low grease water (slaughterhouse)	Washwater from nonproduction areas, finished product chill showers, coolers and freezers, edible and inedible grease, settling and storage tank area, casing stripper water (catch basin effluent), chitterling washwater (catch basin effluent), tripe washers, tripe and tongue scalders
Manure-free, low grease water (cutting rooms, processing and packing)	Washwater from nonproduction areas, green meat boning areas, finished product packaging, sausage manufacture, can filling area, loaf cook water, spice preparation area
Clear water	Storm water, roof drains, cooling water (from compressors, vacuum pumps, air conditioning) steam condenser water (if cooling tower is not used or condensate not returned to boiler feed), ice manufacture, canned product chill water

Source: Ref. 12.

Table 3.5   Wastewater Generation Rate from Meat Processing

Meat type	Slaughterhouse	Packinghouse	Reference
Cattle	• 312–601 gal/10^3 lb LWK (2604–5015 L/tonne)		14
	• 395 gal/animal (1495 L/animal)	• 2189 gal/animal (8286 L/animal)	15
	• 345–390 gal/10^3 lb LWK (2879–3255 L/tonne)	• 835 gal/10^3 lb LWK (6968 L/tonne)	1
	• 185–264 gal/animal (700–1000 L/animal)		11
	• 256 gal/10^3 lb LWK (2136 L/tonne)		16
	• 185–265 gal/animal (700–1003 L/animal)		17
	• 300–4794 gal/10^3 lb (2500–40,000 L/ tonne)	• 240–7190 gal/10^3 lb (2000–60,000 L/ tonne)	13
Hog	• 243–613 gal/10^3 lb LWK (2028–5115 L/tonne)	• 1143 gal/10^3 lb LWK (9539 L/tonne)	1
	• 155 gal/10^3 lb LWK (1294 L/tonne)	• 435–455 gal/10^3 lb LWK (3630–3797 L/tonne)	18
	• 143 gal/animal (541 L/animal)	• 552 gal/animal (1976 L/animal)	15
	• 60–100 gal/animal (227–379 L/animal)		17
	• 42–61 gal/animal 160–230 L/animal)		11
	• 269 gal/10^3 lb LWK (2245 L/tonne)		19
	• 180–1198 gal/10^3 lb (1500–10,000 L/ tonne)		13
Sheep	• 26–40 gal/animal (100–150 L/animal)		11
Mixed	• 359 gal/animal (1359 L/animal)	• 996 gal/animal (3770 L/animal)	15
		• 38–80 gal/animal (144–189 L/animal)	18
		• 1500 gal/10^3 lb LWK (12,518 L/animal)	12
	• 606–6717 L/10^3 lb LWK (1336–14,808 L/tonne)		20
	• 152–1810 gal/animal (575–6852 L/animal)		21
	• 599–1798 gal/10^3 lb (5000–15,000 L/tonne)		9

LWK, live weight kill.

slaughterhouses worldwide; hence the nature of the wastewater and its temperature will show a marked differentiation between the two stages. During the processing stage water use and wastewater generation are relatively constant and at a low temperature compared to the cleanup period. Water use and wastewater generation essentially cease after the cleanup period until processing begins next day.

Wastewater Characteristics

Effluents from slaughterhouses and packing houses are usually heavily loaded with solids, floatable matter (fat), blood, manure, and a variety of organic compounds originating from proteins. As already stated the composition of effluents depends very much on the type of production and facilities. The main sources of water contamination are from lairage, slaughtering, hide or hair removal, paunch handling, carcass washing, rendering, trimming, and cleanup operations. These contain a variety of readily biodegradable organic compounds, primarily fats and proteins, present in both particulate and dissolved forms. The wastewater has a high strength, in terms of biochemical oxygen demand (BOD), chemical oxygen demand (COD), suspended solids (SS), nitrogen and phosphorus, compared to domestic wastewaters. The actual concentration will depend on in-plant control of water use, byproducts recovery, waste separation source and plant management. In general, blood and intestinal contents arising from the killing floor and the gut room, together with manure from stockyard and holding pens, are separated, as best as possible, from the aqueous stream and treated as solid wastes. This can never be 100% successful, however, and these components are the major contributors to the organic load in the wastewater, together with solubilized fat and meat trimmings.

The aqueous pollution load of a slaughterhouse can be expressed in a number of ways. Within the literature reports can be found giving the concentration in wastewater of parameters such as BOD, COD, and SS. These, however, are only useful if the corresponding wastewater flow rates are also given. Even then it is often difficult to relate these to a meaningful figure for general design, as the unit of productivity is often omitted or unclear. These reports do, however, give some indication as to the strength of wastewaters typically encountered, and some of their particular characteristics, which can be useful in making a preliminary assessment of the type of treatment process most applicable. Some of the reported values for typical wastewater characterization parameters are listed along with the source reference in Table 3.6. These values could be averaged, but the value of such an exercise would be limited as the variability between the wastewaters, for the reasons previously mentioned, is considerable. At best it can be concluded that slaughterhouse wastewaters have a pH around neutral, an intermediate strength in terms of COD and BOD, are heavily loaded with solids, and are nutrient-rich.

It is, therefore, clear that for the purposes of design of a treatment facility a much better method of assessing the pollution load is required. For this purpose the typical pollution load resulting from the slaughter of a particular animal could be used, but as animals vary in weight depending upon their age and condition at the time of slaughter, it is better to use the live weight at slaughter as the unit of productivity rather than just animal numbers. Some typical pollution loads per unit of productivity are given in Table 3.7 along with the source references for different types of slaughtering operations.

Very little information is available on where this pollution load arises within the slaughterhouse, as waste audits on individual process streams are not commonly reported. Nemerow and Agardy [15] describe the content of individual process wastes from a slaughterhouse (Table 3.8). It can be seen that the two most contaminated process streams are related to blood and paunch contents. Blood and meat proteins are the most significant sources of nitrogen in the wastewater and rapidly give rise to ammonical nitrogen as breakdown occurs.

The wastewater contains a high density of total coliform, fecal coliform, and fecal streptococcus groups of bacteria due to the presence of manure material and gut contents. Numbers are usually in the range of several million colony forming units (CFU) per 100 mL. It is also likely that the wastewater will contain bacterial pathogens of enteric origin such as *Salmonella* sp. and *Campylobacter jejuni*, gastrointestinal parasites including *Ascaris* sp., *Giardia lamblia*, and *Cryptosporidium parvum*, and enteric viruses [1]. It is, therefore, essential

Table 3.6 Reported Chemical Compositions of Meat Processing Wastewater

Item	Hog	Cattle	Mixed	Reference
		Type of meat		
pH			7.1–7.4	12
			6.5–8.4	9
			7.0	22
			6.3–10.5	23
		6.7–9.3		24
			6.5–7.2	25
	7.3			26
		6.0–7.5		27
			6.7	28
			7.3–8.0	29
COD (mg/L)			960–8290	9
			1200–3000	30
			583	22
		3000–12,873		24
	3015			26
		2100–3190		27
			5100	28
			12,160–18,768	29
BOD (mg/L)	2220	7237		1
			900–2500	12
			600–2720	9
	1030–1045	448–996	635–2240	15
			700–1800	30
			404	22
			950–3490	23
		900–4620		24
			944–2992	25
	1950			26
		975–3330		27
			3100	28
			8833–11,244	29
Suspended solids (SS) (mg/L)	3677	3574		1
			900–3200	12
			300–4200	15
	633–717	467–820	457–929	30
			200–1000	22
			1375	23
			381–3869	24
		865–6090		26
	283		310	28
			10,588–18,768	29
Nitrogen (mg/L)	253	378		1
			22–510	9
	122	154	113–324	15

(*continues*)

Table 3.6 Continued

Item	Type of meat			Reference
	Hog	Cattle	Mixed	
			70–300	30
			152	22
			89–493	23
		93–148		24
			235–309	25
	14.3			26
			405	28
			448–773	29
Phosphorus	154	79		1
(mg/L)		26		24
	5.2			26
			30	28

that slaughterhouse design ensures the complete segregation of process washwater and strict hygiene procedures to prevent cross-contamination. The mineral chemistry of the wastewater is influenced by the chemical composition of the slaughterhouse's treated water supply, waste additions such as blood and manure, which can contribute to the heavy metal load in the form of copper, iron, manganese, arsenic, and zinc, and process plant and pipework, which can contribute to the load of copper, chromium, molybdenum, nickel, titanium, and vanadium.

3.3 WASTEWATER MINIMIZATION

As indicated previously, the overall waste load arising from a slaughterhouse is determined principally by the type and number of animals slaughtered. The partitioning of this load between the solid and aqueous phases will depend very much upon the operational practices adopted, however, and there are measures that can be taken to minimize wastewater generation and the aqueous pollution load.

Minimization can start in the holding pens by reducing the time that the animals remain in these areas through scheduling of delivery times. The incorporation of slatted concrete floors laid to falls of 1 in 60 with drainage to a slurry tank below the floor in the design of the holding pens can also reduce the amount of washdown water required. Alternatively, it is good practice to remove manure and lairage from the holding pens or stockyard in solid form before washing down. In the slaughterhouse itself, cleaning and carcass washing typically account for over 80% of total water use and effluent volumes in the first processing stages. One of the major contributors to organic load is blood, which has a COD of about 400,000 mg/L, and washing down of dispersed blood can be a major cause of high effluent strength. Minimization can be achieved by having efficient blood collection troughs allowing collection from the carcass over several minutes. Likewise the trough should be designed to allow separate drainage to a collection tank of the blood and the first flush of washwater. Only residual blood should enter a second drain for collection of the main portion of the washwater. An efficient blood recovery

Table 3.7 Pollutant Generation per Unit of Production for Meat Processing Wastewater

| Parameter | Type of meat | | | Reference |
	Hog	Cattle	Mixed	
BOD	16.7 lb/10^3 lb or kg/tonne LWK	38.4 lb/10^3 lb or kg/tonne LWK		1
	6.5–9.0 lb/10^3 lb or kg/tonne		1.9–27.6 lb/10^3 lb or kg/tonne	12
			1.1–1.2 lb/hog-unit 2.4–2.6 Kg/hog-unit	18
			8.6–18.0 lb/10^3 lb or kg/tonne	31
Suspended solids	13.3 lb/10^3 lb or kg/tonne	11.1 lb/10^3 lb or kg/tonne		1
			1.2–53.8 lb/10^3 lb or kg/tonne	12
			5.5–15.1 lb/10^3 lb or kg/tonne	31
Total volatile solids (VS)			3.1–56.4 lb/10^3 lb or kg/tonne	12
Grease			0.2–10.2 lb/10^3 lb or kg/tonne	31
Hexane extractables	3.7 lb/10^3 lb or kg/tonne	6.2 lb/10^3 lb or kg/tonne		1
Total Kjeldahl nitrogen	1.3 lb/10^3 lb or kg/tonne	1.2 lb/10^3 lb or kg/tonne		1
Total phosphorus	0.8 lb/10^3 lb or kg/tonne	0.2 lb/10^3 lb or kg/tonne		1
Fecal coliform bacterial	6.2×10^{10} CFU/10^3 lb	2.9×10^{10} CFU/10^3 lb		1
	13.6×10^{10} CFU/tonne	6.4×10^{10} CFU/tonne		

LWK, live weight kill; CFU, colony forming unit.

Table 3.8 Typical Wastewater Properties for a Mixed Kill Slaughterhouse

Source	SS (mg/L)	Organic-N (mg/L)	BOD (mg/L)	pH
Killing floor	220	134	825	6.6
Blood and tank water	3690	5400	32,000	9.0
Scald tank	8360	1290	4600	9.0
Meat cutting	610	33	520	7.4
Gut washer	15,120	643	13,200	6.0
Byproducts	1380	186	2200	6.7

Original data from US Public Health Service and subsequently reported in Refs. 15 and 33.
SS, suspended solids; BOD, biochemical oxygen demand.

system could reduce the aqueous pollution load by as much as 40% compared to a plant of similar size that allows the blood to flow to waste [18].

The second area where high organic loads into the wastewater system can arise is in the gut room. Most cattle and sheep abattoirs clean the paunch (rumen), manyplies (omasum), and reed (abomasum) for tripe production. A common method of preparation is to flush out the gut manure from the punctured organs over a mechanical screen, and allow water to transport the gut manure to the effluent treatment system.

Typically the gut manure has a COD of over 100,000 mg/L, of which 80% dissolves in the washwater. Significant reductions in wastewater strength can be made by adopting a "dry" system for removing and transporting these gut manures. The paunch manure in its undiluted state has enough water present to allow pneumatic transport to a "dry" storage area where a compactor can be used to reduce the volume further if required. The tripe material requires washing before further processing, but with a much reduced volume of water and resulting pollution load.

The small and large intestines are usually squeezed and washed for use in casings. To reduce water, washing can be carried out in two stages: a primary wash in a water bath with continuous water filtration and recirculation, followed by a final rinse in clean potable water. Other measures that can be taken in the gut room to minimize water use and organic loadings to the aqueous stream include ensuring that mechanical equipment, such as the hasher machine, are in good order and maintained regularly.

Within the slaughtering area and cutting rooms, measures should be adopted to minimize meat scraps and fatty tissue entering the floor drains. Once in the drains these break down due to turbulence, pumping, or other mechanical actions (e.g., on screens), leading to an increase in effluent COD. These measures include using fine mesh covers to drains, encouraging operators to use collection receptacles for trimmings, and using well-designed equipment with catch trays. Importantly, a "dry" cleaning of the area to remove solid material, for example using cyclonic vacuum cleaners, should take place before any washdown.

Other methods can also be employed to minimize water usage. These will not in themselves reduce the organic load entering the wastewater treatment system, but will reduce the volume requiring treatment, and possibly influence the choice of treatment system to be employed. For example, high-strength, low-volume wastewaters may be more suited to anaerobic rather than aerobic biological treatment methods. Water use minimization methods include:

- the use of directional spray nozzles in carcass washing, which can reduce water consumption by as much as 20%;
- use of steam condensation systems in place of scald tanks for hair and nail removal;
- fitting washdown hoses with trigger grips;
- appropriate choice of cleaning agents;
- reuse of clear water (e.g., chiller water) for the primary washdown of holding pens.

3.4 WASTEWATER TREATMENT PROCESSES

The degree of wastewater treatment required will depend on the proposed type of discharge. Wastewaters received into the sewer system are likely to need less treatment than those having direct discharge into a watercourse. In the European Union, direct discharges have to comply with the Urban Waste Water Treatment Directive [32] and other water quality directives. In the United States the EPA is proposing effluent limitations guidelines and standards (ELGs) for the

Meat and Poultry Products industries with direct discharge [1]. These proposed ELGs will apply to existing and new meat and poultry products (MPP) facilities and are based on the well-tested concepts of "best practicable control technology currently available" (BPT), the "best conventional pollutant control technology" (BCT), the "best available technology economically achievable" (BAT), and the "best available demonstrated control technology for new source performance standards" (NSPS). In summary, the technologies proposed to meet these requirements use, in the main, a system based on a treatment series comprising flow equalization, dissolved air flotation, and secondary biological treatment for all slaughterhouses; and require nitrification for small installations and additional denitrification for complex slaughterhouses. These regulations will apply to around 6% of an estimated 6770 MPP facilities.

There is some potential, however, for segregation of wastewaters allowing specific individual pretreatments to be undertaken or, in some cases, bypass of less contaminated streams. Depending on local conditions and regulations, water from boiler houses and refrigerating systems may be segregated and discharged directly or used for outside cleaning operations.

3.4.1 Primary and Secondary Treatment

Primary Treatment

Grease removal is a common first stage in slaughterhouse wastewater treatment, with grease traps in some situations being an integral part of the drainage system from the processing areas. Where the option is taken to have a single point of removal, this can be accomplished in one of two ways: by using a baffled tank, or by dissolved air flotation (DAF). A typical grease trap has a minimum detention period of about 30 minutes, but the period need not to be greater than 1 hour [33]. Within the tank, coagulation of fats is brought about by cooling, followed by separation of solid material in baffled chambers through natural flotation of the less dense material, which is then removed by skimming.

In the DAF process, part of the treated water is recycled from a point downstream of the DAF. The recycled flow is retained in a pressure vessel for a few minutes for mixing and air saturation to take place. The recycle stream is then added to the DAF unit where it mixes with the incoming untreated water. As the pressure drops, the air comes out of solution, forming fine bubbles. The fine bubbles attach to globules of fat and oil, causing them to rise to the surface where they collect as a surface layer.

The flotation process is dependent upon the release of sufficient air from the pressurized fluid when the pressure is reduced to atmospheric. The nature of the release is also important, in that the bubbles must be of reasonably constant dimensions (not greater than 130 microns), and in sufficient numbers to provide blanket coverage of the retaining vessel. In practice, the bubble size and uniform coverage give the appearance of white water. The efficiency of the process depends upon bubble size, the concentration of fats and grease to be separated, their specific gravity, the quantity of the pressurized gas, and the geometry of the reaction vessel.

Figure 3.3 shows a schematic diagram of a typical DAF unit. The DAF unit can also be used to remove solids after screening, and in this case it usually incorporates chemical dosing to bring about coagulation and flocculation of the solids. When used for this purpose, the DAF unit will remove the need for a separate sedimentation tank.

Dissolved air flotation has become a well-established unit operation in the treatment of abattoir wastes, primarily as it is effective at removing fats from the aqueous stream within a short retention time (20–30 minutes), thus preventing the development of acidity [18]. Since the 1970s, DAF has been widely used for treating abattoir and meat-processing wastes. Some early

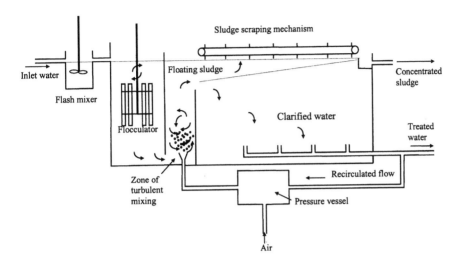

Figure 3.3 Schematic diagram of typical DAF unit.

texts mention the possibility of fat and protein recovery using DAF separation [9,34]. Johns [14] reported, however, that such systems had considerable operating problems, including long retention times and low surface overflow rates, which led to solids settling, large volumes of putrefactive and bulky sludge with difficult dewatering properties, and sensitivity to flow variations.

DAF units are still extensively used within the industry, but primarily now as a treatment option rather than for product recovery. The effectiveness of these units depends on a number of factors and on their position within the series of operations. The efficiency of the process for fat removal can be reduced if the temperature of the water is too hot (>100°F or 38°C); the increase in fat recovery from reducing the wastewater temperature from 104 to 86°F (40 to 30°C) is estimated to be up to 50% [35]. Temperature reduction can be achieved by wastewater segregation or by holding the wastewater stream in a buffer or flow equalization tank. Operated efficiently in this manner the DAF unit can remove 15–30% COD/BOD, 30–60% SS, and 60–90% of the oil and grease without chemical addition. Annual operating costs for DAF treatment remain high, however, indicating that the situation has not altered significantly since Camin [36] concluded from a survey of over 200 meat packing plants in the United States that air flotation was the least efficient treatment in terms of dollars per weight of BOD removed.

Chemical treatment can improve the pollution removal efficiency of a DAF unit, and typically ferric chloride is used to precipitate proteins and polymers used to aid coagulation. The adjustment of pH using sulfuric acid is also reported to be used in some slaughterhouses to aid the precipitation of protein [37]. Travers and Lovett [38] reported enhanced removal of fats when a DAF unit was operated at pH 4.0–4.5 without any further chemical additions. Such a process would require substantial acid addition, however.

A case study in a Swiss slaughterhouse describes the use of a DAF plant to treat wastewater that is previously screened at 0.5 mm (approx 1/50 inch) and pumped to a stirred equalization tank with five times the volumetric capacity of the hourly DAF unit flow rate [39,40]. The wastewater, including press water returns, is chemically conditioned with iron(III) for blood coagulation, and neutralized to pH 6.5 with soda lime to produce an iron hydroxide floc, which is then stabilized by polymer addition. This approach is claimed to give an average of

80% COD removal, between 40 and 60% reduction in total nitrogen, a flotation sludge with 7% dry solids with a volume of 2.5% of the wastewater flow. The flotation sludge can then be dewatered further with other waste fractions such as slurry from vehicle washing and bristles from pig slaughter to give a fraction with around 33% dry solids.

It must be borne in mind that although chemical treatment can be used successfully to reduce pollution load, especially of soluble proteinaceous material, it results in much larger quantities of readily putrescible sludge. It will, however, significantly reduce the nutrient load onto subsequent biological processes.

In many existing plants a conventional train of unit operations is used, in which solids are removed from the wastewater using a combination of screens and settlement. Screening is usually carried out on a fine-mesh screen (1/8 to 1/4 inch aperture, or 0.3–0.6 cm), which can be of a vibrating, rotating, or mechanically cleaned type. The screen is designed to catch coarse materials such as hair, flesh, paunch manure, and floating solids. Removals of 9% of the suspended solids on a 20-mesh screen and 19% on a 30-mesh screen have been reported [15]. The coarser 20-mesh screen gives fewer problems of clogging, but even so the screen must be provided with some type of mechanism to clean it. In practice mechanically cleaned screens using a brush type of cleaner give the best results. Finer settleable solids are removed in a sedimentation tank, which can be of either a rectangular or circular type. The size and design of sedimentation tanks varies widely, but Imhoff tanks with retentions of 1–3 hours have been used in the past in the United States and are reported to remove about 65% of the suspended solids and 35% of BOD [18]. The use of a deep tank can lead to high head loss, or to the need for excavation works to avoid this. For this reason, longitudinal or radial flow sedimentation tanks are now preferred for new installations in Europe. The usual design criteria for these when dealing with slaughterhouse wastewaters is that the surface loading rate should not exceed 1000 gal/ft^2 day (41 m^3/m^2 day).

As discussed above, the nature of operations within a slaughterhouse means that the wastewater characteristics vary considerably throughout the course of a working day or shift. It is, therefore, usually necessary to include a balancing tank to make efficient use of any treatment plant and to avoid operational problems. The balancing tank should be large enough to even out the flow of wastewater over a 24-hour period. To be able to design the smallest, and, therefore, most economical, balancing tank requires a full knowledge of variations in flow and strength throughout the day. This information is often not available, however, and in this case it is usual to provide a balancing tank with a capacity of about two-thirds of the daily flow.

Secondary Treatment

Secondary treatment aims to reduce the BOD of the wastewater by removing the organic matter that remains after primary treatment. This is primarily in a soluble form. Secondary treatment can utilize physical and chemical unit processes, but for the treatment of meat wastes biological treatment is usually favored [41].

Physicochemical Secondary Treatment

Chemical treatment of meat-plant wastes is not a common practice due to the high chemical costs involved and difficulties in disposing of the large volumes of sludge produced. There are, however, instances where it has been used successfully. Nemerow and Agardy [15] report a treatment facility that used $FeCl_3$ to reduce the BOD from 1448 to 188 mg/L (87% reduction) and the suspended solids from 2975 to 167 mg/L (94% reduction), with an operation cost of US$68 per million gallons. Using chlorine and alum in sufficient quantities could also significantly reduce the BOD and color of the wastes, but once again the chemical costs are high.

With this approach the BOD of raw wastewaters ranging from 1500 to 3800 mg/L can be reduced to between 400 and 600 mg/L. Dart [18] reported a 64% reduction in BOD using alumina-ferric as a coagulant with a dosing rate equivalent to 17 mg/L of aluminum. Chemical treatment has also been used to remove phosphates from slaughterhouse wastewater. Aguilar et al. (2002) used $Fe_2(SO_4)_3$, $Al_2(SO_4)_3$, and poly-aluminum chloride (PAC) as coagulants with some inorganic products and synthetic polyelectrolytes to remove approximately 100% orthophosphate and between 98.93 and 99.90% total phosphorus. Ammonia nitrogen removal was very low, however, despite an appreciable removal of albuminoidal nitrogen (73.9–88.77%).

The chemical processes described rely on a physical separation stage such as sedimentation, as illustrated in Figure 3.4, or by using a DAF unit (see "Primary Treatment" section and Fig. 3.3). Using this approach coupled with sludge dewatering equipment it is possible to achieve a good effluent quality and sludge cake with a low water content.

Biological Secondary Treatment

Using biological treatment, more than 90% efficiency can be achieved in pollutant removal from slaughterhouse wastes. Commonly used systems include lagoons (aerobic and anaerobic), conventional activated sludge, extended aeration, oxidation ditches, sequencing batch reactors, and anaerobic digestion. A series of anaerobic biological processes followed by aerobic biological processes is often useful for sequential reduction of the BOD load in the most economic manner, although either process can be used separately. As noted above, slaughterhouse wastewaters vary in strength considerably depending on a number of factors. For a given type of animal, however, this variation is primarily due to the quantity of water used within the abattoir, as the pollution load (as expressed as BOD) is relatively constant on the basis of live weight slaughtered. Hence, the more economical an abattoir is in its use of water, the stronger the effluent will be, and vice versa. The strength of the organic degradable matter in the wastewater is an important consideration in the choice of treatment system. To remove BOD using an aerobic biological process involves supplying oxygen (usually as a component in air) in proportion to the quantity of BOD that has to be removed, an increasingly expensive process as

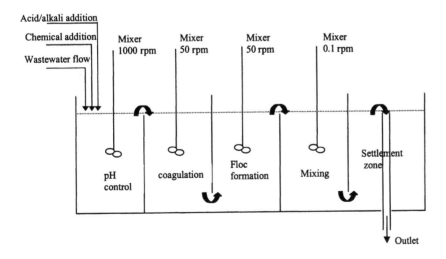

Figure 3.4 Typical chemical treatment and conditioning system.

the BOD increases. On the other hand an anaerobic process does not require oxygen in order to remove BOD as the biodegradable fraction is fermented and then transformed to gaseous endproducts in the form of carbon dioxide (CO_2) and methane (CH_4).

3.4.2 Anaerobic Treatment

Anaerobic digestion is a popular method for treating meat industry wastes. Anaerobic processes operate in the absence of oxygen and the final products are mixed gases of methane and carbon dioxide and a stabilized sludge. Anaerobic digestion of organic materials to methane and carbon dioxide is a complicated biological and chemical process that involves three stages: hydrolysis, acetogenesis, and finally methanogenesis. During the first stage, complex compounds are hydrolyzed to smaller chain intermediates. In the second stage acetogenic bacteria convert these intermediates to organic acids and then ultimately to methane and carbon dioxide via the methanogenesis phase (Fig. 3.5).

In the United States, anaerobic systems using simple lagoons are by far the most common method of treating abattoir wastewater. These are not particularly suitable for use in the heavily populated regions of western Europe due to the land area required and also because of the difficulties of controlling odors in the urban areas where abattoirs are usually located. The extensive use of anaerobic lagoons demonstrates the amenability of abattoir wastewaters to anaerobic stabilization, however, with significant reductions in the BOD at a minimal cost.

The anaerobic lagoon consists of an excavation in the ground, giving a water depth of between 10 and 17 ft (3–5 m), with a retention time of 5–15 days. Common practice is to provide two ponds in series or parallel and sometimes linking these to a third aerobic pond. The pond has no mechanical equipment installed and is unmixed except for some natural mixing brought about by internal gas generation and surface agitation; the latter is minimized where possible to prevent odor formation and re-aeration. Influent wastewater enters near the bottom of the pond and exits near the surface to minimize the chance of short-circuiting. Anaerobic ponds can provide an economic alternative for purification. The BOD reductions vary widely, although

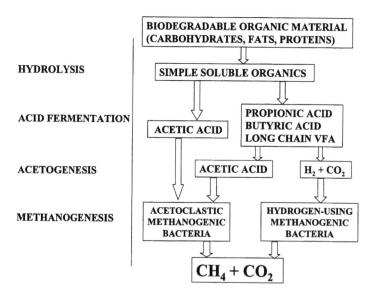

Figure 3.5 The microbial phases of anaerobic digestion.

excellent performance has been reported in some cases, with reductions of up to 97% in BOD, up to 95% in SS, and up to 96% in COD from the influent values [14,20,42]. Table 3.9 summarizes some of the literature data on the performance of anaerobic lagoons for the treatment of slaughterhouse wastes. The use of anaerobic lagoons in New Zealand is reported by Cooper et al. [30].

Anaerobic lagoons are not without potential problems, relating to both their gaseous and aqueous emissions. As a result of breakdown of the wastewater, methane and carbon dioxide are both produced. These escape to the atmosphere, thus contributing to greenhouse gas emissions, with methane being 25 times more potent than carbon dioxide in this respect. Gaseous emissions also include the odoriferous gases, hydrogen sulfide and ammonia. The lagoons generally operate with a layer of grease and scum on the top, which restricts the transfer of oxygen through the liquid surface, retains some of the heat, and helps prevent the emission of odor. Reliance on this should be avoided wherever possible, however, since it is far from a secure means of preventing problems as the oil and grease cap can readily be broken up, for example, under storm water flow conditions. Odor problems due to anaerobic ponds have a long history: even in the 1960s when environmental awareness was lower and public threshold tolerances to pollution were higher, as many as nine out of ten anaerobic lagoons in the United States were reported as giving rise to odor nuisance [43]. A more satisfactory and environmentally sound solution is the use of membrane covers that prevent odor release, while at the same time allowing collection of the biogas that can be used as fuel source within the slaughterhouse. This sort of innovation moves the lagoon one step closer to something that can be recognized as a purpose-built treatment system, and provides the opportunity to reduce plant size and improve performance.

The use of fabricated anaerobic reactors for abattoir wastewater treatment is also well established. To work efficiently these are designed to operate either at mesophilic (around 95°F or 35°C) or thermophilic (around 130°F or 55°C) temperatures. Black et al. [47] reported that the practicality of using anaerobic digestion for abattoir wastewater treatment was established in the 1930s. Their own work concerned the commissioning and monitoring of an anaerobic contact process installed at the Leeds abattoir in the UK. The plant operated with a 24-hour retention time at a loading of 29.3 lb BOD/10^3 gal (3.5 kg BOD/m^3) and showed an 88–93% reduction in BOD, giving a final effluent concentration of around 220 mg/L. Bohm [48] conducted trials using a 106 ft^3 (3 m^3) anaerobic contact process at a loading of 21.7 lb BOD/10^3 gal day (2.6 kg BOD/m^3 day), with a removal efficiency of 80%. An economic evaluation of the process showed savings on effluent disposal charges. The review by Cillie et al. [49] refers to work by Hemens and Shurben [50] showing a 95% BOD reduction from an influent BOD of 2000 mg/L.

Table 3.9 Treatment of Meat Industry Wastes by Anaerobic Lagoon

Loading rate [lb/10^3 gal day (kg BOD/m^3 day)]	Retention time (days)	Depth [feet (m)]	BOD removal (%)	Reference
–	16	6.9 (2.1)	80	43
1.1 (0.13)	7–8	15.1 (4.6)	60	31
1.6 (0.19)	5	14.1 (4.3)	80	31
1.7 (0.20)	–	10.5 (3.2)	86	31
3.4 (0.41)	3.5	15.1 (4.6)	87	27
1.8 (0.21)	1.2	15.1 (4.6)	58	44
1.3 (0.15)	11	8.9 (2.7)	92	45
1.3 (0.16)	–	15.1 (4.6)	65	46

Gas production was only just sufficient to maintain the digester temperature of 91°F (33°C), however. The Albert Lee plant in Minnesota, Unites States, is also mentioned, in which an anaerobic contact digester with vacuum degassing operating at a retention time of 30 hours achieved a 90% reduction in BOD. Work is also described at the Lloyd Maunder Ltd abattoir in Devon, UK, again using an anaerobic contact digester. This achieved 90% BOD removal, but only a low gas production. In the conclusion of their review Cillie et al. [49] state that the most successful anaerobic plants for industrial waste liquids seem to be those dealing with slaughterhouse and meat-packing wastes.

Kostyshyn et al. [24] used both mesophilic and thermophilic anaerobic contact processes as an alternative to physicochemical treatment over an 8-month trial period. At a loading rate of 22.9 lb COD/10^3 gal day (2.75 kg COD/m^3 day) and a retention time of 2.5 days they achieved an average of 93.1% BOD removal and 74.9% COD removal. The process appears to be able to operate successfully at loadings of up 20.9 lb COD/10^3 gal day (2.5 kg COD/m^3 day). This is possible because the anaerobic contact process maintains a high biomass density and long solids retention time (SRT) in the reactor by recirculation of sludge from a separation stage, which usually involves sedimentation. The high biomass density, long SRT, and elevated temperature enable a short hydraulic retention time. As with most anaerobic reactor systems, however, they are expensive to install and require close technical supervision.

Anaerobic filters have also been applied to the treatment of slaughterhouse wastewaters. These maintain a long SRT by providing the microorganisms with a medium that they can colonize as a biofilm. Unlike conventional aerobic filters, the anaerobic filter is operated with the support medium submerged in an upflow mode of operation. Because anaerobic filters contain a support medium, there is potential for the interstitial spaces within the medium to become blocked, and effective pretreatment is essential to remove suspended solids as well as solidifiable oils, fats, and grease.

Andersen and Schmid [51] used an anaerobic filter for treating slaughterhouse wastewater, and encountered problems with grease in the startup period. The problem was solved by introducing dissolved air flotation as a pretreatment for the removal of grease. The filter showed between 62 and 93% removal of COD over a trial period of 22 weeks, but the authors concluded that the process required close supervision and emphasized the need for good pretreatment. Arora and Routh [29] also used an anaerobic filter with a 24-hour retention time and loads of up to 58.4 lb COD/10^3 gal day (7.0 kg COD/m^3 day). Treatment efficiency was up to 90% at loadings up to 45.9 lb COD/10^3 gal day (5.5 kg COD/m^3 day). Festino and Aubart [52,53] used an anaerobic filter for wastewaters containing less than 1% solids, but the main focus of their work was on the high solids fraction of abattoir wastes in complete mix reactors. Generally speaking, a safe operational loading range for a mesophilic anaerobic filter appears to be between 16.7 and 25.0 lb COD/10^3 gal day (2–3 kg COD/m^3 day), and at this loading a COD reduction of between 80 and 85% might conservatively be expected.

The third type of high-rate anaerobic system that can be applied to slaughterhouse wastewaters is the upflow anaerobic sludge blanket reactor (UASB). This is basically an expanded-bed reactor in which the bed comprises anaerobic microorganisms, including methanogens, which have formed dense granules. The mechanisms by which these granules form are still poorly understood, but they are intrinsic to the proper operation of the process. The influent wastewater flows upward through a sludge blanket of these granules, which remain within the reactor as their settling velocity is greater than the upflow velocity of the wastewater. The reactor therefore exhibits a long sludge retention time, high biomass density per unit reactor, and can operate at a short HRT.

UASB reactors overcome the limitations of anaerobic contact plant and anaerobic filters, yet their application to slaughterhouse wastewater appears limited to laboratory- and pilot-scale reactors. The reason for this is the difficulties in trying to form stable granules when dealing with slaughterhouse wastewater, and this may be due to the high fat concentrations [54].

Although anaerobic processes have generally shown good results in the treatment of abattoir wastewaters, some problems have also been reported. Nell and Krige [55] comment in their paper on aerobic composting systems that in the anaerobic process the high organic content leads to a resistance to fermentation and there is a tendency towards scum formation. The work carried out at the Lloyd Maunder Ltd. Plant [49] reports the buildup of scum in the digestion process. Grease was also shown to be a problem in the digester operated by Andersen [51]. Cooper et al. [30], in the paper on abattoir waste treatment in New Zealand, state that the use of anaerobic contact and anaerobic filters is not economic as the energy content in the fat is adsorbed and not really broken down in the anaerobic process. This demonstrates the need for proper pretreatment and for an energy balance as part of the design work.

There is a substantial amount of evidence at laboratory, pilot, and full scale that anaerobic systems are suitable for the treatment of abattoir wastewaters. There is also evidence that with the weaker abattoir wastewaters with BODs around 2000 mg/L, gas production is only just sufficient to maintain reactor temperature as might be predicted from thermodynamics. Table 3.10 summarizes some results achieved using anaerobic reactors of different types applied to slaughterhouse wastewaters.

3.4.3 Aerobic Treatment

Aerobic biological treatment for the treatment of biodegradable wastes has been established for over a hundred years and is accepted as producing a good-quality effluent, reliably reducing influent BOD by 95% or more. Aerobic processes can roughly be divided into two basic types: those that maintain the biomass in suspension (activated sludge and its variants), and those that retain the biomass on a support medium (biological filters and its variants). There is no doubt that either basic type is suitable for the treatment of slaughterhouse wastewater, and their use is well documented in works such as Brolls and Broughton [9], Dart [31], and Kaul [68], where aerobic processes are compared with anaerobic ones. In selecting an aerobic process a number of factors need to be taken into account. These include the land area available, the head of water available, known difficulties associated with certain wastewater types (such as bulking and stable foam formation), energy efficiency, and excess biomass production. It is important to realize that the energy costs of conventional aerobic biological treatment can be substantial due to the requirement to supply air to the process. It is, therefore, usual to only treat to the standard required, as treatment to a higher standard will incur additional cost. For example, in order to convert ammonia to nitrate requires 4.5 moles of oxygen for every mole of ammonia converted. In effect this means that a 1 mg/L concentration of ammonia has an equivalent BOD of 4.5 mg/L. It is, therefore, only usual to aim for the conversion of ammonia to nitrate when this is required.

The most common aerobic biological processes used for the treatment of meat industry wastes are biological filtration, activated sludge plants, waste stabilization ponds, and aerated lagoons.

Waste Stabilization Ponds

A waste stabilization pond (WSP) is the simplest method of aerobic biological treatment and can be regarded as bringing about the natural purification processes occurring in a river in a more restricted time and space. They are often used in countries where plenty of land is available and weather conditions are favorable. In the United States, WSPs with depths of between 1.5 and 9 ft (0.5–2.7 m; typical value 4 ft or 1.2 m) have been used. A typical BOD loading of

Table 3.10　Anaerobic Treatment of Abattoir Wastes

Reactor type	Loading rate [lb COD/ft² day (kg COD/m³ d)]	Retention time	Temperature (°C)	Removal (%)	Gas production	Reference
Lagoon	0.1–0.6 (0.016–0.068)	10–12 days	Ambient	82.6 (BOD)	–	30
Contact	10.0–18.4 (1.2–2.2)	1–1.7 days	35	–	–	56
AF[a]	16.7 (2.0)	–	–	85.0 (COD)	–	6
AF[a]	45.9 (5.5)	1 day	37	90.5 (COD)	–	29
Two stage	–	1 day	30–40		$0.2–0.3$ m³ CH_4/kg COD removed	57
AF[a]	6.7–30.0 (0.8–3.6)	1.4 day	32	62–92 (COD)	–	51
AF[a]	35.9–50.1 (4.3–6.0)	0.71 day	35	49–57 (COD)	$0.8–2.2$ mL CH_4/g COD removed	58
CSTR[b]	7.7 (0.92)	23 days	35	56.6 (COD)	0.2 m³ CH_4/kg COD removed	59
CSTR[b]	24.3–73.0 lb VS/10³ gal day (2.9–8.75 g VS/L-day)	12 days	35–55	45–65 (COD)	$0.30–0.43$ m³ CH_4/kg COD removed	60
Contact	22.9 (2.75)	2.5 days	35	84.5 (COD)	0.28 m³ CH_4/kg COD removed	24
UASB[c]	20.9–162.7 (2.5–19.5)	1.7–9 hours	30	53–67	0.82–5.2	61
	25–100 (3.0–12.0)	5–10 hours	20	40–62 (COD)	1.22–3.2 kg CH_4 – COD/m³ d	
UASB[c]	4.2–167 (0.5–20)	0.5–1.7 days	30	68.4–82.3 (COD)	–	62
Contact	8.3 (1.0)	3.3 days	22	70.0 (COD)	–	63
Contact	133.5 lb TS/10³ gal day (16 kg TS/m³ day)	10 days	55	27.0 (TS)	0.08 m³ CH_4/kg TS added	64
AF[a]	16.7–154.4 (2–18.5)	5–0.5 days	–	27–85 (COD)	–	65
ABR[d]	5.6–39.5 (0.67–4.73)	0.1–1.1 days	25–35	75–90 (COD)	$0.07–0.15$ m³ CH_4/kg COD removed	66
Two stage UASB[c]	125.2 (15)	5.5 hours	18	90.0 (COD)	–	67

[a]AF, anaerobic filter
[b]CSTR, classic continuous stirred tank reactor
[c]UASB, upflow anaerobic sludge blanket
[d]ABR, anaerobic baffle reactor.
VS, volatile solids; TS, total solids

20–30 lb BOD/day acre (22–34 kg BOD/ha day) with a typical retention time of 30 to 120 days has been reported [18]. Such ponds are often used in series and can incorporate an anaerobic pond as the first stage (see Section 3.4.2), followed by a facultative pond and maturation ponds. By using a long total retention and low overall BOD loading a good-quality effluent can be achieved. As a stand-alone system the facultative pond may be expected to give between 60 and 90% BOD/COD reduction and between 10 and 20% reduction in total nitrogen. When coupled with maturation lagoons a further 40–70% reduction in BOD/COD can be achieved, primarily as a result of the settlement and breakdown of biomass generated in the facultative pond. This will result in an overall suspended solids reduction of up to 80% [35].

In both the facultative and maturation ponds the oxygen required for the growth of the aerobic organisms is provided partly by transfer across the air/water interface and partly by algae as a result of photosynthesis. This leads to a very low operating cost as there is no requirement for mechanically induced aeration. Conditions in WSPs are not easily controlled due to the lack of mixing, and organic material can settle out near the inlet of the pond causing anaerobic conditions and offensive smells, especially when treating meat industry wastes that contain grease and fat materials. It is, therefore, not uncommon to find that the facultative pond may also be fitted with a floating surface aerator to aid oxygen transfer and to promote mixing. There is a point, however, when the oxygen input by mechanical means exceeds that naturally occurring by surface diffusion and photosynthesis: at this point the facultative lagoon is best described as an aerated lagoon. The design of a WSP system depends on a number of climatic and other factors: excellent guidance can be found in the USEPA design manual and the work of Mara and Pearson [69,70].

Biological Filters

Biological filters can also be used for treating meat industry wastes. In this process the aerobic microorganisms grow as a slime or film that is supported on the surface of the filter medium. The wastewater is applied to the surface and trickles down while air percolates upwards through the medium and supplies the oxygen required for purification (Fig. 3.6). The treated water along

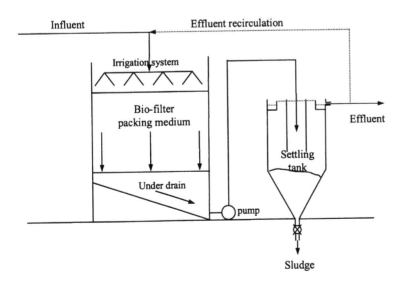

Figure 3.6 Typical biological filtration treatment system.

with any microbial film that breaks away from the support medium collects in an under-drain and passes to a secondary sedimentation tank where the biological solids are separated. Trickling filters require primary treatment for removal of settleable solids and oil and grease to reduce the organic load and prevent the system blocking. Rock or blast furnace slag have traditionally been used as filter media for low-rate and intermediate-rate trickling filters, while high-rate filters tend to use specially fabricated plastic media, either as a loose fill or as a corrugated prefabricated module. The advantage of trickling filters is their low energy requirement, but the disadvantage is the low loading compared to activated sludge, making the plant larger with a consequent higher capital cost. Hydraulic loading rates range from $0.02-0.06$ gal/ft^2 day ($0.001-0.002$ m^3/m^2 day) for low-rate filters to $0.8-3.2$ gal/ft^2 day ($0.03-0.13$ m^3/m^2 day) for high-rate filters. Organic loading rates range from $5-25$ lb BOD/10^3 ft^2 day to $100-500$ lb BOD/10^3 ft^2 day ($0.02-0.12$ kg/m^2 day to $0.49-2.44$ kg/m^2 day). The overall BOD removal efficiency can be as great as 95%, but this is dependent on the loading applied and the mode of operation. A typical performance envelope for biological filters operating with a plastic support medium is given in Figure 3.7.

Because of the relatively high strength of slaughterhouse wastewater, biological filters are more suited to operation with effluent recirculation, which effectively increases surface hydraulic loading without increasing the organic loading. This gives greater control over microbial film thickness. In the United States, high-rate single-stage percolating filters with high recirculation ratios have been used. An overall BOD removal of 92–98% was reported using a high-rate filter with a BOD loading of $2.6-3.8$ lb BOD/10^3 gal media day ($0.31-0.45$ kg BOD/m^3 media day) and a recirculation ratio of about $5:1$ for treating preliminary treated slaughterhouse wastes [71]. Dart [18] reported that a high-quality effluent with 11 mg/L of BOD and 25 mg/L SS could be obtained using alternating double filtration (ADF) at a loading rate of 2.8 lb/10^3 gal day (0.34 kg BOD/m^3 day) for treating screened and settled abattoir waste; the influent was diluted $1:1$ with recirculated effluent. Higher loadings with a BOD of between 17 and 33 lb/10^3 gal ($2-4$ kg BOD/m^3) and a surface hydraulic loading of 884 gal/ft^2 day (1.5 m^3/m^2 hour) and recirculation ratios of $3-4$ are given as a typical French design guideline aimed at providing a roughing treatment in reactors 13.1 ft (4 m) high [14]. Such a design is likely to give a BOD removal of less than 75% (Fig. 3.7) and not to provide any nitrification.

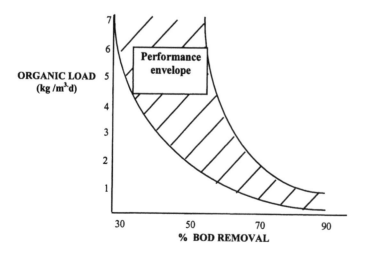

Figure 3.7 Performance envelope for high rate biological filtration.

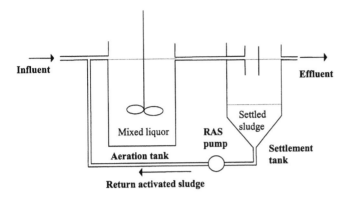

Figure 3.8 Schematic for a completely mixed continuous flow activated sludge plant.

Dart [31] summarized the performance of some high-rate filtration plants treating meat industry wastes (Table 3.11).

Biological filters have not been widely adopted for the treatment of slaughterhouse wastewaters despite the lower operating costs compared with activated sludge systems. Obtaining an effluent with a low BOD and ammonia in a single-reactor system can provide conditions suitable for the proliferation of secondary grazing macro-invertebrate species such as fly larvae, and this may be unacceptable in the vicinity of a slaughterhouse. There is also the need for very good fat removal from the influent wastewater flow, as this will otherwise tend to coat the surface of the biofilm support medium. The use of traditional biological filtration for abattoir wastewater treatment is discussed by Philips [72], and further reviewed by Parker et al. [73].

Rotating Biological Contactors

Rotating biological contactors (RBCs) are also fixed biofilm reactors, which consist of a series of closely spaced circular discs mounted on a longitudinal shaft. The discs are rotated, exposing the attached microbial mass alternately to air and to the wastewater being treated, and allowing the adsorption of organic matter, nutrients, and oxygen. Typical design values for hydraulic and

Table 3.11 Treatment of Meat Industry Wastewaters by High-Rate Biological Filtration

Medium	BOD load		BOD removal (%)
	(lb/10^3 gal day)	(kg/m^3 day)	
Cloisonyle	67.6	8.1	75
Flocor	14.2	1.7	72
Flocor	15.0	1.8	85
Flocor	20.0	2.4	66
Flocor	25.0	3.0	50
Flocor	25.9	3.1	60
Flocor	26.7	3.2	60
Rock	12.5	1.5	61
Unspecified PVC	10.0	1.2	74

Source: Ref. 31.

organic loading rates for secondary treatment are 2–4 gal/ft^2 day (0.08–0.16 m^3/m^2 day) and 2.0–3.5 lb total BOD/10^3 ft^2 day (0.01–0.017 kg BOD/m^2 day) respectively, with effluent BOD concentrations ranging from 15 to 30 mg/L. For secondary treatment combined with nitrification, typical hydraulic and organic loading rate design values are 0.75–2 gal/ft^2 day and 1.5–3.0 lb total BOD/10^3 ft^2 day, respectively (0.03–0.08 m^3/m^2 day and 0.007–0.014 kg BOD/m^2 day), producing effluent BOD concentrations between 7 and 15 mg/L and NH$_3$ concentrations of less than 2 mg/L [74]. The above performance figures are typical of this type of unit, but are not necessarily accurate when applied to the treatment of slaughterhouse wastewaters. Bull et al. [75] and Blanc et al. [76] reported that the performance of RBCs appeared inadequate when compared to activated sludge or high-rate biological filtration. Another report of RBC use in slaughterhouse wastewater treatment is given by Bilstad [77], who describes the upgrading of a plant using one of these systems.

Aerated Filters

These comprise an open tank containing a submerged biofilm support medium, which can be either static or moving. The tank is supplied with air to satisfy the requirements of the bio-oxidation process. There are a number of proprietary designs on the market, but each works on the principle of retaining a high concentration of immobilized biomass within the aerobic reaction tank, thus minimizing the need for secondary sedimentation and sludge recycle. The major differences between the processes are the type of biomass support medium, the mechanism of biofilm control, and whether or not the support medium is fixed or acts as an expanded or moving bed. As an example of the use of such a process, a Wisconsin slaughterhouse installed a moving-bed biofilm reactor (MBBR) to treat a wastewater flow of 168,000 USgal/day, with surge capabilities to 280,000 gal/day (636 and 1060 m^3/day). Average influent soluble BOD and soluble COD concentrations were 1367 mg/L, and 1989 mg/L, respectively. The Waterlink, Inc., process selected used a small polyethylene support element that occupied 50% of the 9357 ft^3 (265 m^3) volume provided by two reactors in series to give 10 hours hydraulic retention time at average flows and 6 hours at peak hydraulic flow [78]. Effluent from the second MBBR was sent to a dissolved air flotation unit, which removed 70–90% of the solids generated. The average effluent soluble BOD and COD were 59 mg/L and 226 mg/L, respectively.

Activated Sludge

The activated sludge process has been successfully used for the treatment of wastewaters from the meat industry for many decades. It generally has a lower capital cost than standard-rate percolating filters and occupies substantially less space than lagoon or pond systems. In the activated sludge process the wastewaters are mixed with a suspension of aerobic microorganisms (activated sludge) and aerated. After aeration, the mixed liquor passes to a settlement tank where the activated sludge settles and is returned to the plant inlet to treat the incoming waste. The supernatant liquid in the settlement tank is discharged as plant effluent. Air can be supplied to the plant by a variety of means, including blowing air into the mixed liquor through diffusers; mechanical surface aeration; and floor-mounted sparge pipes. All the methods are satisfactory provided that they are properly designed to meet the required concentration of dissolved oxygen in the mixed liquor (greater than 0.5 mg/L) and to maintain the sludge in suspension; for nitrification to occur it may be necessary to maintain dissolved oxygen concentrations above 2.0 mg/L.

The activated sludge process can be designed to meet a number of different requirements, including the available land area, the technical expertise of the operator, the availability of sludge disposal routes, and capital available for construction. Excellent descriptions of the process can be found in many texts: Metcalf & Eddy provides many good examples [74]. The first step in the design of an activated sludge system is to select the loading rate, which is usually defined as the mass ratio of substrate inflow to the mass of activated sludge (on a dry weight basis); this is commonly referred to as the food to microorganism (F : M) ratio and is usually reported as lb BOD/lb MLSS day (kg BOD/kg MLSS day). For conventional operation the range is 0.2–0.6; the use of higher values tends to produce a dispersed or nonflocculent sludge and lower values require additional oxygen input due to high endogenous respiration rates. Systems with F : M ratios above 0.6 are sometimes referred to as high rate, while those below 0.2 are known as extended aeration systems (Table 3.12). The latter, despite their higher capital and operating costs are commonly chosen for small installations because of their stability, low sludge production, and reliable nitrification. Because of the stoichiometric relationship between F : M ratio and mean cell residence time (MCRT), high-rate plants will have an MCRT of less than 4 days and extended aeration plants of greater than 13 days. Because of the low growth rates of the nitrifying bacteria, which are also influenced markedly by temperature, the oxidation of ammonia to nitrates (nitrification) will only occur at F : M ratios less than 0.1. It is also sometimes useful to consider the nitrogen loading rate, which for effective nitrification should be in the range 0.03–0.08 lb N/lb MLSS-day (kg N/kg MLSS day).

Conventional plants can be used where nitrification is not critical, for example, as a pretreatment before sewer discharge. One of the main drawbacks of the conventional activated sludge process, however, is its poor buffering capability when dealing with shock loads. This problem can be overcome by the installation of an equalization tank upstream of the process, or by using an extended aeration activated sludge system. In the extended aeration process, the aeration basin provides a 24–30 hour (or even longer) retention time with complete mixing of tank contents by mechanical or diffused aeration. The large volume combined with a high air input results in a stable process that can accept intermittent loadings. A further disadvantage of using a conventional activated sludge process is the generation of a considerable amount of surplus sludge, which usually requires further treatment before disposal. Some early work suggested the possible recovery of the biomass as a source of protein [30,79], but concerns over the possible transmission of exotic animal diseases would make this unacceptable in Europe [80]. The use of extended aeration activated sludge or aerated lagoons minimizes biosolids production because of the endogenous nature of the reactions. The size of the plant and the additional aeration required for sludge stabilization does, however, lead to increased capital and operating costs. Considering the high concentrations of nitrogen present in slaughterhouse wastewater, ammonia removal is often regarded as essential from a regulatory standpoint for direct discharge, and increasingly there is a requirement for nutrient removal. It is therefore not surprising that most modern day designs are of an extended aeration type so as to promote

Table 3.12 Classification of Activated Sludge Types Based on the F : M Ratio Showing Appropriate Retention Times and Anticipated Sludge Yields

Mode of operation	F : M ratio	Retention time (hours)	Sludge yield [lb SS/ lb BOD (kg SS/kg BOD)]	BOD removal efficiency (%)
High rate	0.6–0.35	1	1.0	60–70
Conventional	0.2–0.6	6–10	0.5	90–95
Extended aeration	0.03–0.2	24+	0.2	90–95

Typical values derived from a wide range of sources.

reliable nitrification as well as to minimize sludge production. Efficient designs will also attempt to recover the chemically bound oxygen in nitrate through the process of denitrification, thus reducing treatment costs and lowering nitrate concentrations in the effluent.

Design criteria and loadings for activated sludge treatment have been widely reported and reliable data can be found in a number of reports [9,34,81–83].

In recent years, a great deal of interest has been shown in the use of sequencing batch reactors (SBRs) for food-processing wastewaters, as these provide a minimum guaranteed retention time and produce a high-quality effluent. A batch process also often fits well with the intermittent discharge of an industrial process working on one or two shifts. Advantages are an ideal plug flow that maximizes reaction rates, ideal quiescent sedimentation, and flow equalization inherent in the design. Decanting can be achieved using floating outlets and adjustable weirs, floating aera-tors are commonly employed, and an anoxic fill overcomes problems of effluent turbidity [84] as well as providing ideal conditions for denitrification reactions.

Hadjinicolaou [85] described using a pilot-scale SBR activated sludge system for the treatment of cattle slaughterhouse wastewaters. The system was operated on a 24-hour cycle and 97.8% of COD removal was achieved with an influent COD concentration of 3512 mg/L. A case study showing the use of an SBR in conjunction with an anaerobic lagoon has shown the potential of the system for both organic load reduction and nutrient removal [86]. The main effluent stream from the slaughterhouse containing some blood, fat, and protein enters a save-all for separation of fat and settleable solids. The flow is then equally split, one portion going to the anaerobic lagoon, which also receives clarified effluent from settling pits that are used to collect manure and paunch material, and the other to the SBR; the effluent from the lagoon subsequently also enters the SBR. The average ratio of BOD : total N entering the SBR is 3 : 1, which provides sufficient carbon to achieve complete nitrification and denitrification. The SBR has a cycle during which filling takes place over an 11-hour period corresponding to the daily operation and cleaning cycle of the slaughterhouse. The aeration period is 12 hours, settlement 1 hour and decanting to a storage lagoon over 3 hours. The total volume of the plant is 0.33 Mgal (1250 m^3), with a daily inflow of 66,000 gal (250 m^3) at a BOD of 600–800 mg/L, total N of 200 mg/L, and total P of 40 mg/L. The plant is reported to achieve a final effluent with values less than 2 mg/L NH$_3$-N, 10 mg/L NO$_3$-N, total P 20 mg/L, BOD 20 mg/L, and SS 20 mg/L. Additionally, all surplus activated sludge from the SBR is returned to the anaerobic lagoon.

3.5 SOLID WASTES

If good operational practice is followed in the slaughterhouse, the solids and organic loading entering the aqueous phase can be minimized. The separated solids still require treatment prior to disposal, however, and traditional rendering of some of these fractions is uneconomic because of the high water and low fat content. These fractions are the gut manures, the manure and bedding material from holding pens, material from the wastewater screens and traps on surface drains, sedimentation or DAF sludge, and possibly hair where no market exists for this material. Other high-protein and fat-containing residues such as trimmings, nonedible offal, and skeletal material can be rendered to extract tallow and then dried to produce meat and bone meal. The traditional rendering process is not within the scope of the present chapter, but consideration is given to the disposal of the other fractions as these may appear in the form of a wastewater sludge, although in an efficient slaughterhouse they would be "dry" separated.

Manures from stockyards and holding pens are likely to be similar in composition to the animal slurries that are generated on a farm. Typical characteristics of these are reported by Gendebien et al. [5] and are shown in Table 3.13. It is important that as much of the material as

Table 3.13 Comparison of Farm Collected Animal Slurries to Manure Washed from a Stockyard Cleaning Operation

Characteristics	Concentration (g/L)		
	Stockyard[a]	Cattle slurry[b]	Pig slurry[b]
Total suspended solids	0.173	10–180	10–180
Volatile suspended solids	0.132	10–107	34–70
Organic nitrogen	0.011	2–18	2–16
Ammonia nitrogen	0.08	0.6–2.2	2.1–3.6
BOD	0.64	27[c]	37[c]

[a]Derived from Nemerow and Agardy [15].
Source: Refs. 5, 15, 87
[b]Derived from Gendebien et al. [5].
[c]Derived from American Society of Agricultural Engineers [87].
Source: Refs. 5, 15, 87

possible is removed for further processing before the holding pen area is washed down as, otherwise, this will result in a high-volume, high water content waste flow that can only be handled in the wastewater treatment system. For example, results from a survey of Chicago stockyards (Table 3.13) by Nemerow and Agardy [15] showed the effluent to be weak in strength compared to animal slurries collected on farms for land spreading.

Gut manures that can be dry separated for separate processing also have a very high pollutant load and solids content, as indicated in Table 3.14.

3.5.1 Land Disposal

Land disposal of abattoir solid wastes, either by land spreading or landfill, has been a common practice for many years, but concern over the transmission of exotic animal diseases has already seen a decline in this practice in Europe over the past several years. The EU Animal By-products Regulations [80] now prohibit land disposal of all animal wastes with the exception of manures and digestive tract contents, and these only when "the competent authority does not consider them to present a risk of spreading any serious transmissible disease." The only restriction on digestive tract contents is that spreading is limited to nonpasture land. In the EU operators will

Table 3.14 Analysis of Paunch Contents of Ruminant Animals and Mixed Gut Material

Type of animal	Total solids (%)	COD [lb/10³ lb (g/kg)]	BOD (mg/L)	Reference
Cattle			50,000	7
	14.2	134		88
	12.7	134		88
	14			8
Sheep			30,000	7
Lamb	9.3	165		88
Pig	26			8
Mixed kill	4.7–9.7			7
	22		50,000	9
	2.4–21		6000–41,000	5

also no longer be able to spread untreated blood from abattoirs onto land or dispose of it down wastewater drainage systems for treatment by water companies. The blood will need to be treated in an approved rendering, biogas, or composting plant before it can be land-spread. The regulations will vary from country to country, but it should be noted that land-spreading of any abattoir waste is liable to cause public nuisance through odor and environmental concerns, and has potential for disease transmission. It is, therefore, beneficial to treat the waste by a stabilization process before land application, and where this is not possible, it is imperative that land application be undertaken with great care. The rate of application of the waste should be based on the level of plant nutrients present.

Where there are no country-specific regulations, as a general rule, all abattoir wastes should be injected into the soil to reduce odor and avoid any potential pathogen transmission, and should not be surface-spread on pasture land or forage crops. If these materials are surface-spread on arable land, they should be incorporated immediately by plowing. Injection into grassland should be followed by a minimum interval of 3 weeks before the grass is used for grazing or conservation. Storage time for the materials should be kept to a minimum to avoid further development of odors [5].

The regulations currently in force in the United States for the land application of slaughterhouse-derived biosolids are given in the USEPA's Guide to Field Storage of Biosolids [89].

3.5.2 Composting

Composting can be defined as the biological degradation of organic materials under aerobic conditions into relatively stable products, brought about by the action of a variety of microorganisms such as fungi, bacteria, and protozoa. The process of composting may be divided into two main stages: stabilization and maturation. During stabilization, three phases may be distinguished: first a phase of rising temperature, secondly the thermophilic phase where high temperature is maintained, and, thirdly, the mesophilic phase where the temperature gradually decreases to ambient. During the first phase, a vigorous multiplication of mesophilic bacteria is observed, and a transformation of easily oxidized carbon compounds, rich in energy, to compounds of lower molecular structure takes place. Excess energy results in a rapid rise in temperature and leads to the thermophilic phase when less easily degraded organic compounds are attacked. When the energy source is depleted the temperature decreases gradually to ambient. Actinomycetes and fungi become active in the mesophilic phase, during which biological degradation of the remaining organic compounds is slowly continued. At this stage the composting material is stabilized but not yet matured. During maturation, mineralization of organic matter continues at a relatively slow rate until a carbon : nitrogen (C : N) ratio of 10 : 1 is reached and the volatile matter content falls below 50%. Only then is the compost completely matured.

Composting of dewatered sedimentation tank solids from a slaughterhouse in mechanically turned open windrows was described by Supapong [90]. The material was kept in windrows for 40 days, and the temperature fluctuated between 149 and 158°F (65–70°C) for up to 3 weeks. The resulting product was a friable, odorless, and microbiologically satisfactory material whose bulk was only one-fifth of the original volume. It contained 0.5 and 3% by weight of phosphorus and nitrogen, respectively, and was an acceptable soil conditioner. Nell and Krige [55] conducted an in-vessel composting process for treating solid wastes mainly consisting of paunch and pen manure. The reactor was an insulated rotating stainless steel drum of 10 ft^3 (280 L) capacity. After 4 days retention in the reactor, the waste reached the stabilization stage, and after a further 50 days the composting was completed. The final product had the following characteristics: pH 8.6, 65.1% moisture content, 55.3% of volatile matter, 2.1% of nitrogen, and 17.9% of carbon. The Australian Environmental Protection Authority [91] suggests paunch contents can be efficiently and economically disposed of by composting as long as offensive odors are not generated, and state that the most suitable composting techniques are turned windrows and aerated static piles.

3.5.3 Anaerobic Digestion

Anaerobic digestion of abattoir solid wastes is not common in the United States, UK, or elsewhere, despite the potential for stabilization of the solid residues with the added bonus of fuel gas production. Cooper et al. [30] looked at the potential in New Zealand for production of methane from both the solid and liquid fraction of abattoir wastes. Based on tests carried out by Buswell and Hatfield in 1939, they concluded that paunch contents and fecal matter would not give an economic return. In these very early tests it was reported that a retention time of 38–40 days might be required and that the expected gas yield would be 2500 ft^3/lb solids added (156 m^3/kg). In the UK the first of a new generation of well-mixed digestion plants to treat slaughterhouse wastes was installed in 1984 to treat all the paunch wastes, blood, and settlement tank solids produced by a small abattoir in Shropshire. The operation and performance of a 3531 ft^3 (100 m^3) demonstration-scale anaerobic digester treating cattle and lamb paunch contents, blood, and process wastewaters from a slaughterhouse was described by Banks [4]. Anaerobic digestion of the solid fraction of abattoir wastes suffers from low methane production and solid reduction as well as requires a longer retention time compared to sewage and food processing wastes [30]. Steiner et al. [60] reported the failure of a digester when treating a mixture of abattoir wastes. The mixture contained 13% of rumen and intestine contents, 25% of manure from animal buildings, 44% of surplus sludge from an aerobic sewage treatment plant, and 19% fat derived from the fat separator, and exhibited a COD of 165 g/L, a BOD of 112 g/L, a dry weight of 120 g/L, and a volatile solids concentration of 105 g/L consisting of 25% fat and 23% protein. The experiment was carried out in a cylindrical completely mixed reactor with a capacity of 0.07 ft^3 (2 L). When the organic loading rate was raised to more than 73 lb VS/10^3 gal day (8.75 g VS/L day), digestion failure occurred and was caused by enrichment of volatile acids in the digester. In his paper, Banks [4] also mentioned serious problems associated with the accumulation of ammonia concentration in the process. Several other authors also indicate that where blood and fat form a significant proportion of the feedstock it is found to be digestible in only limited quantities due to an inhibitory effect on methanogenesis, thought to be caused by accumulation of toxic intermediates produced by the hydrolysis/acidification stages [57,92,93].

Using a two-stage anaerobic process, Banks and Wang [94] successfully overcame the toxicity problems associated with the accumulation of ammonia and volatile fatty acids when treating a mixture of cattle paunch contents and cattle blood. The first-stage reactor was operated in a hydraulic flush mode to maintain a significantly shorter liquid retention time than the solids retention time of the fibrous components in the feedstock. The first-stage reactor was run in this mode using solids retention times of 5, 10, 15, 20, and 30 days with liquid retention of between 2 and 5 days. Up to 87% solid reductions were achieved compared to a maximum of 50% when the control reactor was operated in single-pass mode with solids and liquid retentions of equal duration. The liquid effluent from the first stage hydrolysis reactor was treated by a second-stage completely mixed immobilized-cell digester. Operated at a retention time of between 2 and 10 days with loading rates in the range of 36–437 lb/10^3 ft^3 day (0.58–7.0 kg COD/m^3 day), the second stage reactor achieved a COD removal of 65–78% with a methane conversion efficiency between 2 and 4 ft^3 CH$_4$/lb COD removed (0.12–0.25 m^3 CH$_4$/kg COD removed).

Other than these few reports there has been little research on the anaerobic digestion of the solid waste fraction and it is clear that certain conditions and waste types lead to operational instability. Early work questions the economic viability of the digestion process when used only for the treatment of paunch content and intestinal fecal material and it may be necessary to look at the codigestion of slaughterhouse waste fractions with other waste materials. One successful operation is the Kristianstad biogas plant in Sweden, which coprocesses organic household waste, animal manure, gastrointestinal waste from two slaughterhouses, biosludge from a distillery, and some vegetable processing waste [95]. The slaughterhouse waste fraction is 24,600 tonnes per annum of a total throughput of 71,200 tonnes which is treated in the 1.2 Mgal (4500 m^3) digester. The plant biogas production was equivalent to 20,000 MWh and the digester residue is returned to the land as a fertilizer. The plant represents an environmentally friendly method of waste treatment and appears to have overcome the problems of trying to digest slaughterhouse solid wastes in isolation.

REFERENCES

1. USEPA. *Development document for the proposed effluent limitations guidelines and standards* for the meat and poultry products industry. Office of Water (4303T), United States Environmental Protection Agency (USEPA), EPA-821-B-01–007, 2002.
2. Council of the European Communities. Council Directive on health problems affecting the production and marketing of meat products and certain other products of animal origin. *Official J. Eur. Comm.* **1977**, *L 026* (31 October 1977), 0085–0100.
3. Warris, P.D. *Meat Science: An Introductory Text*; CABI Publishing: New York, 2000.
4. Banks, C.J. Anaerobic digestion of solid and high nitrogen content fractions of slaughterhouse wastes. In *Environmentally Responsible Food Processing*; Niranjan, K., Okos, M.R., Rankowitz, M., Eds.; Vol. AIChE Symposium Series. American Institute of Chemical Engineers: New York, 1994; 103–109.
5. European Commission. *Survey of wastes spread on land – Final report*; Gendebien, A., Ferguson, R., Brink, J., Horth, H., Sullivan, M., Davis, R., Brunet, H., Dalimier, F., Landrea, B., Krack, D., Perot, J., and Orsi, C; Report No. CO 4953–2; Directorate-General for Environment, 2001.
6. Denmead, C.F. *Methane production from meat industry wastes and other potential methods for their utilization*, Publication no. 602; Meat Industry Research Institute; Hamilton, New Zealand, 1977.
7. Fernando, T. Utilization of Paunch Content Material by ultrafiltration. *Process Biochem.* **1980**, *15*, 7–9.
8. Pollack, H. Biological waste disposal from slaughterhouses. In *Anaerobic Digestion and Carbohydrate Hydrolysis of Wastes*; Ferrero, G.L., Ferranti, M.P. Naveau, H., Eds.; Elsevier Applied Science Publishers: London and New York, 1984: 323–330.
9. Brolls, E.K.; Broughton, M. The treatment of effluents arising from the animal by-products industry. In *Food Industry Wastes: Disposal and Recovery*; Herzka, A., Booth, R.G., Eds.; Applied Science Publishers: London and New Jersey, 1981; 184–203.
10. Grady, C.P.L.; Lim, H.C. *Biological Wastewater Treatment: Theory and Applications*; Marcel Dekker Inc: New York, 1980.
11. AEA Technology. *Environmental Technology Best Practice Programme: Reducing Water and Effluent Costs in Red Meat Abattoirs*, Report no. GG234; AEA Technology: Harwell, UK, 2000.
12. Johnson, A.S. Meat. In *Industrial Wastewater Control*; Gurnham, C.F., Ed.; Academic Press: New York and London, 1965.
13. World Bank. Meat processing and rendering. In *Pollution Prevention and Abatement Handbook*; World Bank: Washington DC, 1988: 336–340.
14. Johns, M.R. Developments in wastewater treatment in the meat processing industry: a review. *Biores. Technol.* **1995**, *54*, 203–216.

15. Nemerow, N.L.; Agardy, F.J. Origin and characteristics of meat-packing wastes. In *Strategies of Industrial and Hazardous Waste Management*; Agardy, F.J., Ed., Van Nostrand Reinhold: New York, 1998; 427–432.

16. Carawan, R.E.; Pilkington, D.H. Reduction in load from a meat-processing plant – beef. Randolph Packing Company/North Carolina Agricultural Extension Service, 1986.

17. UNEP. Cleaner production in meat processing. COWI Consulting Engineers/UNEP/Danish Environmental Protection Agency, 2000.

18. Dart, M.C. Treatment of meat trade effluents. In *Practical Waste Treatment and Disposal*; Dickinson, D., Ed.; Applied Science Publishers Ltd: London, **1974**; 75–86.

19. Denker, D.O.; Grothman, D.L.; Berthouex, P.M.; Scully, L.J.P.; Kerrigan, J.E.O. *Characterization and potential methods for reducing wastewater from in-plant hog slaughtering operations*; Interim Report to Mayer and Company/University of Wisconsin, 1973.

20. USEPA. *Development document for effluent limitation guidelines and new source performance standards for red meat segment of the meat product and rendering processing point source category*; Effluent Guidelines Division, Office of Air and Water Programs, USEPA, EPA-440/1–74–012a, 1974.

21. Macon, J.A.; Cote, D.N. *Study of meat packing wastes in North Carolina*; Industrial Extension Service, School of Engineering, North Carolina State College, 1961.

22. Millamena, S.M. Ozone treatment of slaughterhouse and laboratory wastewaters. *Aquacult. Eng.* **1992**, *11*, 23–31.

23. Lloyd, R.; Ware, G.C. Anaerobic digestion of waste waters from slaughterhouses. *Food Manuf.* **1956**, *31*, 511–515.

24. Kostyshyn, C.R.; Bonkoski, W.A.; Sointio, J.E. Anaerobic treatment of a beef processing plant wastewater: a case history. *Proceedings of 42nd Industrial Waste Conference*, Purdue University, Lafayette, IN, 1987, 673–692.

25. Jayangoudar, I.; Thanekar, A.; Krishnamoorthi, K.P.; Satyanarayana, S. Growth potentials of algae in anaerobically treated slaughterhouse waste. *Indian J. Environ. Health* **1983**, *25*, 209–213.

26. Gariepy, S.; Tyagi, R.D.; Couillard, D.; Tran, F. Thermophilic process for protein recovery as an alternative to slaughterhouse wastewater treatment. *Biol. Waste.* **1989**, *29*, 93–105.

27. Enders, K.E.; Hammer, M.; Weber, C.L.; Anaerobic lagoon treatment of slaughterhouse waste. *Water Sewage Works* **1968**; *115*, 283–288.

28. Borja, R.; Banks, C.J.; Wang, Z. Performance and kinetics of an Upflow Anaerobic Sludge Blanket (UASB) reactor treating slaughterhouse wastewater. *J. Environ. Sci. Heal. A* **1994**, *29*, 2063–2085.

29. Arora, H.C.; Routh, T. Treatments of slaughterhouse effluents by anaerobic contact filter. *Indian Assoc. Water Pollut. Control* **1980**, *16*, 67–78.

30. Cooper, R.N.; Heddle, J.F.; Russell, J.M. Characteristics and treatment of slaughterhouse effluents in New Zealand. *Prog. Wat. Treat.* **1979**, *11*, 55–68.

31. Dart, M.C. Treatment of waste waters from the meat industry. *Process Biochem.* **1974**, *9*, 11–14.

32. Council of the European Communities. Urban Waste Water Treatment Directive 91/271/EEC. *Off. J. Eur. Comm.*, **1991**, *L135/40-52*, (30 May 1991).

33. Eldridge, E.F. Meat-packing and slaughterhouse wastes. In *Industrial Waste Treatment Practice*; McGraw-Hill, London and New York, 1942.

34. Hopwood, D. Effluent treatment in meat and poultry processing industries. *Process Biochem.* **1977**, *12*, 5–8.

35. Meat and Livestock Australia Ltd. *Eco-Efficiency Manual for Meat Processing*, ABN 39 081 678 364 (MLA), 2002.

36. Camin, K.Q. Cost of waste treatment in the meat packing industry. In *Proceedings of 25th Purdue Industrial Waste Conference*, Purdue University Lafayette, IN, 1970; 193–202.

37. Masse, D. I.; Masse, L. Characterization of wastewater from hog slaughterhouses in Eastern Canada and evaluation of their in-plant wastewater treatment systems. *Can. Agr. Eng.* **2000**, *42*, 131–137.

38. Travers, S.M.; Lovett, D.A. Pressure flotation of slaughterhouse wastewaters using carbon dioxide. *Water Res.* **1985**, *19*, 1479–1482.

39. Hans Huber. *Wastewater treatment in slaughterhouses and meat processing factories*; Technical brochure; Hans Huber AG, Maschinen-und-Andagenbau: Berching, Germany, 2002.

40. Hans Huber. *Slaughterhouse wastewater treatment in combination with flotate sludge dewatering*, practice report/application info; Hans Huber AG, Maschinen-und-Andagenbau: Berching, Germany, 2002.

41. Peavy, H.S.; Rowe, D.R.; Tchobanoglous, G. *Environmental Engineering*; McGraw-Hill: New York, 1986.

42. USEPA. *Development document for effluent limitation guidelines and new source performance standards for the poultry segment of the meat product and rendering processing point source category*; USEPA, EPA-440/1-75-031b, 1975.

43. Steffen, A.J. Stabilisation ponds for meat packing wastes. *J. Wat. Pollut. Control Fed.* **1963**; *35*, 440–444.

44. Rollag, D.A.; Dornbush, J.N. Design and performance evaluation of an anaerobic stabilization pond system for meat-processing wastes. *J. Wat. Pollut. Control Fed.* **1966**, *38*, 1805–1812.

45. Witherow, J.L. Small meat-packers waste treatment systems. In *Proceedings of 28th Industrial Waste Conference*, Purdue University, Lafayette, IN, **1973**; 994–1009.

46. Wymore, A.H.; White, J.E. Treatment of slaughterhouse waste using anaerobic and aerated lagoons. *Water Sewage Works* **1986**; *115*, 492–498.

47. Black, M.G.; Brown, J.M.; Kaye, E. Operational experiences with an abattoir waste digester plant at Leeds. *Water Pollut. Control*, **1974**, *(73)*, 532–537.

48. Bohm, J.L. Digestion anaerobie des effluents d'abattoirs dans une unite pilote de 3000 litres epuration et production d'energie. *Entropie* **1986**, 130–131, 83–87.

49. Cillie, G.G.; Henzen, M.R.; Stander, G.J.; Baillie, R.D. Anaerobic digestion IV. The application of the process in waste purification. *Water Res.* **1969**, *3*, 623–643.

50. Hemens, J.; Shurben, D.G. Anaerobic digestion of wastewaters from a slaughterhouse. *Food Trade Rev.* **1959**, *29*, 2–7.

51. Andersen, D.R.; Schmid, L.A. Pilot plant study of an anaerobic filter for treating wastes from a complex slaughterhouse. In *Proceedings of 40th Purdue Industrial Waste Conference*, Purdue University, Lafayette, IN, 1985; 87–98.

52. Festino, C.; Aubart, C. Optimisation of anaerobic digestion of slaughterhouse wastes and mixtures of animal wastes with sewage sludges and slaughterhouse wastes. *Entropie* **1986**, 130–131, 20.

53. Narbonne, C.; Fromantin-Souli. Epuration d'effluents liquides et valorisation energetique de dechets solides d'abattoir par voie anaerobie. *Entropie* **1986**, 130–131, 57–60.

54. Rajeshwari, K.V.; Balakrishnan, M.; Kansal, A.; Lata, K.; Kishore, V.V.N. State-of-the-art of anaerobic digestion technology for industrial wastewater treatment. *Renew. Sust. Energy. Rev.* **2000**, *4*, 135–156.

55. Nell, J.H.; Krige, P.R. The disposal of solid abattoir waste by composting. *Water Res.* **2000**, *5*, 1177–1189.

56. Meat Industry Research Institute. The anaerobic treatment of effluent from meat processing operations; Denmead, C.F. Publication no. 405, 1974.

57. Vollmer, H.; Scholz, W. Recycle international In *4th International Recycling Congress*; Thome-Kozmiensky, K.J., Ed. Berlin, **1984**; 667–671.

58. Wheatley, A.D.; Cassell, L. Effluent treatment by anaerobic biofiltration. *Water Pollut. Control* **1985**, *84*, 10–22.

59. Campagna, R.; Del Medico, G.; Pieroni, M. Methane from biological anaerobic treatment of industrial organic wastes. In *3rd EC Conference on Energy from Biomass*; Elsevier Applied Science Publishers: London & New York, 1985.

60. Steiner, A.E.; Wildenauer, F.X.; Kandler, O. Anaerobic digestion and methane production from slaughterhouse wastes. In *3rd Energy Conservation Conference*; Elsevier Applied Science Publishers: London & New York, 1985.

61. Sayed, S.; van Campen, L.; Lettinga, G. Anaerobic treatment of slaughterhouse waste using a granular sludge UASB reactor. *Biol. Waste.* **1987**; *21*, 11–28.

62. Sayed, S.; De Zeeuw, W. The performance of a continuously operated flocculent sludge UASB reactor with slaughterhouse wastewater. *Biol. Waste.* **1988**, *24*, 213–226.

63. Ross, W.R. Anaerobic treatment of industrial effluents in South Africa. *Water SA* **1989**, *15*, 231–246.

64. Marchaim, U.; Levanon, D.; Danai, O.; Musaphy, S. A suggested solution for slaughterhouse wastes: uses of the residual materials after anaerobic digestion. *Biores. Technol.* **1991**, *37*, 127–134.

65. Tritt, W.P. The anaerobic treatment of slaughterhouse wastewater in fixed-bed reactors. *Biores. Technol.* **1992**, *41*, 201–207.

66. Polprasert, C.; Kemmadamrong, P.; Tran, F.T. Anaerobic Baffle Reactor (ABR) process for treating a slaughterhouse wastewater. *Environ. Technol.* **1992**; *13*, 857–865.

67. Sayed, S.K.I.; van der Spoel, H.; Truijen, G.J.P. A complete treatment of slaughterhouse wastewater combined with sludge stabilization using two stage high rate UASB process. *Water Sci. Technol.* **1993**, *27*, 83–90.

68. Kaul, S.N. Biogas from industrial wastewaters. *J. IPHE India* **1986**, *3*, 5–17.

69. Mara, D.D.; Pearson, H.W. *Design Manual for Waste Stabilisation Ponds in Mediterranean Countries*; Lagoon Technology International: Leeds, 1998.

70. USEPA. *Design Manual: Municipal Wastewater Stabilization Ponds*; USEPA, EPA 625/1-83-016, 1983.

71. US Department of Health Education and Welfare. *An Industrial Waste Guide to the Meat Industry*; USDHEW. US Public Health Service Publication No. 386, 1958.

72. Philips, S.A. Wastewater treatment handles cattle killing waste. *Water Sewage Works* **1975** *122*, 50–51.

73. Parker, D.S.; Lutz, M.P.; Pratt, A.M. New trickling filter applications in the USA. *Water Sci. Technol.* **1990**, *22*, 215–226.

74. Tchobanoglous, G.; Burton, F.L. *Wastewater Engineering: Metcalf & Eddy*; McGraw-Hill: New York, 1991.

75. Bull, M.A.; Sterritt, R.M.; Lester, J.N. The treatment of wastewaters from the meat industry: a review. *Environ. Tech. Lett.* **1982**, *3*, 117–126

76. Blanc, F.C.; O'Shaughnessy, J.C.; Corr, S.H. Treatment of soft drink bottling wastewater from bench-scale treatability to full-scale operation. In *Proceedings of 38th Purdue Industrial Waste Conference*; Purdue University, Lafayette, In, 1984; 243–256.

77. Bilstad, T. Upgrading slaughterhouse effluent with rotating biological contactors. In *Proceedings of the 1st International Conference on Fixed-Film Biological Processes*; Wu, Y.C., Smith, E.D. Miller, R.D., Eds. Kings Island: Ohio, USA, 1982; 892–912.

78. Bibby, J. *Innovative Technology Reduces Costs for a Meat Processing Plant*; Waterlink, Inc. 2002 (www.waterlink.com).

79. Gariepy, S.; Jyagi, R.D.; Couillard, D.; Tran, F. Thermophylic process for protein recovery as an alternative to slaughterhouse wastewater treatment. *Biol. Waste.* **1989**, *29*, 93–105.

80. Council of the European Communities. Regulation (EC) no. 1774/2002 of the European Parliament and of the Council of 3 October 2002. Laying down health rules concerning animal by-products not intended for human consumption. *Off. J. Eur. Comm.*, **2002**. *L 273/1-95* (10 October 2002).

81. Heddle, J.H. Activated sludge treatment of slaughterhouse wastes with protein recovery. *Water Res.* **1979**, *13*, 581–584.

82. Lovett, D.A.; Travers, S.M.; Davey, K.R. Activated sludge treatment of slaughterhouse wastewater – I. Influence of sludge age and feeding pattern. *Water Res.* **1984**, *18*, 429–434.

83. Travers, S.M.; Lovett, D.A. Activated sludge treatment of slaughterhouse wastewater – II. Influence of dissolved oxygen concentration. *Water Res.* **1984**, *18*, 435–439.

84. Irvine, R.L.; Miler, G.; Bhamrah, A.S. Sequencing batch treatment of wastewater in rural areas. *J. Wat. Pollut. Control Fed.* **1979**, *51*, 244.

85. Hadjinicolaou, J. Evaluation of a controlled condition in a sequencing batch reactor pilot plant operation for treatment of slaughterhouse wastewaters. *Can. Agr. Eng.* **1989**, *31*, 249–264.

86. EIDN. *Nutrient Removal System for Treatment of Abattoir Wastewater. Environmental Case Studies Directory*; Environmental Industries Development Network, www.eidn.co.au/Poowon.htm, 2002.

87. American Society of Agricultural Engineers. *Manure Production and Characteristics*; Standard ASAE D384.1, 1999.

88. Wang, Z. *Evaluation of a Two Stage Anaerobic Digestion System for the Treatment of Mixed Abattoir Wastes*. PhD Thesis, University of Manchester Institute of Science and Technology (UMIST), UK, 1996.

89. USEPA. *Guide to Field Storage of Biosolids*; Office of Wastewater, EPA/832-B-00-007, United States Environmental Protection Agency: Washington, DC, 2000.

90. Meat Industry Research Institute. *Stabilization of Save-All Bottom Solids Including Paunch Material, by Composting*; Supapong, B. Publication no. 336, 1973.

91. Australian Environment Protection Authority. Authorised officers manual, EPA 95/89, 1995.

92. Rooke, C.D. *Pilot Scale Studies on the Anaerobic Digestion of Abattoir Waste*; BSc Industrial Biology Dissertation, Southbank Polytechnic: Borough Rd, London, UK, 1988.

93. Cox, D.J.; Banks, C.J.; Hamilton, I.D.; Rooke, C.D. *The Anaerobic Digestion of Abattoir Waste at Bishops Castle Meat Co. Part I and II*. Meat and Livestock Commission Report, MLC: Milton Keynes, UK, 1987.

94. Banks, C.J.; Wang, Z. Development of a two phase anaerobic digester for the treatment of mixed abattoir wastes. *Water Sci. Technol.* **1999**, *40*, 67–76.

95. Centre for Renewable Energy, Environmental Technology Support Unit (ETSU). *Co-Digestion of Manure with Industrial and Household Waste*, Technical Brochure No. 118. AEA Technology: Harwell, UK, 2000.

4

Treatment of Palm Oil Wastewaters

Mohd Ali Hassan and Shahrakbah Yacob
University Putra Malaysia, Serdang, Malaysia

Yoshihito Shirai
Kyushu Institute of Technology, Kitakyushu, Japan

Yung-Tse Hung
Cleveland State University, Cleveland, Ohio, U.S.A.

4.1 INTRODUCTION

This chapter discusses the palm oil extraction process, wastewater treatment systems, and future technologies and applications for the palm oil industry. Crude palm oil (CPO) is extracted from the mesocarp of the fruitlets while palm kernel oil is obtained from the kernel (Fig. 4.1). The oil contents originating from mesocarp and kernel are 20 and 4%, respectively. Palm oil is a semisolid oil, rich in vitamins and several major fatty acids: oleic, palmitic, and linoleic. To produce palm oil, a considerable amount of water is needed, which in turn generates a large volume of wastewater. Palm oil mills and palm oil refineries are two main sources of palm oil wastewater; however, the first is the larger source of pollution and effluent known as palm oil mill effluent (POME). An estimated 30 million tons of palm oil mill effluent (POME) are produced annually from more than 300 palm oil mills in Malaysia. Owing to the high pollution load and environmental significance of POME, this chapter shall place emphasis on its treatment system.

4.1.1 Production of Crude Palm Oil (CPO)

It is important to note that no chemicals are added in the extraction of oil from the oil palm fruits, therefore, making all generated wastes nontoxic to the environment. The extraction of crude palm oil involves mainly mechanical and heating processes, and is illustrated in several steps below (Fig. 4.2).

Sterilization

To ensure the quality and the productivity of palm oil mill, the fresh fruit bunches (FFB) must be processed within 24 hours of harvesting. Thus, most of the palm oil mills are located in close

Figure 4.1 (A) Cross-section cutting of oil palm fruit showing shell, mesocarp, and kernel sections; (B) manual harvesting of fresh fruit bunches using sickle. (Courtesy of Malaysian Palm Oil Board.)

proximity to the oil palm plantation. During sterilization, the FFB is subjected to three cycles of pressures (30, 35, and 40 psi) for a total holding time of 90 minutes. There are four objectives of the FFB sterilization: (a) to remove external impurities, (b) to soften and loosen the fruitlets from the bunches, (c) to detach the kernels from the shells, and, most importantly, (d) to deactivate the enzymes responsible for the buildup of free fatty acids. The sterilization process acts as the first contributor to the accumulation of POME in the form of sterilizer condensate.

Bunch Stripping

Upon completion of the sterilization, the "cooked" FFB will be subjected to mechanical threshing to detach the fruitlets from the bunch. At this stage the loose fruitlets are transferred to the next process while the empty fruit bunches (EFB) can be recycled to the plantation for mulching or as organic fertilizer.

Digestion and Pressing

The digester consists of a cylindrical vessel equipped with stirrer and expeller arms mainly to digest and press the fruitlets. Steam is introduced to facilitate the oil extraction from the digested mesocarp. At the end of the process, oil and pressed cake comprising nuts and fiber are produced. The extracted oil will then be purified and clarified in the next stage. At the same time the fiber and nuts are separated in the depericarper column. The waste fiber is then burnt for energy generation inside the boiler.

Oil Clarification and Purification

As the name of this process implies, the extracted oil is clarified and purified to produce CPO. Dirt and other impurities are removed from the oil by centrifugation. Before the CPO is transferred to the storage tank, it is subjected to high temperatures to reduce the moisture content in the CPO. This is to control the rate of oil deterioration during storage prior to processing at the palm oil refinery. The sludge, which is the byproduct of clarification and purification procedures, is the main source of POME in terms of pollution strength and quantity.

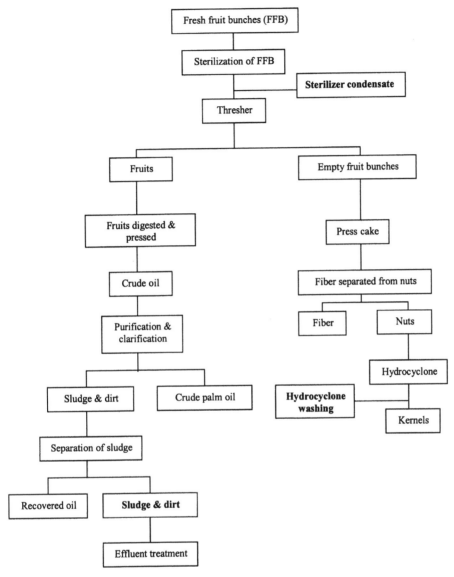

Figure 4.2   Flow diagram of crude palm oil extraction processes and sources of POME.

Nut Cracking

At this point, the nuts from the digestion and pressing processes are polished (to remove remnants of fiber) before being sent to the nut-cracking machine or ripple mill. The cracked mixture of kernels and shells is then separated in a winnowing column using upwards suction (hydrocyclone) and a clay bath. The third source of POME is the washing water of the hydrocyclone. The kernel produced is then stored before being transferred to palm kernel mill for oil extraction. Shell wastes will join the fiber at the boiler for steam and power generation.

4.1.2   Production of Refined Bleached Deodorized Palm Oil (RBDPO)

The refining of CPO employs physical/steam refining in which steam distillation is used to separate free fatty acids under high temperature and vacuum (Fig. 4.3). It consists of two main processes as follows.

Pretreatment

Before the actual refining process is carried out, the CPO is pretreated with phosphoric acid to eliminate impurities such as gums and trace metals. A bleaching technique is then used to remove phosphoric acid and its content under vacuum, followed by a filtration method. Solid waste in the form of sludge is disposed and buried in a landfill.

Deodorization

At this stage, steam is introduced under a vacuum condition to strip the pretreated oil of volatile free fatty acids, odoriferous compounds, and unstable pigments. The distillate for the deodorization process will form the main source of palm oil refinery effluent (PORE). The distillate has a free fatty acid content of approximately 80–90%. After the refining process, the oil is known as refined, bleached and deodorized palm oil (RBDPO). Further process such as fractionation of RBDPO will separate palm olein and stearin based on the different melting points of each component.

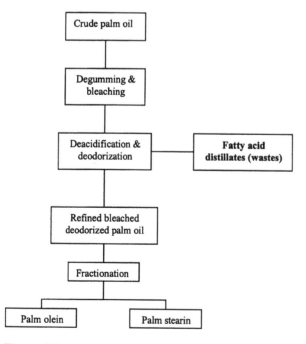

Figure 4.3   Flow diagram of physical refining process of crude palm oil and source of PORE.

4.2 PALM OIL MILL EFFLUENT (POME)

Palm oil mill effluent originates from two main processes: sterilization and clarification stages, as the condensate and clarification sludge, respectively (Fig. 4.2). The clarification sludge shows higher level of solid residues compared to the sterilizer condensate. Both contain some level of unrecovered oils and fats. The final POME would of course include hydrocyclone washing and cleaning up processes in the mill [1]. Approximately 1–1.5 tons of water are required to process 1 ton of FFB.

4.2.1 Properties of POME

Based on the process of oil extraction and the properties of FFB, POME is made up of about 95–96% water, 0.6–0.7% oil, and 4–5% total solid, including 2–4% suspended solids, which are mainly debris from palm mesocarp [2]. No chemicals are added during the production of palm oil; thus it is a nontoxic waste. Upon discharge from the mill, POME is in the form of a highly concentrated dark brown colloidal slurry of water, oil, and fine cellulosic materials. Due to the introduction of heat (from the sterilization stage) and vigorous mechanical processes, the discharge temperature of POME is approximately 80–90°C. The chemical properties of POME vary widely and depend on the operation and quality control of individual mills [3]. The general properties of POME are indicated in Table 4.1.

Apart from the organic composition, POME is also rich in mineral content, particularly phosphorus (18 mg/L), potassium (2270 mg/L), magnesium (615 mg/L) and calcium (439 mg/L) [2]. Thus most of the dewatered POME dried sludge (the solid endproduct of the POME treatment system) can be recycled or returned to the plantation as fertilizer.

4.2.2 Biological Treatment

Owing to its chemical properties, POME can be easily treated using a biological approach. With high organic and mineral content, POME is a suitable environment in which microorganisms can thrive. Hence, it could harbor a consortium of microorganisms that will consume or break down the wastes or pollutants, turning them into harmless byproducts. In some cases, these byproducts have high economic value and can be used as potential renewable sources or energy. In order to achieve such a goal, a suitable mixed population of microorganisms must be introduced and the

Table 4.1 Chemical Properties of Palm Oil Mill Effluent (POME)

Chemical property	Average	Range
pH	4.2	3.4–5.2
BOD (mg/L)	25,000	10,250–43,750
COD (mg/L)	50,000	15,000–100,000
Oil and grease (mg/L)	6000	150–18,000
Ammoniacal nitrogen (mg/ L)	35	4–80
Total nitrogen (mg/L)	750	180–1400
Suspended solid (mg/L)	18,000	5000–54000
Total solid (mg/L)	40,000	11,500–78,000

Source: Refs. 3, 4.

process should be optimized. Three biological processes are currently employed by the industry as a series of anaerobic, facultative anaerobic, and aerobic treatments. However, the major reduction of POME polluting strength — up to 95% of its original BOD — occurs in the first stage, that is, during the anaerobic treatment [4].

The anaerobic process involves three main stages; hydrolytic, acidogenic, and methanogenic. In the first stage, hydrolytic microorganisms secrete extracellular enzymes to hydrolyze the complex organic complexes into simpler compounds such as triglycerides, fatty acids, amino acids, and sugars. These compounds are then subjected to fermentative microorganisms that are responsible for their conversion into short-chain volatile fatty acids — mostly acetic, propionic, butyric acids, and alcohols. In the final stage, there are two separate biological transformations: first, the conversion of acetic acid into methane and carbon dioxide by methanogens; secondly, the conversion of propionic and butyric acids into acetic acid and hydrogen gas before being consumed by the methanogens. The endproducts of the anaerobic degradation are methane and carbon dioxide. Traces of hydrogen sulfide are also detected as the result of the activity of sulfate-reducing bacteria in the anaerobic treatment. The biochemical oxygen demand (BOD) at the first two stages remains at the same level as when it entered the anaerobic treatment, because only the breakdown of the complex compounds to a simpler mixture of organic materials has occurred. Only after the methanogenic stage will the BOD be reduced significantly.

4.2.3 Wastewater Treatment Systems for POME

The choice of POME wastewater treatment systems is largely influenced by the cost of operation and maintenance, availability of land, and location of the mill. The first factor plays a bigger role in the selection of the treatment systems. In Malaysia, the final discharge of the treated POME must follow the standards set by the Department of Environment (DOE) of Malaysia, which is 100 mg/L of BOD or less (Table 4.2) regardless of which treatment system is being utilized.

Pretreatment

Prior to the primary treatment, the mixed raw effluent (MRE, a mixture of wastewater from sterilization, clarification, and other sources) will undergo a pretreatment process that includes the removal of oil and grease, followed by a stabilization process. The excess oil and grease is extracted from the oil recovery pit using an oil skimmer. In this process, steam is continuously

Table 4.2 Environmental Regulations for Watercourse Discharge for Palm Oil Mill Effluent (POME)

Parameters	Level
BOD (mg/L)	100
Suspended solids (mg/L)	400
Oil and grease (mg/L)	50
Ammoniacal nitrogen (mg/L)	150
Total nitrogen (mg/L)	200
pH	5–9

Source: Ref. 5.

supplied to the MRE to aid the separation between oil and liquid sludge. The recovered oil is then reintroduced to the purification stage. The process will prevent excessive scum formation during the primary treatment and increase oil production. The MRE is then pumped into the cooling and mixing ponds for stabilization before primary treatment. No biological treatment occurs in these ponds. However, sedimentation of abrasive particles such as sand will ensure that all the pumping equipment is protected. The retention time of MRE in the cooling and mixing ponds is between 1 and 2 days.

Ponding System

The ponding system is comprised of a series of anaerobic, facultative, and algae (aerobic) ponds. These systems require less energy due to the absence of mechanical mixing, operation control, or monitoring. Mixing is very limited and achieved through the bubbling of gases; generally this is confined to anaerobic ponds and partly facultative ponds. On the other hand, the ponding system requires a vast area to accommodate a series of ponds in order to achieve the desired characteristics for discharge. For example, in the Serting Hilir Palm Oil Mill, the total length of the wastewater treatment system is about 2 km, with each pond about the size of a soccer field (Fig. 4.4). Only a clay lining of the ponds is needed, and they are constructed by excavating the earth. Hence, the ponding system is widely favored by the palm oil industry due to its marginal cost.

In constructing the ponds, the depth is crucial for determining the type of biological process. The length and width differ based on the availability of land. For anaerobic ponds, the optimum depth ranges from 5–7 m, while facultative anaerobic ponds are 1–1.5 m deep. The effective hydraulic retention time (HRT) of anaerobic and facultative anaerobic systems is 45 and 20 days, respectively. A shallower depth of approximately 0.5–1 m is required for aerobic ponds, with an HRT of 14 days. The POME is pumped at a very low rate of 0.2 to 0.35 kg $BOD/m^3 \cdot$ day of organic loading. In between the different stages of the ponding system, no pumping is required, as the treated POME will flow using gravity or a sideways tee-type subsurface draw-off system. Under these optimum conditions, the system is able to meet the requirement of DOE. The number of ponds will depend on the production capacity of each palm oil mill.

One problem faced by pond operators is the formation of scum, which occurs as the bubbles rise to the surface, taking with them fine suspended solids (Fig. 4.5). This results from the presence of oil and grease in the POME, which are not effectively removed during the pretreatment stage. Another disadvantage of the ponding system is the accumulation of solid sludge at the bottom of the ponds (Fig. 4.6). Eventually the sludge and scum will clump together inside the pond, lowering the effectiveness of the pond by reducing the volumetric capacity and HRT. When this happens, the sludge may be removed by either using submersible pumps or excavators. The removed sludge is dewatered and dried before being used as fertilizer. The cleanup is normally carried out every 5 years or when the capacity of the pond is significantly reduced.

Open Digester and Ponding Systems

This system is a combination of an open digester tank and a series of ponding systems (Fig. 4.7). The anaerobic digestion is carried out in the digester, then in the facultative anaerobic and algae ponds. It has been shown that by using an open digester, a better reduction of BOD can be achieved in a shorter time. Digesters are constructed of mild steel at various volumetric capacities ranging from 600 up to 3600 m^3. The treatment of treated POME from the digester

Figure 4.4   A series of ponds for POME treatment occupying a large land area. (Courtesy of Felda Palm Industries.)

will start at the facultative ponds, followed by the algae ponds. A description of the ponding systems is outlined in the previous section "Pretreatment."

The HRT of the digester is only 20–25 days and has a higher organic loading of $0.8–1.0$ BOD kg/m$^3 \cdot$ day compared to anaerobic ponds. With minimal financial input from the operators, no mechanical mixing equipment is installed in the digesters. Using the same principle as anaerobic ponds, mixing of POME is achieved via bubbling of biogas. Occasionally, the mixing is also achieved when the digester is being recharged with fresh POME. The treated POME is then overflowed into the ponding system for further treatment.

Although the digester system has been proven to be superior to anaerobic ponds, it also has similar problems of scum formation and solid sludge accumulation. Another serious problem is

Figure 4.5   Active bubbling of gases leading to the formation of scum.

the corrosion of the steel structures due to long exposure to hydrogen sulfide. Incidents such as burst and collapsed digesters have been recorded. Accumulated solids could be easily removed using the sludge pipe located at the bottom of the digester. The dewatered and dried sludge can then be disposed for land application.

Extended Aeration

To complement the previous systems, mechanical surface aerators can be introduced at the aerobic ponds (Fig. 4.8). This effectively reduces the BOD through aerobic processes. The aerators are normally installed at the end of the ponding system before discharge. However, this happens only where land area is a constraint and does not permit extensive wastewater treatment. Otherwise, aerators must be provided to meet DOE regulations.

4.3   PALM OIL REFINERY EFFLUENT (PORE)

Following the production of CPO from the palm oil mill, the CPO is then subjected to further refining before it can be categorized as edible oil. Even after the clarification and purification processes, there are still large amounts of impurities such as gums, pigments, trace of metals, and soluble fats that cause unpleasant taste, odor, and color. There are three common types of operation in the palm oil refineries: (a) physical refining and dry fractionation, (b) physical refining and detergent fractionation, and (c) physical and chemical refining with dry/detergent fractionation [6].

Figure 4.6 Formation of islets of sludge in the middle of the pond. (Courtesy of Felda Palm Industries.)

4.3.1 Chemical Properties of PORE

The main sources of PORE are water from the deodorization process and cleaning operations within the mill (Fig. 4.3). The characteristics of PORE are very much dependent on the type of process employed. The main chemical properties of PORE are as described in Table 4.3.

Figure 4.7 A series of $3600\,m^3$ open digesters for POME treatment. (Courtesy of Felda Palm Industries.)

Figure 4.8 An aerator system installed to accelerate BOD reduction at the aerobic pond. (Courtesy of Malaysian Palm Oil Board.)

In comparison with POME, PORE is less polluting. This is largely due to the absence of oil and grease, and its low organic load. From Table 4.3, it is obvious why most of the palm oil refineries in Malaysia have adopted physical refining and dry fractionation to produce edible oil. Not only does the system reduce the effluent problem, but higher yield and oil purity with lower operating costs are obtained.

Table 4.3 Chemical Properties of Palm Oil Refinery Effluent (PORE) Based on Different Operations

Chemical properties	Type of refinery processes		
	Physical refining and dry fractionation	Physical refining and detergent fractionation	Physical and chemical refining with dry/detergent fractionation
Temperature (°C)	35	42	57
pH	5.3	4.9	3.0
BOD (mg/L)	530	2640	4180
COD (mg/L)	890	5730	7700
Total solids (mg/L)	330	1170	2070
Suspended solids (mg/L)	50	12	6
Phosphorus (mg/L)	4	1	12
Total fatty matter (mg/L)	220	1580	3550

Source: Ref. 3.

4.3.2 Wastewater Treatment Systems for PORE

Unlike POME wastewater treatment systems, the PORE system is more systematic and predictable. Most PORE systems involve biological processes, with some using physical and chemical methods such as sedimentation, dissolved air flotation after coagulation and flocculation using lime, alum and polyelectrolytes [8].

Pretreatment of PORE

The first step in ensuring satisfactory performance of a PORE treatment plant is to remove oil and fat from the MRE. The separation is carried out using several methods such as fat traps, tilted-plate separators, and dissolved flotation units. Beside physical separations, the addition of chemical flocculants and coagulants also helps in reducing the total fatty matter and other suspended solids. Before the commencement of the biological treatment, the pH of the PORE is adjusted to the desired level as pH plays an important role in the optimum biodegradation of PORE.

Activated Sludge System

Many palm oil refineries use activated sludge systems to treat PORE. This is because of land constraints (for ponding systems) and effective removal of BOD in a short HRT [6]. This system can be very effective if the level of total fatty matter is kept low after the pretreatment stage. The presence of fatty matter in the activated sludge systems will contribute not only higher BOD but the formation of scum. This leads to poor performance of the system.

The treatment is carried out by bringing PORE into contact with a mixed population of aerobic microorganisms in the controlled environment of the activated sludge system. In this process, oxygen is supplied via aeration or vigorous agitation for the oxidation of organic wastes to carbon dioxide. After the treatment, the suspended biomass is separated in the clarifier. The biomass is reintroduced back into the activated sludge systems as "return activated sludge." This is to ensure the density of microorganisms is maintained at an optimum level for maximum removal of BOD. The supernatant from the clarifier can then be safely discharged into the watercourse. The HRT of PORE and cell residence time are 1–2 days and 3–10 days, respectively. Using this system, a final BOD and suspended solids of 20 and 30 mg/L, respectively, can be obtained with 1500–2000 mg/L of mixed liquor suspended solids. Apart from the energy requirement to operate the treatment facilities, additional nutrients are normally added to the effluent. This is because the effluent from the palm oil refinery is low in nitrogen content, which is essential for the growth of aerobic micro-organisms. A ratio of BOD:N of 100:5 is kept constant throughout the process.

4.4 POTENTIAL TECHNOLOGIES AND COMMERCIAL APPLICATIONS OF PALM OIL WASTES

4.4.1 POME Treatment

Evaporation Technology

In one study, a 200 L single-effect evaporator was constructed to test the evaporation technique in POME treatment [8]. It used the principle of rapid heating to vaporize water at 600 mmHg and 80°C using a plate heat exchanger. Staggered feeding of fresh POME was introduced into the evaporator when the liquor dropped by half of the initial volume. The feeding was carried out until the accumulated solid sludge reached the pre-set level of 30%. The solid was then discharged before the new cycle began. The single-effect evaporator was able to recover 85% of water from POME with a good quality distillate of 20 mg/L BOD. The distillate could be recycled as process water or feedwater for the boiler in the mill. Even though the system promises a significant reduction of liquid waste (and thus less dependence on vast land area for ponding systems), the energy required for heating may impose financial constraints for the mill operator. Moreover, the mill may have to make a big investment in equipment, skilled operators, and maintenance. Further studies are being carried out to produce cost-effective systems such as utilizing excess organic biomass from the mill as an energy source.

High-Technology Bioreactor Design

There have been numerous studies to optimize the anaerobic treatment of POME using various designs of bioreactor. Laboratory-scale studies have been carried out to evaluate the effectiveness of anaerobic filters (AF) and a fluidized-bed reactor (FBR) in treating POME [9]. About 90% of the fed COD was effectively removed by both reactor systems. However, when the COD loading was increased, a significant reduction in terms of COD removal was recorded in the FBR system, while clogging of the filter was evident in the AF reactor. A higher COD removal efficiency was reported [10] when using a modified anaerobic baffled reactor (MABR). The system also demonstrated a short retention time of 3 days. Despite the good potential of the bioreactor systems for POME treatment, none has been implemented at a larger scale.

Power Generation: Closed Digester

The composition of biogas emitted from an open digester tank and the lagoon was lower than that reported for laboratory studies [2]. The biogas composition was 40% methane and 60% carbon dioxide for the open digester tank, and 55% methane and 45% carbon dioxide in anaerobic lagoons. In terms of energy value, it is comparable to commercially available gas fuels as shown in Table 4.4. The potential energy that could be generated from 1 m^3 of biogas is 1.8 kWh [11].

A closed digesting system was tested to improve the anaerobic digestion of POME, leading to the production of biogas. Using the same design of open digester, a fixed or floating cover is included, equipped with the other facilities such as gas collector, safety valves, and monitoring facilities.

Compost

Based on our research, dewatered POME sludge can be composted with domestic wastes and bulking agents such as shredded wood and sawdust. A modified composter from a cement mixer with insulated drum was used as a reactor to run the composting process. Experimental

Table 4.4 Comparisons Between Methane Derived from Anaerobic Digestion of POME and Other Gas Fuels

Chemical properties	Methane	Natural gas	Propane
Gross calorific value (kcal/kg)	4740–6150	907	24,000
Specific gravity	0.847–1.002	0.584	1.5
Ignition temperature (°C)	650–750	650–750	450–500
Inflammable limits (%)	7.5–21.0	5–15	2–10
Combustion air required (m^3/m^3)	9.6	9.6	13.8

Source: Ref. 11.

parameters such as aeration, pH, temperature, C/N ratio, and moisture content were controlled and monitored during the fermentation phase of the composting process. It took about 40 days to completely convert the POME sludge into compost via the solid substrate fermentation process with mixed microbial inoculum. The carbon content decreased towards the end of the composting process, which resulted in a decrease of the C/N ratio from 30 to 20. The low C/N ratio of the final compost product was very important as an indicator of maturity. The characteristics of the final compost products for POME sludge were similar to commercial composts and complied with US Environmental Protection Agency (EPA) standards, especially in heavy metal content and total coliforms. Planting out tests with leafy vegetables showed satisfactory performance [12].

Organic Acids

Two-stage fermentation was carried out in a study where POME was used as substrate for volatile fatty acids (VFA) production by continuous anaerobic treatment using a locally fabricated 50 L continuous stirred tank reactor (CSTR). The highest VFA obtained was at 15 g/L at pH 6.5, 30°C, 100 rpm, sludge to POME ratio 1 : 1, HRT 4 days, without sludge recycle. The highest BOD removal corresponded with the high production of organic acids. The organic acids produced from POME were then recovered and purified using acidification and evaporation techniques. A clarified concentrated VFA comprised of 45 g/L acetic, 20 g/L propionic and 22 g/L butyric acids were obtained with a recovery yield of 76% [13].

Production of Polyhydroxyalkanoates

The organic acids from treated POME can be used to biologically synthesize polyhydroxyalkanoates (PHA), a bacterial bioplastic. The concentrated organic acids obtained were used in a fed-batch culture of *Alcaligenes eutrophus* for the production of PHA. About 45% PHA content in the dry cells could be obtained, corresponding to a yield of 0.32 from acetic acid. The overall volumetric productivity of PHA is estimated at 0.09 g PHA/L hour. This indicates that the application of a high-density cell culture to produce bioplastic from POME can be achieved [14].

Biological Hydrogen

Another potential application of POME as a renewable resource of energy is the production of biological hydrogen via a fermentation process. The main purpose of producing biological hydrogen is to offer an alternative source of energy to fossil fuels. The major advantage of biological hydrogen is the lack of polluting emission since the utilization of hydrogen, either via combustion or fuel cells, results in pure water [15]. Currently, two proposed systems produce biological hydrogen using photoheterotrophic and heterotrophic bacteria. However, the latter is most suitable for POME due to limited light penetration caused by the sludge particles as experienced during the production of PHA by phototrophic *Rhodobacter sphaeroides* [16]. Moreover, it would be costly to construct and maintain a photobioreactor at a commercial-scale operation.

In the anaerobic degradation of POME, complex organic matter is converted into a mixture of methane and carbon dioxide in a network of syntrophic bacteria. Prior to this, fermentative and acetogenic bacteria first convert organic matter into a mixture of VFA and hydrogen before being consumed by methanogenic bacteria. Based on the metabolic activities of these microorganisms in POME degradation, a system combining the organic acids and biological hydrogen production is suggested. However, the utilization of biological hydrogen from POME is still at the planning stage. Major development in terms of selection of suitable microorganisms and optimization of process conditions is required for cost-effective production of hydrogen. Nevertheless, this technology promises a means to conserve the environment by generating clean energy.

4.4.2 PORE Treatment

Sequential Batch Reactor System

A new technology using the sequential batch reactor (SBR) technique has been shown to provide an effective treatment of PORE [7] as shown in Figure 4.9. Among the advantages of SBR over the conventional activated sludge are an automated control system, more versatility, stability, and the ability to handle high fluctuations in organic loading. A consistent output of BOD below 50 mg/L was observed. With this system, the hydraulic retention time and solid sludge content could be controlled, thus eliminating the need for clarifier and sludge recycling facilities.

4.5 FUTURE TRENDS

From the preceding section, several potential and emerging technologies for POME wastewater treatment system can be integrated into the palm oil mill operation (Fig. 4.10). The strategy is to combine the existing wastewater treatment system with the production of appropriate bioproducts, towards zero discharge for the palm oil industry [17]. In anaerobic treatment, methanogenic activity will be suppressed or inhibited in order to extract the organic acids produced. This, in turn, shall lower the greenhouse gases (methane and carbon dioxide) emissions from the anaerobic digestion, thus reducing the effects of global warming. Further separation and purification processes are needed before organic acids can be utilized as a

Figure 4.9 A pilot plant sequential bioreactor system tested for POME treatment. (Courtesy of Malaysian Palm Oil Board.)

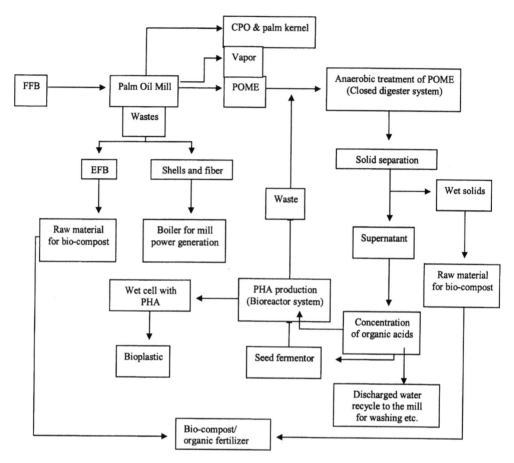

Figure 4.10 Proposed integrated palm oil production and POME wastewater treatment system (from Ref. 17).

substrate for PHA-producing microorganisms. The solid wastes (sludges) generated from the wastewater treatment system will be used as a mixture with EFB to form biocompost.

Wastes generated from the palm oil mill contain a high percentage of degradable organic material and can be converted into value-added products and chemicals. It is expected that changes in the technologies in POME treatment could lead to a substantial reduction in terms of waste discharged. On the other hand, the palm oil industry will experience a sustainable growth by addressing the excessive pollution issue through development of biowastes as alternative sources of renewable energy and valued chemicals. This in turn shall generate additional revenue for the industry. Finally, better-integrated waste management is associated with other environmental benefits such as reduction of surface waterbody and groundwater contamination, less waste of land and resources, lower air pollution, and a reduction of accelerating climate changes.

REFERENCES

1. Agamuthu, P. Palm oil mill effluent — treatment and utilization. In *Waste Treatment Plant*; Sastry, C.A., Hashim, M.A., Agamuthu, P., Eds.; Narosa Publishing House: New Delhi, India, 1995; 338–360.

2. Ma, A.N. Treatment of palm oil mill effluent. In *Oil Palm and the Environment — A Malaysian Perspective*; Singh, G., Lim, K.H., Leng, T., David, L.K., Eds.; Malaysian Oil Palm Growers Council: Selangor, Malaysia, 1999; 113–126.

3. Basiron, Y.; Darus, A. The oil palm industry — from pollution to zero waste. *The Planter* 1995, *72* (*840*), 141–165.

4. Ma, A.N. Environment management for the palm oil industry. *Palm Oil Develop.* 1999, *30*, 1–10.

5. Md. Noor, M. Environmental legislation: Environmental quality act, 1974. In *Oil Palm and the Environment — A Malaysian Perspective*; Singh, G., Lim, K.H., Leng, T., David, L.K., Eds.; Malaysian Oil Palm Growers Council: Selangor, Malaysia, 1999; 261–264.

6. Ma, A.N. Treatment of palm oil refinery effluent. In *Oil Palm and the Environment — A Malaysian Perspective*; Singh, G., Lim, K.H., Leng, T., David, L.K., Eds.; Malaysian Oil Palm Growers Council: Selangor, Malaysia, 1999; 127–136.

7. Sastry, C.A. Waste treatment case studies. In *Waste Treatment Plant*; Sastry, C.A., Hashim, M.A., Agamuthu, P., Eds.; Narosa Publishing House: New Delhi, 1995; 361–379.

8. Ma, A.N. Evaporation technology for pollution abatement in palm oil mills. In *Proceedings of the National Seminar on Palm Oil Milling, Refining Technology and Quality*; Chang, A.K.C., Ed.; PORIM: Selangor, Malaysia, 1997; 167–171.

9. Borja, R.; Banks, C.J. Comparison of an anaerobic filter and an anaerobic fluidized bed reactor treating palm oil mill effluent. *Process Biochem.* 1995, *30* (*6*), 511–521.

10. Faisal, M; Unno, H. Kinetic analysis of palm oil mill wastewater treatment by a modified anaerobic baffled reactor. *Biochem. Eng. J.* 2001, *9*, 25–31.

11. Ma, A.N.; Toh, T.S.; Chua, N.S. Renewable energy from oil palm industry. In *Oil Palm and the Environment — A Malaysian Perspective*; Singh, G., Lim, K.H., Leng, T., David, L.K., Eds.; Malaysian Oil Palm Growers Council: Selangor, Malaysia, 1999; 253–260.

12. Abdul Rahman, A.R.; Baharum, Z.; Hassan, M.A.; Idris, A. Bioreactor composting of selected organic sludges. *Proceedings of the 13th National Biotechnology Seminar*, Penang, Malaysia, 2001.

13. Noraini, A.R.; Hassan, M.A.; Shirai, Y.; Karim, M.I.A. Production of organic acids from palm oil mill effluent during continuous anaerobic treatment. *Asia-Pac. J. Mol. Biol.* 1999, *7* (*2*), 179–184.

14. Hassan, M.A.; Shirai, Y.; Umeki, H.; Yamazumi, H.; Jin, S.; Yamamoto, S.; Abdul Karim, M.I.; Nakanishi, K.; Hashimoto, K. Acetic acid separation from anaerobically treated palm oil mill effluent by ion exchange resin for the production of polyhydroxyalkanoate by *Alcaligenes eutrophus. Biosci. Biotech. Biochem.* 1997, *61* (*9*), 1465–1468.

15. Claassen, P.A.M; van Lier, J.B.; Lopez Contreras, A.M.; van Niel, E.W.J.; Sijtsma, L.; Stams, A.J.M.; de Vries, S.S.; Weusthuis, R.A. Utilization of biomass for the supply of energy carrier. *Appl. Microbiol. Biot.* 1999, *52*, 741–755.

16. Hassan, M.A.; Shirai, Y.; Kusubayashi, N.; Abdul Karim, M.I.; Nakanishi, K.; Hashimoto, K. The production of polyhydroxyalkanoate from anaerobically treated palm oil mill effluent by *Rhodobacter sphaeroides. J. Ferment. Bioeng.* 1997, *83* (*3*), 485–488.

17. Hassan, M.A.; Nawata, O.; Shirai, Y.; Noraini, A.R.; Yee, P.L.; Karim, M.I.A. A proposal for zero emission from palm oil industry incorporating the production of polyhydroxyalkanoates from palm oil mill effluent. *J. Chem. Eng. Jpn.* 2002, *35* (*1*), 9–14.

18. Malaysia Palm Oil Board. The processing sector, In *Oil Palm Industry in Malaysia*; Salleh, M., Ed.; Malaysia Palm Oil Board: Selangor, Malaysia, 2000; 10–17.

5
Olive Oil Waste Treatment

Adel Awad and Hana Salman
Tishreen University, Lattakia, Syria

Yung-Tse Hung
Cleveland State University, Cleveland, Ohio, U.S.A.

5.1 INTRODUCTION

The extraction and use of olive oil has been linked to Mediterranean culture and history since 4000 BC. Several terms used today are reminders of this ancient heritage. For example, the Latin words *olea* (oil) and *olivum* (olive) were derived from the Greek word *elaia*. As a dietary note, olive oil is high in nutrition, and appears to have positive effects in the prevention and reduction of vascular problems, high blood pressure, arteriosclerosis, thrombosis, and even some types of cancer [1].

The social and economic importance of the olive production sector may be observed by considering some representative data. In the European Union (EU), there are about 2 million companies related to olives and olive oil. Worldwide olive oil production is about 2.6 million tons per year, 78% (about 2.03 million tons) of which are produced in the EU (main producers: Spain, Greece, and Italy). Other main producers are Turkey (190,000 tons), Tunisia (170,000 tons), Syria (110,000 tons), and Morocco (70,000 tons). More than 95% of the world's olives are harvested in the Mediterranean region. In Spain alone, more than 200 million olive trees out of the total world number of 800 million are cultivated on an area of approximately 8.5 million ha. Within Spain, 130 million olive trees are found in Andalusia, where about 15% of the total arable land is used for olive cultivation [2].

According to the FAOSTAT database [3], the total waste generated by olive oil production worldwide in 1998 was 7.3 million tons, 80% of which was generated in the EU and 20% generated in other countries. In Spain, the top olive oil producer, the generated waste in 1998 alone was 2.6 million tons, or about 36% of the waste generated worldwide.

Approximately 20 million tons of fresh water are required for olive oil production in the Mediterranean area, resulting in up to 30 million tons of solid–liquid waste (*orujo* and *alpeorujo*) per year. By comparison, the annual amount of sewage sludge in Germany is 55 million m^3, with 5% dry solid matter content [4].

5.2 OLIVE OIL MILL TECHNOLOGY

The olive oil extraction industry is principally located around the Mediterranean, Aegean, and Marmara seas, and employs a very simple technology (Fig. 5.1). First, the olives are washed to remove physical impurities such as leaves, pieces of wood, as well as any pesticides. Afterwards, the olives are ground and mixed into paste. Although a large variety of extracting systems are available, two methods are generally employed: traditional pressing and modern centrifuging. Pressing is a method that has evolved since ancient times, while centrifuging is a relatively new technology. Figures 5.2 and 5.3 are schematic drawings of the two systems. Figure 5.2 represents the traditional discontinuous press of olive oil mills, while Figure 5.3 represents more recent continuous solid/liquid decanting system (three-phase decanting mills). Both systems (traditional and three-phase decanter) generate one stream of olive oil and two streams of wastes, an aqueous waste called *alpechin* (black water) and a wet solid called *orujo*. A new method of two-phase decanting, extensively adopted in Spain and growing in popularity in Italy and Greece, produces one stream of olive oil and a single stream of waste formed of a very wet solid called *alpeorujo*.

Looking at milling systems employed worldwide, a greater percentage of centrifuge systems are being used compared to pressing systems. Because of the higher productivity of the more modern centrifuge systems, they are capable of processing olives in less time, which is a requisite for a final quality product [5].

Furthermore, in contrast to the three-phase decanter process, the two-phase decanter does not require the addition of water to the ground olives. The three-phase decanter requires up to 50 kg water for 100 kg olive pulp in order to separate the latter into three phases: oil, water, and solid suspension [6]. This is necessary, since a layer of water must be formed with no bonds to the oil and solid phase inside the decanter. Thus, up to 60 kg of alpechin may be produced from 100 kg olives. Alpechin is a wastewater rich in polyphenols, color, and soluble stuffs such as sugar and salt [7].

In the two-phase decanter, there must be no traces of water inside the decanter to prevent water flowing out with the oil and reducing the paste viscosity, which leads to improved oil extraction [8]. The two-phase decanter process is considered more ecological, not only because it reduces pollution in terms of the alpechin, but since it requires less water for processing [9]. Depending on the preparation steps (ripeness, milling, malaxing time, temperature, using enzymes or talcum, etc.), the oil yield using the two-phase decanter may be higher than that using the three-phase decanter [10]. The oil quality is also different in each process. In the case

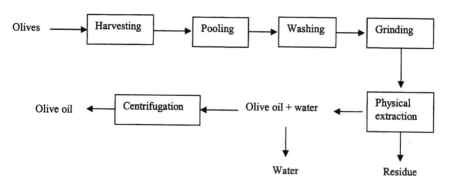

Figure 5.1 Technology generally used to produce olive oil (from Ref. 5).

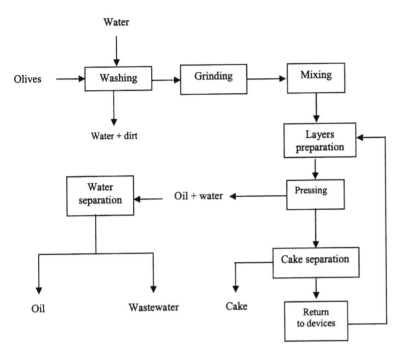

Figure 5.2 Traditional pressing for olive oil production (from Ref. 5).

of the three-phase decanter, the main part of the polyphenols will be washed out in the alpechin phase. These chemicals, which also provide antioxidation protection, are sustained in the oil phase using the two-phase decanter; the results are better conditions for a long oil shelf life as well as a more typical fruit taste [11].

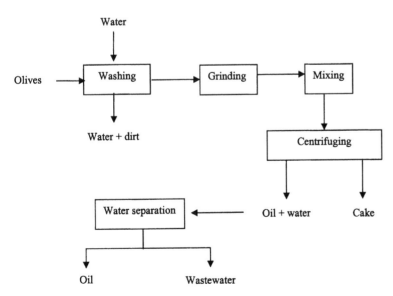

Figure 5.3 Modern centrifuging for olive oil production (three-phase decanter) (from Ref. 5).

The alpeorujo (solid/liquid waste) has a moisture content of 60–65% at the decanter output while the moisture content of the solid waste using the three-phase decanter is about 50%, and by traditional pressing is about 25%. One drawback is that two-phase alpeorujo is more difficult to store due to its humidity. Comparing the three different solids (orujo press cake, three-phase decanter orujo, and two-phase decanter alpeorujo), the two-phase decanter alpeorujo is the best residue to be reprocessed for oil [9].

5.3 OLIVE OIL WASTEWATER CHARACTERISTICS

The olive consists of flesh (75–85% by weight), stone (13–23% by weight) and seed (2–3% by weight) [12].The chemical composition of the olive is shown in Table 5.1. The quantities and composition of olive mill waste (OMW) vary considerably, owing to geographical and climatic conditions, tree age, olive type, extraction technology used, use of pesticides and fertilizers, harvest time, and stage of maturity.

In waste generated by olive oil mills, the only constituents found are produced either from the olive or its vegetation water, or from the production process itself. Auxiliary agents, which are hardly used in production, may be influenced and controlled by process management. Therefore, they are not important to the composition of wastewater. However, the composition of the olive and its vegetation wastewater cannot be influenced; thus, the constituents of vegetation wastewater are decisive for the expected pollution load. Table 5.2 summarizes some literature data concerning the constituents of olive oil wastewater [13–25]. The variations of maximum and minimum concentrations of olive oil wastewater resulting from both methods (traditional presses and decanter centrifuge) are also presented, according to the International Olive Oil Council (IOOC) in Madrid [26], in Table 5.3.

Wastewater from olive oil production is characterized by the following special features and components [27]:

- color ranging from intensive violet–dark brown to black;
- strong olive oil odor;
- high degree of organic pollution (COD values up to 220 g/L, and in some cases reaching 400 g/L) at a COD/BOD_5 ratio between 1.4 and 2.5 and sometimes reaching 5 (difficult to be degraded);

Table 5.1 Composition of Olives

Constituents	Pulp	Stone	Seed
Water	50–60	9.3	30
Oil	15–30	0.7	27.3
Constituents containing nitrogen	2–5	3.4	10.2
Sugar	3–7.5	41	26.6
Cellulose	3–6	38	1.9
Minerals	1–2	4.1	1.5
Polyphenol (aromatic substances)	2–2.25	0.1	0.5–1
Others	–	3.4	2.4

Note: Values in percent by weight (%).
Source: Ref. 12.

Table 5.2 Summary of the Constituents of Olive Oil Wastewater (Alpechin) According to Different Literature Data

Parameter	Pompei[13] (1974)	Fiestas[14] (1981)	Garcia[18] (1989)[a]	Steegmans[15] (1992)	Hamdi[16] (1993)	Borja[25] (1995)	Beccari[23] (1996)[f]	Ubay[22] (1997)[e]	Zouari[24] (1998)[c]	Andreozzi[17] (1998)	Beltran-Heredia[21] (2000)[d]	Kissi[20] (2001)[b]	Rivas[19] (2001)[a]
pH	–	4.7	–	5.3	3–5.9	5.2	5.06	4.7	–	5.09	13.6	4.2	12.9
Chemical oxygen demand, COD (g/L)	195	–	15–40	108.6	40–220	60	90 (filtered 63)	115–120	225	121.8	6.7	50	24.45
Biochemical oxygen demand in 5 days, BOD$_5$ (g/L)	38.44	–	9–20	41.3	23–100	–	–	–	58	–	4.3	–	14.8
Total solids, TS (g/L)	–	1–3	–	19.2	1–20	48.6	51.5	8.5–9 (SS)	–	102.5	22.9	4 (SS)	–
Organic total solids (g/L)	–	–	–	16.7	–	41.9	37.2	–	190	81.6	4.6	–	–
Fats (g/L)	–	3–8	–	2.33	1–23	–	–	7.7	–	9.8	–	–	–
Polyphenols (g/L)	17.5	–	0.5	0.002	5–80	0.3	3.3	–	–	6.2	0.12	12	0.833
Volatile organic acids (g/L)	–	5–10	–	0.78	0.8–10	0.64	15.25	–	–	0.96	–	–	–
Total nitrogen (g/L)	0.81	0.3–0.6	–	0.6	0.3–1.2	0.16 (N-NH$_4$)	0.84	0.18	1.2	0.95	–	–	–

[a]Wastewater generated in the table olive processing industries during different stages including washing of fruits, debittering of green olives (addition of sodium hydroxide), fermentation and packing.

[b]Other parameters were measured such as: color (A$_{395}$) = 16; Cl$^-$ = 11.9 g/L; K$^+$ = 2.5 g/L; NH$_4^+$ = 0.15 g/L.

[c]Since the dark color of olive oil mill effluent was difficult to determine quantitatively, the optical value (OD) at 390 nm was measured; this value was 8.5.

[d]Represents wastewater generated in table olive processing plant (black olives). Aromatic compounds (A) = 17 were determined by measuring the absorbance of the samples at 250 nm (the maximum absorbance wavelength of these organic compounds).

[e]Represents concentrated black water from a traditional olive oil mill plant. Other parameters were measured such as SS = 8.5–9 g/L, Total P = 1.2 g/L.

[f]Other parameters were measured such as TC = 25.5 g/L, Total P = 0.58 g/L, Lipids = 8.6 g/L.

Source: Refs. 13–25.

Table 5.3 Maximum and Minimum Concentration Values of Olive Oil Wastewater According to Applied Type of Technology

	Technology type	
Parameters	Centrifuge	Traditional presses
pH	4.55–5.89	4.73–5.73
Dry matter (g/L)	9.5–161.2	15.5–266
Specific weight	1.007–1.046	1.02–1.09
Oil (g/L)	0.41–29.8	0.12–11.5
Reducing sugars (g/L)	1.6–34.7	9.7–67.1
Total polyphenols (g/L)	0.4–7.1	1.4–14.3
O-diphenols (g/L)	0.3–6	0.9–13.3
Hydroxytyrosol (mg/L)	43–426	71–937
Ash (g/L)	0.4–12.5	4–42.6
COD (g/L)	15.2–199.2	42.1–389.5
Organic nitrogen (mg/L)	140–966	154–1106
Total phosphorus (mg/L)	42–495	157–915
Sodium (mg/L)	18–124	38–285
Potassium (mg/L)	630–2500	1500–5000
Calcium (mg/L)	47–200	58–408
Magnesium (mg/L)	60–180	90–337
Iron (mg/L)	8.8–31.5	16.4–86.4
Copper (mg/L)	1.16–3.42	1.10–4.75
Zinc (mg/L)	1.42–4.48	1.6–6.50
Manganese (mg/L)	0.87–5.20	2.16–8.90
Nickel (mg/L)	0.29–1.44	0.44–1.58
Cobalt (mg/L)	0.12–0.48	0.18–0.96
Lead (mg/L)	0.35–0.72	0.40–1.85

Source: Ref. 26.

- pH between 3 and 5.9 (slightly acid);
- high content of polyphenols, up to 80 g/L; other references up to 10 g/L [28];
- high content of solid matter (total solids up to 102.5 g/L);
- high content of oil (up to 30 g/L).

Table 5.4 compares the composition values of olive oil mill wastewater (A and B) with those of municipal wastewater (C). While the ratio COD/BOD_5 in both types of wastewater is rather close (between 1.5 and 2.5), there is a big difference between the two for the ratio (BOD:N:P); olive oil wastewater (100:1:0.35) highly deviates from that in municipal wastewater (100:20:5).

Based on Tables 5.2 and 5.3, the phenols and the organic substances responsible for the high COD value must be considered as problematic for treatment of this wastewater, and the presence of inhibitory or toxic substances may seriously affect the overall treatment system. Therefore, the chemical oxygen demand (COD), the total aromatic content (A), and the total phenolic content (TPh) are mostly selected as representative parameters to follow the overall purification process [19,21,29].

The terms and definitions for the waste resulting from the different oil extraction processes are neither standardized nor country specific [30]. Table 5.5 shows the nominations found in the Mediterranean countries, while Table 5.6 shows the most common terminology used in these countries with descriptions.

Table 5.4 Comparison of Composition Values of Olive Oil Wastewater from a Small Mill (A) and a Big Mill (B) with Municipal Wastewater (C)

Parameter	Source of liquid waste		
	A	B	C
pH	4.5–5.3	5.3–5.7	7–8
BOD_5 (g/L)	15–65	17–41	0.1–0.4
COD (g/L)	37–150	30–80	0.15–1
Total solids (g/L)	24–115	19–75	0.35–1.2
Volatile solids (g/L)	20–97	17–68	0.18–0.6
Suspended solids (g/L)	5.7–14	0.7–26	0.1–0.35
Fats and oils (g/L)	0.046–0.76	0.1–8.2	0.05–0.1
Total nitrogen (g/L)	0.27–0.51	0.3–0.48	0.02–0.08
Total phosphorus (g/L)	0.1–0.19	0.075–0.12	0.006–0.02
COD/BOD_5	2.3–2.5	1.8–2	1.5–2.5
$BOD_5:N:P$	100:0.98:0.37	100:1.3:0.34	100:20:5

Between 400 and 600 L of liquid waste are generated per ton of processed olives from the traditional presses used for olive oil extraction, which are operated discontinuously. Depending on its size, the capacity of such an olive oil mill is about 10–20 ton of olives/day. With a capacity of 20 ton of olives/day and a process-specific wastewater volume of 0.5 m^3/ton of olives, the daily wastewater can range up to 10 m^3/day.

Compared to the traditional presses, twice the quantity of wastewater (from 750 to 1200 L per ton of olives) is produced with the three-phase decanting method. Depending on their size, the capacities of the olive oil mills are also between 10 and 20 ton of olives/day. With a capacity of 20 ton of olives/day and a process specific wastewater volume of about 1 m^3/ton of olives, the daily wastewater volume from a continuous process is up to 20 m^3/day. The concentration of the constituents in wastewater from traditional presses is therefore twice as high as in the wastewater resulting from three-phase decanting. In general, the organic pollution

Table 5.5 Nominations of Waste Resulting from Different Oil Extraction Processes as Found in the Mediterranean Area

	Pressing	Three-phase decanting	Two-phase decanting
Solid	Orujo (Sp) Pirina (Gr, Tk) Hask (It, Tu) Grignons (Fr)	Orujo (Sp) Grignons (Fr) Pirina (Gr, Tk) Hask (It, Tu) Orujillo (Sp) after de-oiling of solid waste	Alpeorujo (in two- phase decanting mainly alpeorujo is produced)
Wastewater	Alpechin (Sp) Margine (Gr) Jamila (It)	Alpechin (Sp) Margine (Gr) Jamila (It)	Alpechin

Oil (from
  de-oiling of
  solid waste)
	–	Orujooil	Orujooil

Note: Sp, Spanish; Gr, Greek; It, Italian; Tu, Tunisian; Tk, Turkish; Fr, French.
Source: Ref. 30.

Table 5.6 Terminology of the Olive Oil Sector Related with Waste

Name	Description
Flesh, pulp (En)	Soft, fleshy part of the olive fruit
Pit, husk, stone (En)	Nut, hard part of the olive
Kernel, seed (En)	Softer, inner part of the olive
Alpeorujo, orujo de dos fases, alperujo (Sp)	Very wet solid waste from the two-phase decanters
Orujo, orujo de tres fases (Sp)	
Pirina (Gr/Tk)	
Pomace (It)	Wet solid waste from the three-phase decanters and presses
Grignons (Fr)	
Husks (It/Tu)	
Orujillo (Sp)	De-oiled orujo, de-oiled alpeorujo
Alpechin (Sp)	Liquid waste from the three-phase decanters and presses
Margine (Gr)	
Jamila (It)	
Alpechin-2 (Sp)	
Margine-2 (Gr)	Liquid fraction from secondary alpeorujo treatment (second decanting, repaso, etc.)
Jamila-2 (It)	

Note: En, English; Sp, Spain; Gr, Greek; It, Italian; Tu, Tunisian; Tk, Turkish; Fr, French.
Source: Ref. 1.

load in wastewater from olive oil extraction processes is practically independent of the processing method and amounts to 45–55 kg BOD_5 per ton of olives [31].

The input–output analysis of material and energy flows of the three production processes (press, two-phase, and three-phase decanting) is shown in Table 5.7. The basis of reference is one metric ton of processed olives.

5.3.1 Design Example 1

What is the population equivalent (pop. equ.) of the effluents discharged from a medium-sized oil mill processing about 15 ton (33,000 lb) of olives/day by using the two systems of traditional pressing or continuous centrifuging?

Solution

Traditional pressing of olives results in a wastewater volume of approximately 600 L (159 gal) per ton of olives; thus wastewater flow rate $= 15\,T \times 0.6\,m^3/T = 9\,m^3/day$ (2378 gal/day). Assuming a BOD_5 concentration of 40 g/L (0.34 lb/gal), the resulting total BOD_5 discharged per day $= 9\,m^3/day \times 40\,kg/m^3 = 360\,kg\,BOD_5/day$ (792 lb/day).

$$BOD_5 \text{ per person} = 54 - 60\,g/p.day\ (0.119 - 0.137\,lb/p.day)$$

then

$$\text{Pop. equ.} = \frac{360}{0.06} = 6000 \text{ persons}$$

Continuous centrifuging (three-phase decanting) of olives results in a wastewater volume of approximately 1000 L (264.2 gal) per ton of olives, thus wastewater flow rate =

Table 5.7 An Input–Output Analysis of Material and Energy Flows of the Production Processes Related to One Ton of Processed Olives

Production process	Input	Amount of input	Output	Amount of output
Traditional pressing process	Olives	1000 kg	Oil	200 kg
	Washing water	0.1–0.12 m^3	Solid waste (25% water + 6% oil)	400 kg
	Energy	40–63 kWh	Wastewater (88% water)	600 L[a]
Three-phase decanters	Olives	1000 kg	Oil	200 kg
	Washing water	0.1–0.12 m^3	Solid waste (50% water + 4% oil)	500–600 kg
	Fresh water for decanter	0.5–1 m^3	Wastewater (94% water +1% oil)	1000–1200 L[b]
	Water to polish the impure oil	10 kg		
	Energy	90–117 kWh		
Two-phase decanter	Olives	1000 kg	Oil	200 kg
	Washing water	0.1–0.12 m^3	Solid waste (60% water +3% oil)	800–950 kg
	Energy	<90–117 kWh		

[a]According to International Olive Oil Council: (400–550 L/ton processed olives)
[b]According to International Olive Oil Council: (850–1200 L/ton processed olives)
Source: Ref. 1.

15 T × 1 m^3/T = 15 m^3/day (3963 gal/day). Assuming a BOD$_5$ concentration of about 23 g BOD$_5$/L (0.192 lb/gal), the resulting total BOD$_5$ discharged per day is:

$$15 \, m^3/day \times 23 \, kg/m^3 = 345 \, kg/day \; (759 \, lb/day)$$

then

$$Pop. \; equ. = \frac{345}{0.06} = 5750 \; persons$$

5.4 ENVIRONMENTAL RISKS

Olive oil mill wastewaters (OMW) are a major environmental problem, in particular in Mediterranean countries, which are the main manufacturers of olive oil, green and black table olives. In these countries, the extraction and manufacture of olive oil are carried out in numerous small plants that operate seasonally and generate more than 30 million tons of liquid effluents (black water) [16], called "olive oil mill wastewaters" (OMW) each year. These effluents can cause considerable pollution if they are dumped into the environment because of their high organic load, which includes sugar, tannins, polyphenols, polyalcohols, pectins, lipids, and so on. Seasonal operation, which requires storage, is often impossible in small plants [32]. In fact, 2.5 L of waste are released per liter of oil produced [28].

Olive oil mill wastewaters contain large concentrations of highly toxic phenol compounds (can exceed 10 g/L) [33]. Much of the color of OMW is due to the aromatic compounds present, which have phytotoxic and antibacterial effects [34,35].

Despite existing laws and regulations, disposal of untreated liquid waste into the environment is uncontrolled in most cases. When it is treated, the most frequent method used is to retain the effluent in evaporation ponds. However, this procedure causes bad odors and risks polluting surface waters and aquifers. Therefore, this process presents an important environmental problem. Table 5.8 displays the risks that arise from direct disposal of olive oil mill wastewater (OMW) in the environment (soil, rivers, ground water). Examples of the risks [2] are described in the following sections.

5.4.1 Discoloring of Natural Waters

This is one of the most visible effects of the pollution. Tannins that come from the olive skin remain in the wastewater from the olive oil mill. Although tannins are not harmful to people, animals, or plants, they dye the water coming into contact with them dark black-brown. This undesired effect can be clearly observed in the Mediterranean countries [2].

5.4.2 Degradability of Carbon Compounds

For the degradation of the carbon compounds (BOD_5), the bacteria mainly need nitrogen and phosphorus besides some trace elements. The $BOD_5 : N : P$ ratio should be $100 : 5 : 1$. The optimal ratio is not always given and thus an excess of phosphorus may occur [36].

5.4.3 Threat to Aquatic Life

Wastewater has a considerable content of reduced sugar, which, if discharged directly into natural waters, would increase the number of microorganisms that would use this as a source of

Table 5.8 The Environmental Risks Resulting from the Direct Disposal of the Olive Oil Mill Liquid Water Without Treatment

Pollutants	Medium/environment	Effects
Acids	Soil	Destroys the cation exchange capacity of soil
Oil		Reduction of soil fertility
Suspended solids		Bad odors
Organics	Water	Consumption of dissolved oxygen
Oil		Eutrophication phenomena
Suspended solids		Impenetrable film
		Aesthetic damage
Acids	Municipal wastewater sewerage	Corrosion of concrete and metal canals/pipes
Suspended solids		Flow hindrance
		Anaerobic fermentation
Acids	Municipal wastewater treatment plants	Corrosion of concrete and metal canals/pipes
Oil		Sudden and long shocks to activated sludge and trickling filter systems
Organics		
Nutrient imbalance		Shock to sludge digester

Source: Refs. 2 and 15.

substrate. The effect of this is reduction of the amount of oxygen available for other living organisms, which may cause an imbalance of the whole ecosystem.

Another similar process can result from the high phosphorus content. Phosphorus encourages and accelerates the growth of algae and increases the chances of eutrophication, destroying the ecological balance in natural waters. In contrast to nitrogen and carbon compounds, which escape as carbon dioxide and atmospheric nitrogen after degradation, phosphorus cannot be degraded but only deposited. This means that phosphorus is taken up only to a small extent via the food chain: plant → invertebrates → fish → prehensile birds.

The presence of such a large quantity of nutrients in the wastewater provides a perfect medium for pathogens to multiply and infect waters. This can have severe effects on the local aquatic life and humans that may come into contact with the water [2].

5.4.4 Impenetrable Film

The lipids in the wastewater may form an impenetrable film on the surface of rivers, their banks, and surrounding farmlands. This film blocks out sunlight and oxygen to microorganisms in the water, leading to reduced plant growth in the soils and river banks and in turn erosion [2].

5.4.5 Soil Quality

The waste contains many acids, minerals, and organics that could destroy the cation exchange capacity of the soil. This would lead to destruction of microorganisms, the soil–air and the air–water balance, and, therefore, a reduction of the soil fertility [15].

5.4.6 Phytotoxicity

Phenolic compounds and organic acid can cause phytotoxic effects on olive trees. This is of dire importance since wastewater can come into contact with crops due to possible flooding during the winter. The phenols, organic, and inorganic compounds can hinder the natural disinfection process in rivers and creeks [2].

5.4.7 Odors

Anaerobic fermentation of the wastewater causes methane and other gases (hydrogen sulfide, etc.) to emanate from natural waters and pond evaporation plants. This leads to considerable pollution by odors even at great distances [2].

Other risks could be referred to in this respect, such as agricultural-specific problems arising from pesticides and other chemicals, although their effect in olive cultivation is less pronounced than other fields of agriculture. The main problem is soil erosion caused by rainwater, which results in steeper slopes and increases difficulty in ploughing. Soil quality and structure also influence erosion caused by rain. At present, protective measures such as planting of soil-covering species or abstention from ploughing are hardly used.

5.5 LIQUID WASTE TREATMENT METHODS

Disposal and management of highly contaminated wastewater constitute a serious environmental problem due to the biorecalcitrant nature of these types of effluents, in most cases. Generally, biological treatment (mainly aerobic) is the preferred option for dealing with urban and industrial effluents because of its relative cost-effectiveness and applicability for treating a wide variety of hazardous substances [19]. Nevertheless, some drawbacks may be found when applying this technology. For instance, some chemical structures, when present at high concentrations, are difficult to biodegrade because of their refractory nature or even toxicity toward microorganisms. Thus, several substances have been found to present inhibitory effects when undergoing biological oxidation. Among them, phenolic compounds constitute one of the most important groups of pollutants present in numerous industrial effluents [37]. Owing to the increasing restrictions in quality control of public river courses, development of suitable technologies and procedures are needed to reduce the pollutant load of discharges, increase the biodegradability of effluent, and minimize the environmental impact to the biota.

Industries that generate nonbiodegradable wastewater showing high concentrations of refractory substances (chiefly phenol-type compounds) include the pharmaceutical industry, refineries, coal-processing plants, and food-stuff manufacturing. The olive oil industry (a common activity in Mediterranean countries), in particular, generates highly contaminated effluents during different stages of mill olive oil production (washing and vegetation waters).

Therefore, most treatment processes used for high-strength industrial wastewaters have been applied to olive oil mill effluents (OME). Yet, OME treatment difficulties are mainly associated with: (a) high organic load (OME are among the strongest industrial effluents, with COD up to 220 g/L and sometimes reaching 400 g/L); (b) seasonal operation, which requires storage (often impossible in small mills); (c) high territorial scattering; and (d) presence of organic compounds that are difficult to degrade by microorganisms (long-chain fatty acids and phenolic compounds of the C-7 and C-9 phenylpropanoic family) [23].

Furthermore, a great variety of components found in liquid waste (alpachin) and solid waste (orujo and alpeorujo) require different technologies to eliminate those with harmful effects on the environment. Most used methods for the treatment of liquid waste from olive oil production are presented in Table 5.9. They correspond to the current state-of-art-technologies and are economically feasible. These methods are designed to eliminate organic components and to reduce the mass. In some cases, substances belonging to other categories are also partly removed. In practice, these processes are often combined since their effects differ widely [1]. Therefore, methods should be used in combination with each other.

The following key treatment methods are mainly applied to liquid waste. Some of these methods can also be used in the treatment of liquid–solid waste (alpeorujo), for example, treatment by fungi, evaporation/drying, composting, and livestock feeding. However, those methods tested at laboratory scale must be critically examined before applying them at industrial or full-scale, in order to meet the local environmental and economical conditions.

Regarding the olive oil industry, it should always be considered that complicated treatment methods that lack profitable use of the final product are not useful, and all methods should have a control system for the material flows [38].

5.5.1 Low-Cost Primitive Methods

These methods are mostly applied in the developing countries producing olive, due to their simplicity and low costs. Of these methods, the most important are:

- Drainage of olive oil mill liquid waste in some types of soils, with rates up to 50 $m^3/$ ha-year (in the case of traditional mills) and up to 80 m^3/ha-year (in the case of decanting-based methods), or to apply the olive oil mill liquid wastes to the irrigation water for a rate of less than 3%. These processes are risky because they decrease the fertility of the soil. This calls for greater care and scientific research into these methods prior to agronomic application.

Table 5.9 Treatment Methods for the Liquid and Solid Waste from Olive Oil Production

Treatment method of (alpechin)
Low-cost primitive methods
• Drainage in soil
• Simple disposal in evaporation ponds
• Mixing with solid waste in sanitary landfills
Aerobic treatment
Anaerobic treatment
Combined biological treatment methods
Wet air oxidation and ozonation
Fungal treatment
Decolorization
Precipitation/flocculation
Adsorption
Filtration (biofiltration, ultrafiltration)
Evaporation/drying
Electrolysis
Bioremedation and composting[a]
Livestock feeding[a]
Submarine outfall

[a] These recycling methods can be used for liquid as well as solid waste from olive oil production. Products resulting from treatment may be reused, for instance, as fertilizer or fodder in agriculture. For all methods, waste that is not suited for reuse can be disposed at landfills.

- Simple disposal and retention in evaporation ponds (large surface and small depth ponds), preferably in distant regions, to be dried by solar radiation and other climatic factors. This method does not require energy or highly trained personnel. Drawbacks are associated with the evaporation process, which generates odors and additional risks for the aquatic system of the area (filtration phenomena, surface water contamination, etc.). In addition, the disadvantages include: the need for large areas for drying in selected regions with impermeable (clay) soil distant from populated areas; the requirement, in most cases, for taking necessary precautions to prevent pollutants reaching the groundwater through placement of impermeable layers in the ground and walls of ponds; ineffective in higher rainfall regions; emergence of air pollutants caused by decomposition of organic substances (ammonia-hydrocarbon volatile compounds). This method is being applied in many countries of the Mediterranean area. In Spain alone, there are about 1000 evaporation ponds, which improve the water quality, but the ponds themselves caused serious negative environmental impacts. Dried sludge from corporation ponds can be used as fertilizer, either directly or composted with other agricultural byproducts (e.g., grape seed residues, cotton wastes, bean straw) [39].
- Mixing the olive oil mill liquid wastes with municipal solid wastes in sanitary landfills leads to increased organic load on site. Consideration should be made regarding the pollutants that may reach the groundwater, in addition to the risks of combustion due to generation of combustible hydrocarbon gases. These factors should be taken into account in designing and establishing landfills, not forgetting the necessity to collect

and treat the drainage wastewater resulted from applying this method. This method is cost-effective and is suitable for final disposal of the wastes, with the property of obtaining energy from the generated gases. Nevertheless, there are drawbacks such as the air pollution caused by the decomposition, the need for advanced treatment for the highly polluted collected drainage wastewater, and the need for using large areas of land and particular specifications.

5.5.2 Aerobic Treatment

When biodegradable organic pollutants in olive oil mill wastewater (alpechin) are eliminated by oxygen-consuming microorganisms in water to produce energy, the oxygen concentration decreases and the natural balance in the water body is disturbed. To counteract an overloading of the oxygen balance, the largest part of these oxygen-consuming substances (defined as BOD_5) must be removed before being discharged into the water body. Wastewater treatment processes have, therefore, been developed with the aim of reducing the BOD_5 concentration as well as eliminating eutrophying inorganic salts, that is, phosphorus and nitrogen compounds, ammonium compounds, nonbiodegradable compounds that are analyzed as part of the COD, and organic and inorganic suspended solids [38].

In aerobic biological wastewater treatment plants, the natural purification processes taking place in rivers are simulated under optimized technical conditions. Bacteria and monocellular organisms (microorganisms) degrade the organic substances dissolved in water and transform them into carbonic acid, water, and cell mass. The microorganisms that are best suited for the purification of a certain wastewater develop in the wastewater independently of external influences and adapt to the respective substrate composition (enzymatic adaptation). Owing to the oxidative degradation processes, oxygen is required for wastewater treatment. The oxygen demand corresponds to the load of the wastewater.

Two types of microorganisms live in waters: suspended organisms, floating in the water, and sessile organisms, which often settle on the surface of stones and form biofilms. Biofilm processes such as fixed-bed or trickling filter processes are examples of the technical application of these natural processes [38].

Treatment of Olive Oil Mill Wastewaters in Municipal Plants

Municipal wastewater is unique in that a major portion of the organics are present in suspended or colloidal form. Typically, the BOD in municipal sewage consists of 50% suspended, 10% colloidal, and 40% soluble parts. By contrast, most industrial wastewaters are almost 100% soluble. In an activated sludge plant-treating municipal wastewater, the suspended organics are rapidly enmeshed in the flocs, the colloids are adsorbed on the flocs, and a portion of the soluble organics are absorbed. These reactions occur in the first few minutes of aeration contact. By contrast, for readily degradable wastewaters, that is, food processing, a portion of the BOD is rapidly sorbed and the remainder removed as a function of time and biological solids concentration. Very little sorption occurs in refractory wastewaters. The kinetics of the activated sludge process will, therefore, vary depending on the percentage and type of industrial wastewater discharged to the municipal plant and must be considered in the design calculations [40].

The percentage of biological solids in the aeration basin will also vary with the amount and nature of the industrial wastewater. Increasing the sludge age increases the biomass percentage as volatile suspended solids undergo degradation and synthesis. Soluble industrial wastewater will increase the biomass percentage in the activated sludge.

A number of factors should be considered when discharging industrial wastewaters, including olive oil mill effluents, into municipal plants [40]:

- *Effect on effluent quality.* Soluble industrial wastewaters will affect the reaction rate K. Refractory wastewaters such as olive oil mills, tannery, and chemical will reduce K, while readily degradable wastewaters such as food processing and brewery will increase K.
- *Effect on sludge quality.* Readily degradable wastewaters will stimulate filamentous bulking, depending on basin configuration, while refractory wastewaters will suppress filamentous bulking.
- *Effect of temperature.* An increased industrial wastewater input, that is, soluble organics, will increase the temperature coefficient θ, thereby decreasing efficiency at reduced operating temperatures.
- *Sludge handling.* An increase in soluble organics will increase the percentage of biological sludge in the waste sludge mixture. This will generally decrease dewaterability, decrease cake solids, and increase conditioning chemical requirements. One exception is pulp and paper-mill wastewaters in which pulp and fiber serve as a sludge conditioner and enhances dewatering rates.

It is worth pointing out that certain threshold concentrations for inhibiting agent and toxic substances must not be exceeded. Moreover, it should be noted that most industrial wastewaters are nutrient deficient, that is, they lack nitrogen and phosphorus. Municipal wastewater with a surplus of these nutrients will provide the required nutrient balance.

The objective of the activated sludge process is to remove soluble and insoluble organics from a wastewater stream and to convert this material into a flocculent microbial suspension that is readily settleable and permits the use of gravitational solids liquid separation techniques. A number of different modifications or variants of the activated sludge process have been developed since the original experiments of Arden and Lockett in 1914 [40]. These variants, to a large extent, have been developed out of necessity or to suit particular circumstances that have arisen. For the treatment of industrial wastewater, the common generic flow sheet is shown in Figure 5.4.

The activated sludge process is a biological wastewater treatment technique in which a mixture of wastewater and biological sludge (microorganisms) is agitated and aerated. The biological solids are subsequently separated from the treated wastewater and returned to the aeration process as needed. The activated sludge process derives its name from the biological mass formed when air is continuously injected into the wastewater. Under such conditions, microorganisms are mixed thoroughly with the organics under conditions that stimulate their growth through use of the organics as food. As the microorganisms grow and are mixed by the agitation of the air, the individual organisms clump together (flocculate) to form an active mass of microbes (biologic floc) called activated sludge [41].

In practice, wastewater flows continuously into an aeration tank where air is injected to mix the activated sludge with the wastewater and to supply the oxygen needed for the organisms to break down the organics. The mixture of activated sludge and wastewater in the aeration tank is called mixed liquor. The mixed liquor flows from the aeration tank to a secondary clarifier where the activated sludge is settled out. Most of the settled sludge is returned to the aeration tank (return sludge) to maintain a high population of microbes to permit rapid breakdown of the organics. Because more activated sludge is produced than is desirable in the process, some of the return sludge is diverted or wasted to the sludge handling system for treatment and disposal.

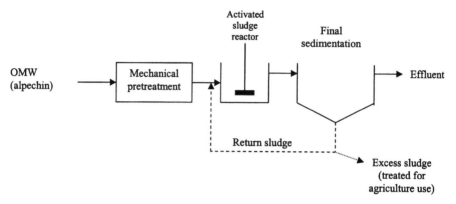

Figure 5.4 Aerobic treatment (activated sludge plant).

Biofilm processes are used when the goal is very far-reaching retention and concentration of the biomass in a system. This is especially the case with slowly reproducing microorganisms in aerobic or anaerobic environments. The growth of sessile microorganisms on a carrier is called biofilm. The filling material (e.g., in a trickling filter stones, lava slag, or plastic bodies) or the filter material (e.g., in a biofilter) serve as carrier. The diffusion processes in biofilm plants are more important than in activated sludge plants because unlike activated sludge flocs the biofilms are shaped approximately two-dimensionally. On the one hand, diffusion is necessary to supply the biofilm with substrate and oxygen; on the other hand, the final metabolic products (e.g., CO_2 and nitrate) must be removed from the biofilm.

For treatment of industrial wastewater, trickling filters are often used. A trickling filter is a container filled completely with filling material, such as stones, slats, or plastic materials (media), over which wastewater is applied. Trickling filters are a popular biological treatment process [42]. The most widely used design for many years was simply a bed of stones, 1–3 m deep, through which the wastewater passed. The wastewater is typically distributed over the surface of the rocks by a rotating arm. Rock filter diameters may range up to 60 m. As wastewater trickles through the bed, a microbial growth establishes itself on the surface of the stone or packing in a fixed film. The wastewater passes over the stationary microbial population, providing contact between the microorganisms and the organics. The biomass is supplied with oxygen using outside air, most of the time without additional technical measures. If the wastewater is not free of solid matter (as in the case of alpechin), it should be prescreened to reduce the risk of obstructions.

Excess growths of microorganisms wash from the rock media and would cause undesirably high levels of suspended solids in the plant effluent if not removed. Thus, the flow from the filter is passed through a sedimentation basin to allow these solids to settle out. This sedimentation basin is referred to as a secondary clarifier, or final clarifier, to differentiate it from the sedimentation basin used for primary settling. An important element in trickling filter design is the provision for return of a portion of the effluent (recirculation) to flow through the filter. Owing to seasonal production of wastewater and to the rather slow growth rates of the microorganisms, these processes are less suited for the treatment of alpechin, compared to the activated sludge process.

Another worthwhile aerobic treatment method developed by Balis and his colleagues [38] is the bioremediation process, based on the intrinsic property of an *Azotobacter vinelandii* strain (strain A) to proliferate on limed olive oil mill wastewater. More specifically, the olive mill wastewater is pretreated with lime to pH 7–8 and then is fed into an aerobic bioreactor equipped with a rotating wheel-type air conductor. The reactor is operated in a repeated fed batch culture fashion with a cycle time of 3 days. During each cycle, the Azotobacter population proliferates and fixes molecular nitrogen. It concomitantly produces copious amounts of slime and plant growth promoting substances. The endproduct is a thick, yellow-brown liquid. It has a pH of about 7.5–8.0, it is nonphytotoxic, soluble in water, and can be used as liquid fertilizer over a wide range of cultivated plants (olives, grapes, citrus, vegetables, and ornamentals). Moreover, there is good evidence that the biofertilizer induces soil suppressiveness against root pathogenic fungi, and improves soil structure. A medium-scale pilot plant of 25 m^3 capacity has been constructed in Greece by the Olive Cooperative of Peta near Arta with the financial support of the General Secretariat of Science and Technology of Greece. The plant has been operating since 1997. The local farmers use the liquid biofertilizer that is produced to treat their olive and citrus groves.

In short, it has been demonstrated that free-living N$_2$-fixing bacteria of Azotobacter grow well in olive mill wastewater and transform the wastes into a useful organic fertilizer and soil conditioner. For further details in this regard, refer to Section 5.5.17 (Bioremediation and Composting).

The following case study explains the influence of aerobic treatments for already fermented olive oil mill wastewater (OMW), on the anaerobic digestion of this waste.

Case Study

This kinetic study [25] allows intercomparison of the effects of different aerobic pre-treatments on the anaerobic digestion of OMW, previously fermented with three microorganisms (*Geotrichum condidum*, *Azotobacter chroococcum*, and *Aspergillus terreus*). The OMW used was obtained from a continuous olive-processing operation. The bioreactor used was batch fed and contained sepiolite as support for the mediating bacteria. The results of the microtox toxicity test expressed as toxic units (TU) for both pretreated and untreated OMW are as follows:

- prior to inoculation (untreated OMW): TU = 156;
- after fermentation with Geotrichum: TU = 64;
- after fermentation with Azotobacter: TU = 32;
- after fermentation with Aspergillus: TU = 20.

The influence of the different aerobic pretreatments on the percentages of elimination of COD and total phenol contents are indicated in Table 5.10.

Table 5.10 Influence of Different Aerobic Pretreatments on the Percentages of Elimination of COD and Total Phenol Contents

Pretreatment	Elimination COD %	Elimination phenols %
Geotrichum	63.3	65.6
Azotobacter	74.5	90.0
Aspergillus	74.0	94.3

Source: Ref. 25.

A kinetic model was developed for the estimation of methane production (G) against time (t), represented in the following equation:

$$G = G_M\left[1 - \exp\left(-\frac{AXt}{S_0}\right)\right], \quad \text{over the COD range studied } (3.9 - 14.5\,\text{g/L})$$

where G_M is the maximum methane volume obtained at the end of digestion time, S_0 is the initial substrate concentration, X is the microorganism concentration, and A is the kinetic constant of the process, which was calculated using a nonlinear regression. This kinetic parameter was found to be influenced by the pretreatment carried out, and was 4.6, 4.1, and 2.3 times higher for Aspergillus-, Azotobacter-, and Geotrichum-pretreated OMWs than that obtained in the anaerobic digestion of untreated OMW. The kinetic constant increased as the phenolic compound content and biotoxicity of the pretreated OMWs decreased.

The final conclusion that can be drawn from this work is that aerobic pretreatment of the OMW with different microorganisms (Geotrichum, Azotobacter, and Aspergillus) considerably reduces the COD and the total phenolic compound concentration of waste that is responsible for its biotoxicity. This fact is shown through enhancement of the kinetic constant for the anaerobic digestion process, and a simultaneous increase in the yield coefficient of methane production.

Case studies regarding the role and importance of the aerobic treatment process combined with chemical oxidation such as wet air oxidation (WAO) are found in Section 5.5.9.

5.5.3 Design Example 2

An olive oil mill is to treat its wastewater in an extended aeration activated sludge plant. The final effluent should have a maximum soluble BOD$_5$ of 20 mg/L during the olive mill operation season. This plant is to be designed under the following conditions: $Q = 60\,\text{m}^3/\text{day}$ (15,850 gal/day); S_0 (diluted) $= 800\,\text{mg/L}$; $S_e = 20\,\text{mg/L}$; $X_v = 3000\,\text{mg/L}$; $a = 0.50$; $a' = 0.6$; $b = 0.10$ at $20°C$; $\theta = 1.065$; $K = 6.0/\text{day}$ at $20°C$; and $b' = 0.12/\text{day}$.

Solution

$$t = \frac{S_0(S_0 - S_e)}{KS_eX_v}$$

$$t = \frac{800(800 - 20)}{6(20)(3000)} = 1.73\,\text{days}$$

$$\frac{F}{M} = \frac{S_0}{X_vt} = \frac{800}{3000 \times 1.73} = 0.154$$

The degradable fraction is determined by:

$$X_d = \frac{0.8}{1 + 0.26\theta c}$$

Assuming $\theta c = 25$ day (SRT)

$$X_d = \frac{0.8}{1 + 0.2 \times 0.1 \times 25} = 0.53$$

The aeration basin volume is: $60 \, \text{m}^3/\text{day} \times 1.73 \, \text{day} = 104 \, \text{m}^3$ (27,421 gal). The sludge yield can be computed as:

$$\Delta X_v = aS_r - bX_dX_vt$$
$$\Delta X_v = 0.5 \times 780 \, \text{mg/L} - 0.10 \times 0.53 \times 3000 \, \text{mg/L} \times 1.73$$
$$\Delta X_v = 115 \, \text{mg/L}$$
$$\Delta X_v = 115 \, \text{mg/L} \times 60 \, \text{m}^3/\text{day} \times 10^{-3}$$
$$= 7.0 \, \text{kg/day} (15.4 \, \text{lb/day})$$

Check the sludge age:

$$\theta_c = \frac{\forall X_v}{\Delta X_v} = \frac{104 \times 3000}{7 \times 1000} = 45 \, \text{day}$$

or

$$\theta_c = \frac{27,421 \, \text{gal} \times 8.34 \times 10^{-6} \times 3000}{15.4} = 45 \, \text{day}$$

Compute the oxygen required:

$$O_2/\text{day} = a'S_rQ + b'X_dX_v\forall$$
$$O_2/\text{day} = (0.6 \times 780 \times 60 + 0.12 \times 0.53 \times 3000 \times 104)10^{-3}$$
$$O_2/\text{day} = 48 \, \text{kg/day} = 2 \, \text{kg/hour} (4.4 \, \text{lb/hour})$$

The oxygen needed can also be calculated directly from the approximate relation:

$$2.0 - 2.5 \, \text{kg} \, O_2/\text{kg BOD}_5$$
$$O_2/\text{day} = 60 \, \text{m}^3/\text{day} \times 800 \, \text{g BOD}_5/\text{m}^3 \times 10^{-3} \times 2 \, \text{kg} \, O_2/\text{kg BOD}_5$$
$$O_2/\text{day} = 96 \, \text{kg} \, O_2/\text{day} (4 \, \text{kg/hour}) (8.8 \, \text{lb/hour})$$

Compute the effluent quality at 15°C:

$$K_{15^\circ} = 6 \times 1.065^{(15-20)} = 4.38/\text{day}$$
$$S_e = \frac{S_0^2}{KX_vt + S_0} = \frac{800^2}{4.38 \times 3000 \times 1.73 + 800}$$
$$S_e = 27 \, \text{mg/L}$$

The effluent quality at 10°C:

$$K_{10^\circ} = 6 \times 1.065^{(10-20)} = 3.19/\text{day}$$
$$S_e = \frac{(800)^2}{3.19 \times 3000 \times 1.73 + 800}$$
$$S_e = 37 \, \text{mg/L}$$

5.5.4 Anaerobic Treatment

Anaerobic processes are increasingly used for the treatment of industrial wastewaters. They have distinct advantages including energy and chemical efficiency and low biological sludge yield, in addition to the possibility of treating organically high-loaded wastewater (COD > 1500 mg/L), with the requirement of only a small reactor volume.

Anaerobic processes can break down a variety of aromatic compounds. It is known that anaerobic breakdown of the benzene nucleus can occur by two different pathways, namely, photometabolism and methanogenic fermentation. It has been shown that benzoate, phenyl-acetate, phenylpropionate, and annamate were completely degraded to CO_2 and CH_4. While long acclimation periods were required to initiate gas production, the time required could be reduced by adapting the bacteria to an acetic acid and substrate before adapting them to the aromatic.

Chmielowski et al. [43] showed that phenol, p-cresol, and resorcinol yielded complete conversion to CH_4 and CO_2.

Principle of Anaerobic Fermentation

In anaerobic fermentation, roughly four groups of microorganisms sequentially degrade organic matter. Hydrolytic microorganisms degrade polymer-type material such as polysaccharides and proteins to monomers. This reduction results in no reduction of COD. The monomers are then converted into fatty acids (VFA) with a small amount of H_2. The principal organic acids are acetic, propionic, and butyric with small quantities of valeric. In the acidification stage, there is minimal reduction of COD. Should a large amount of H_2 occur, some COD reduction will result, seldom exceeding 10%. All formed acids are converted into acetate and H_2 by acetogenic microorganisms. The breakdown of organic acids to CH_4 and CO_2 is shown in Figure 5.5. Acetic acid and H_2 are converted to CH_4 by methanogenic organisms [40].

The specific biomass loading of typical anaerobic processes treating soluble industrial wastewaters is approximately 1 kg COD utilized/(kg biomass-day). There are two classes of methanogenes that convert acetate to methane, namely, *Methanothrix* and *Methanosarcina*. Methanothrix has a low specific activity that allows it to predominate in systems with a low steady-state acetate concentration. In highly loaded systems, Methanosarcina will predominate with a higher specific activity (3 to 5 times as high as Methanothrix) if trace nutrients are

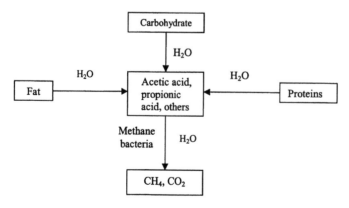

Figure 5.5 Anaerobic degradation of organics (from Ref. 46).

available. At standard temperature and pressure, 1 kg of COD or ultimate BOD removed in the process will yield 0.35 m^3 of methane [40].

The quantity of cells produced during methane fermentation will depend on the strength and character of the waste, and the retention of the cells in the system.

In comparing anaerobic processes and aerobic processes, which require high energy and high capital cost and produce large amounts of secondary biological sludge, the quantity of excess sludge produced is 20 times lower in anaerobic processes. This can be explained by the fact that with the same organic load under oxygen exchange about 20 times less metabolic energy is available for the microorganisms. Anaerobic wastewater treatment methods are mainly used for rather high-loaded wastewaters with a COD of 5000 up to 40,000 mg/L from the food and chemical industry [2]. Unfortunately, these methods are normally employed strictly as pretreatment measures. Aerobic follow-up treatment, for example, in a downstream-arranged activated sludge plant, is possible and recommended (Fig. 5.6).

Factors Affecting Anaerobic Process Operation

The anaerobic process functions effectively over two temperature ranges: the mesophilic range of 85–100°F (29–38°C) and the thermophilic range of 120–135°F (49–57°C). Although the rates of reaction are much greater in the thermophilic range, the maintenance of higher temperatures is usually not economically justifiable.

Methane organisms function over a pH range of 6.6–7.6 with an optimum near pH 7.0. When the rate of acid formation exceeds the rate of breakdown to methane, a process imbalance results in which the pH decreases, gas production falls off, and the CO_2 content increases [40]. pH control is therefore essential to ensure a high rate of methane production. According to German literature, the tolerable pH range for anaerobic microorganisms is between 6.8 and 7.5. This means that the anaerobic biocenosis is very pH-specific [38].

With regard to the influence of initial concentration on anaerobic degradation, preliminary laboratory and pilot-scale experimentation on diluted olive oil mill effluents (OME) [44] showed that the anaerobic contact process was able to provide high organic removal efficiency (80–85%) at 35°C and at an organic load lower than 4 kg COD/m^3/day; however, in particular at high feed concentration, the process proved unstable due to the inhibitory effects of substances

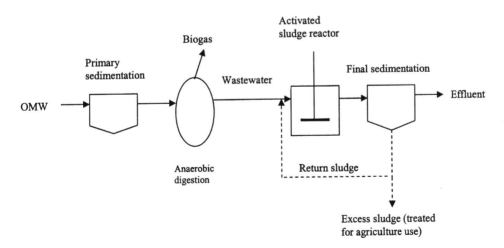

Figure 5.6 Anaerobic–aerobic treatment method.

such as polyphenols. Moreover, additions of alkalinity to neutralize acidity and ammonia to furnish nitrogen for cellular biosynthesis were required.

To overcome these difficulties and improve process efficiency and stability, there are basically two methods that may be adopted [23]: (a) the treatment of combined OME and sewage sludge in contact bioreactors; and (b) operation with more diluted OME in high-rate bioreactors (such as UASB reactors and fixed-bed filters).

In the first method, conventional digesters can be overloaded with concentrated soluble wastes such as OME, and still operate satisfactorily. Moreover, nutrients such as ammonia and buffers are provided by degradation of proteineous substances from sludge. On this basis, laboratory-scale experimentation [45] has shown that removal efficiencies of 65 and 37% in terms of COD and VSS, respectively, were obtained at 35°C and at an organic load of 4.2 kg $COD/m^3/day$ (66% from sewage sludge, 34% from OME). Higher OME additions led to process imbalance due to the inhibitory effects of polyphenols. This method, based on anaerobic contact digestion of combined OME and sewage sludge, seems to be suitable only for those locations where the polluting load due to the OME is lower than the domestic wastewater load. In this regard it is worth considering that during the olive oil milling season, OME pollution largely exceeds that from domestic wastewater [23].

With regard to the second method, based on the use of high-rate bioreactors, experimentation on UASB reactors [46,47] showed that COD removal efficiencies of about 70–75% were obtained at 37°C and at an organic load in the range 12–18 kg $COD/m^3/day$ by adopting a dilution ratio in the range of 1:8 to 1:5 (OME: tap water; diluted OME initial concentration in the range 11–19 g COD/L). Slightly less satisfactory results were obtained by using anaerobic filters filled with macroreticulated polyurethane foam [45].

It is important to note that immobilization of methanogenic bacteria may decrease the toxicity of phenolic compounds. Another pilot-scale anaerobic–aerobic treatment of OME mixed with settled domestic wastewater [48] produced a final COD concentration of about 160 mg/L, provided that a dilution ratio of 1:60 to 1:100 was adopted, corresponding to a COD load ratio equal to 3:1 for OME and domestic wastewater, respectively. This ratio is typical for those locations with a high density of olive oil mills. However, in addition to the high value required for the dilution ratio, the final effluent did not comply with legal requirements in terms of color and nitrogen [23].

The aforementioned data clearly show that in the treatment of OME, even when carried out with the use of most appropriate technology, that is, anaerobic digestion, it was difficult to reach the treatment efficiencies required by national regulations throughout the Mediterranean area. In particular, methanogenesis, which represents the limiting step in the anaerobic digestion of soluble compounds, is severely hindered by the inhibition caused by the buildup of volatile fatty acids (VFAs) and/or the presence of a high concentration of phenolic compounds and/or oleic acid in the OME. As for phenol, 1.25 g/L leads to 50% activity reduction of acetate-utilizing methanogens [49]. As for oleic acid, it is reported that 5 mM is toxic to methanogenic bacteria [50].

The reader may refer to the following Case Study V to better understand the mechanism of biodegradation of the main compounds contained in the OME in relation to pH, temperature, and initial concentration of effluents, and in particular the mutual coherence of the two successive partial stages occurring in anaerobic digestion of OME, acidogenesis, and methanogenesis.

Anaerobic Treatment Systems of Wastewater

Seasonal operation of olive oil mills is not a disadvantage for anaerobic treatment systems because anaerobic digesters can be easily restarted after several months of mill shutdown [51].

At present there are no large-scale plants. However, the anaerobic contact reactors and upflow sludge-blanket reactors have been mainly studied using several pilot tests (Fig. 5.7), besides other tested reactors such as anaerobic filters and fluidized-bed reactors.

Sludge retention is decisive for the load capacity and, thus, the field of application of an anaerobic reactor. In the UASB reactor, favorable sludge retention is realized in a simple way. Wastewater flows into the active space of the reactor, passing from the bottom to the top of the reactor. Owing to the favorable flocculation characteristics of the anaerobic-activated sludge, which in higher-loaded reactors normally leads to the development of activated sludge grains and to its favorable sedimentation capacity, a sludge bed is formed at the reactor bottom with a sludge blanket developing above it. To avoid sludge removal from the reactor and to collect the biogas, a gas-sludge separator (also called a three-phase separator) is fitted into the upper part of the reactor. Through openings in the bottom of this sedimentation unit, the separated sludge returns into the active space of the reactor. Because of this special construction, the UASB reactor has a very high load capacity. In contrast to the contact sludge process, no additional sedimentation tank is necessary, which would require return sludge flow for the anaerobic activated sludge, resulting in a reduction of the effective reactor volume. Several studies on anaerobic treatment of olive oil wastewaters have been carried out, and data from different publications are listed in Table 5.11.

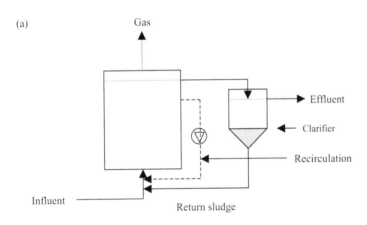

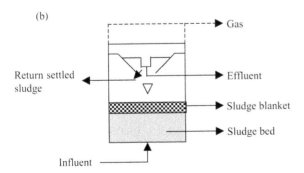

Figure 5.7   Anaerobic treatment processes: (a) Contact sludge reactor; (b) UASB reactor.

Table 5.11 Summary of the Data from Different Publications Related to Anaerobic Treatment of Olive Oil Wastewater

	Fiestas (1981)[14]	FIW[38]	Aveni (1984)[44][a]	FIW[38]	FIW[38]	FIW[38]	Steegmans (1992)[15]	Ubay (1997)[22]
Treatment process	Contact process	UASB reactor	Contact process	Conventional reactor	UASB reactor	Packed-bed reactor	UASB reactor	UASB reactor
Influent	33–42 g BOD_5/L	4–6 g COD/L	–	20–65 g COD/L	5–15 g COD/L	45–50 g COD/L	26.7 g COD/L	5–22.6 g COD/L
Volumetric loading	1.2–1.5 kg BOD/(m^3*day)	15–20 kg COD/(m^3*day)	4 kg COD/(m^3*day)	20–65 kg COD/(m^3*day)	5–21 kg COD/(m^3*day)	–	1.59 kg COD/(m^3*day)	5–18 kg COD/(m^3*day)
Purification efficiency	80–85% BOD	70% COD	80–85% COD	80–85% COD	70–80% COD	45–55% COD	55.9% COD	70–75% COD
Gas production	700 L/kg BOD_{elim}	–	–	550 L/kg COD_{elim}	8000 L/(m_r^3*day)	300–600 L/kg COD_{elim}	50–100 L CH_4/kg COD_{elim}	350 L CH_4/kg COD_{elim}
Methane content	70%	–	–	50–70%	70–80%	84%	70%	–

[a] Based on laboratory and pilot experimentation on diluted olive oil mill effluents.
Source: Refs. 14, 15, 22, 38, 44.

Case Studies

Many anaerobic pilot plants have been applied successfully in treating OMW in various parts of the world. The following describe some of these pilot plants and tests.

Case Study I. The search for an economic treatment process for wastewater from an olive oil extraction plant in Kandano (region of Chania, Crete) led to the concept of a pilot plant. The goal was to study the efficiency of separate anaerobic treatment of the settled sludge and of the sludge liquor from the settling tank (Fig. 5.8) [38].

Description of the plant:

- delivery, storage container;
- settling tank with a capacity of 650 m^3;
- anaerobic digester (volume: 16 m^3) for the sludge;
- UASB (upflow anaerobic sludge blanket) reactor (volume: 18 m^3) for the sludge liquor.

The plant can receive one-sixth of the total wastewater volume produced. The daily influent is 30 m^3. The wastewater is collected in a storage container where its quality and quantity are analyzed. The raw wastewater is then retained for 10 days in the settling tank where the particular substances settle.

Two separate zones are formed:

- the supernatant zone;
- the thickening and scraping zone.

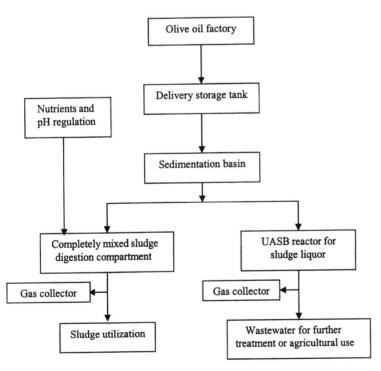

Figure 5.8 Pilot plant for treatment of wastewater from olive oil extraction in Kandano (a region of Chania, Crete) (from Ref. 38).

Both the preclarified sludge liquor and the primary sludge withdrawn are anaerobically treated in parallel. There is the risk of scum layer formation in the settling tank, which may lead to strong odors. This problem can be solved by covering the tank or using a scraper bridge.

The preclarified sludge liquor is preheated and fed into the UASB reactor. The biogas obtained is withdrawn from the upper part of the reactor and conducted to the gas storage room. The liquid phase is submitted to sedimentation, then stored in a container.

After the addition of nutrients and pH regulation, the primary sludge, showing a high water content (65–80%), is fed into a completely mixed digester. The biogas is again withdrawn from the upper part of the digester and conducted to the gas storage room. The treated liquid phase is conducted to the settling tank and then to the collecting container. At this point, the biogas is incinerated.

To build a plant that treats 30 m^3/day, a surface of at least 1 ha is necessary, at the cost of about 150,000 Euro. This sum does not include the construction costs for a soil filter or an irrigation system because these strongly depend on the location of the plant. At least 50% of the staff should be skilled workers, including a chemical engineer who is in charge of plant operation. Because of its high realization costs, this method is suited for industrial-scale oil mills, or as a central treatment facility for several oil mills.

The biogas may be used by the plant itself, or it may be fed into the public supply grid. The liquid phase, designated to be spread on agricultural land, is stored in an open pit. After drying, the solids can be sold as soil-improving material or as humus after having been mixed with vegetable residues. There are no odor nuisances from escaping liquids from the digesters, and maintenance costs are moderate. If the treated wastewater is additionally submitted to soil filtration and then used for irrigation or as fertilizer, the water cycle is closed, thus solving the problem of olive oil waste.

Case Study II. A pilot plant was operated between January 1993 and April 1994 to treat the wastewater from an oil mill in the region of Kalyvia/Attica (Fig. 5.9) [38].

Description of the plant:

- delivery, storage tank with a volume of 20 m^3 for the total quantity of margines produced;
- settling tank with a volume of 4 m^3;
- UASB reactor with a working volume of 2 m^3, additionally equipped with a high-performance heat exchanger to maintain the temperature during the mesophile phase;
- fixed-bed reactor with a working volume of 2 m^3, a high-performance heat exchanger, and recirculation system;
- gas storage room;
- seven tests (mesophile phase) have been carried out under varying operational conditions.

The organic load was degraded by 88–89%. During the fourth test, the phenol content was reduced by 74–75%, while the biogas production was 21–23 L gas per liter of bioreactor volume.

Foregoing the addition of CaO and expensive processing equipment facilitates the treatment for wastewater from oil mills. Plant investments can be quickly amortized by methane production.

Case Study III. A pilot test has been carried out in Tunisia with a sludge-bed reactor and an anaerobic contact reactor, followed by a two-stage aerobic treatment [15,38]. To compare the two different anaerobic processes, the semitechnical pilot plant was designed with parallel streams. The goal was not only to determine parameters and values for design and operation of optimal anaerobic–aerobic treatment, dependent on the achievable purification

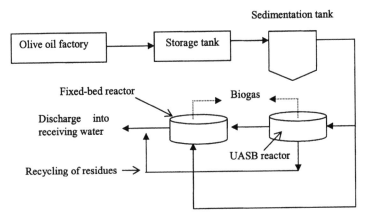

Figure 5.9 Pilot plant for treatment of wastewater from an olive oil mill in the region of Kalyvia, Attica (from Ref. 38).

capacity, but also to examine, modify, and further develop the process technology with regard to optimizing the purification capacity of the single stages, the total purification capacity, and process stability.

The tests determined that both anaerobic–aerobic procedures proved successful in the treatment of liquid waste from olive oil production. Comparing the anaerobic contact process with the bed process, neither is clearly favored. Both procedures lead to nearly the same results with regard to pretreatment of liquid waste from olive oil production.

Case Study IV. The anaerobic treatability of olive mill effluent was investigated using a laboratory-scale UASB reactor (with active volume of 10.35 L) operating for about 6 months. The black water collected from a traditional olive oil extraction plant in Gemlik village (Turkey) was used as the feed [22].

Active anaerobic sludge retained in the UASB reactor after a previous study was used as the seed. During the startup, pH was maintained in the range 6.8–8.0 and the average temperature was kept at mesophilic operating conditions (34°C) in the reactor. NaOH solution was added directly to the reactor to maintain the required pH levels when it was necessary. Urea was added to the feed to provide COD : N : P ratio of 350 : 5 : 1 in the system due to N deficiency of the feed.

In the first part of this study, the reactor was operated with feed COD concentrations from 5000 to 19,000 mg/L and a retention time of 1 day, giving organic loading rates (OLR) of 5–18 kg COD/m³/day. Soluble COD removal was around 75% under these conditions. In the second part of the study, feed COD was varied from 15,000 to 22,600 mg/L while retention times ranged from 0.83 to 2 days; soluble COD removal was around 70%. A methane conversion rate of 0.35 m³/kg COD removed was achieved during the study. The average volatile solids or biomass (VS) concentration in the reactor had increased from 12.75 g/L to 60 g/L by the end of the study. Sludge volume index (SVI) determinations performed to evaluate the settling characteristics of the anaerobic sludge in the reactor indicating excellent settleability with SVI values of generally less than 20 mL/g. Active sludge granules ranging from 3–8 mm in diameter were produced in the reactor.

In short, it may be concluded that anaerobic treatment may be a very feasible alternative for olive mill effluents, but additional posttreatment, such as aerobic treatment, would be needed to satisfy discharge standards required for receiving waters (river, lake).

Case Study V. This experiment aimed at gaining better insight into the degradation of the main compounds contained in the OME, in particular, the interaction between the two successive stages occurring in the anaerobic digestion: acidogenesis and methanogenesis [23].

Fresh OME was obtained from the olive oil continuous centrifuge processing plant of Montelibretti (Rome). The tests were carried out in 500 mL glass bottles with perforated screw tops with latex underneath, which served to ensure that the bottles were airtight. These bottles were filled with OME diluted in distilled water to obtain the required concentration (in the range of 10–60 g COD/L). The inoculum was obtained from a sludge anaerobic digester at the East Rome wastewater treatment plant. The main results that can be drawn from this study are as follows.

Under the most favorable conditions (pH 8.5, 35°C, initial concentration 10 g COD/L, acclimatized inoculum) the OME were degraded with a high conversion yield (70–80%), both in acidogenic and methanogenic tests. Most of the lipids were degraded both in acidogenesis and methanogenesis tests. On the other hand, polyphenol-like substances were not degraded at all in acidogenic conditions, whereas they were partially removed in methanogenic conditions. Such a difference has been observed both in OME and synthetic solutions. A little methanogenic activity, established in acidogenic conditions because of the partial degradation of the chemical inhibitor, seems to be the key factor determining lipids degradation, even in acidogenesis tests.

It was also experimentally reported that polyphenol degradation is directly related to the presence of an intense methanogenic activity. In addition, bioconversion yields of OME in acidogenesis are remarkably less sensitive to the effect of pH and substrate concentrations than in methanogenesis. This result might lead to adoption of two-phase anaerobic digestion of OME as a suitable process for optimizing its performance. It is our recommendation that further research be conducted in this scope.

5.5.5 Design Example 3

The design of an anaerobic contact reactor to achieve 90% removal of COD from a wastewater flow 180 m^3/day (47,600 gal/day) resulted from a group of neighboring olive mills. The following conditions apply: total influent COD = 13,000 mg/L; nonremovable COD = 2500 mg/L; removable COD (COD$_R$) = 10,500 mg/L; and COD to be removed = 90%. The process parameters are: sludge age (SRT) = 15 days (minimum); temperature = 35°C; a = 0.14 mg VSS/mg COD$_R$; b = 0.02 mg VSS/mg VSS-day; K' = 0.0005 L/mg-day; X_v = 5000 mg/L.

Solution

(a) The digester volume from the kinetic relationship:

$$\text{Detention time, } t = \frac{S_r}{X_v K' S} = \frac{(10,500)(0.9)}{(5000)(0.0005)(1050)} = 3.6 \, \text{day}$$

The digester volume is therefore:

$$\forall = (3.6 \, \text{day})(180 \, \text{m}^3/\text{day}) = 648 \, \text{m}^3 \, (0.1712 \, \text{MG})$$

Check SRT from the equation:

$$SRT = \frac{X_v t}{\Delta X_v} = \frac{X_v t}{aS_r - bX_v t}$$

$$= \frac{(5000)(3.6)}{(0.14)(9450) - (0.02)(5000)(3.6)} = 18.7\,\text{day}$$

This is in excess of the recommended SRT of 15 days to ensure the growth of methane formers.
(b) The sludge yield from the process is:

$$\Delta X_v = aS_r - bX_v t$$
$$= (0.14)(9450) - (0.02)(5000)(3.6) = 963\,\text{mg/L}$$

$$\Delta X_v = 963\,\text{mg/L} \times 180\,\text{m}^3/\text{day}$$
$$= 173.34\,\text{kg/day}\ (381.35\,\text{lb/day})$$

(c) Gas production:

$$G = 0.351(S_r - 1.42\Delta X_v),$$

where $G = \text{m}^3$ of CH_4 produced/day

$$G = 0.351[(9.450)(180) - (1.42)(173.34)]$$

$$= 0.351(1701 - 246.14) = 511\,\text{m}^3\ CH_4/\text{day}$$

or

$$G = 5.62(S_r - 1.42\Delta X_v),$$

where $G = \text{ft}^3$ of CH_4 produced/day

$$G = 5.62[(9450)(0.0476\,\text{MG/day})(8.34) - (1.42)(381.35)]$$

$$= 18{,}040\,\text{ft}^3/\text{day}\ (511\,\text{m}^3/\text{day})$$

Gas production can be also determined by using the approximate estimation, which is 1 kg COD_{elim} yields about 0.3–0.5 m³ of methane. Therefore, total gas production:

$$G = 9.45\,\text{kg COD/m}^3 \times 180\,\text{m}^3/\text{day} \times 0.3\,\text{m}^3\ CH_4/\text{kg COD}$$

$$= 510\,\text{m}^3\ CH_4/\text{day}$$

(d) Heat required can be estimated by calculating the energy required to raise the influent wastewater temperature to 35°C (95°F) and allowing 1°F (0.56°C) heat loss per day of detention time. Average wastewater temperature = 24°C (75.2°F) and heat transfer efficiency = 50%.

$$Btu_{req} = \frac{W(T_i - T_e)}{E} \times (\text{specific heat})$$

$$= \frac{(47{,}600\,\text{gal/day})(8.34\,\text{lb/gal})(95°F + 3.6°F - 75.2°F)}{0.5} \times \left(\frac{1\,\text{Btu}}{1\,\text{lb°F}}\right)$$

$$= 18{,}600{,}000\,\text{Btu}\ (19{,}625{,}000\,\text{kJ})$$

The heat available from gas production is:

$$\text{Btu}_{available} = (18{,}040\,\text{ft}^3\,\text{CH}_4/\text{day})(960\,\text{Btu}/\text{ft}^3\,\text{CH}_4)$$
$$= 17{,}320{,}000\,\text{Btu}/\text{day}\ (18{,}300{,}000\,\text{kJ}/\text{day})$$

External heat of $18{,}600{,}000 - 17{,}320{,}000 = 1{,}280{,}000\,\text{Btu}/\text{day}$

$1{,}325{,}000\,\text{kJ}/\text{day}$ should be supplied to maintain the reactor at $35°\text{C}$ ($95°\text{F}$).
(e) Nutrient required as nitrogen is:

$$N = 0.12\Delta X_v = 0.12 \times 173.34\,\text{kg}/\text{day}$$
$$= 20.80\,\text{kg}/\text{day}\ (45.8\,\text{lb}/\text{day})$$

The phosphorus required is:

$$P = 0.025\Delta X_v = 0.025 \times 173.34\,\text{kg}/\text{day}$$
$$= 4.33\,\text{kg}/\text{day}\ (9.534\,\text{lb}/\text{day})$$

Remarks

1. The effluent from the anaerobic plant does not achieve the national quality criteria of the water resources because of the high values of residual COD_R ($10\% = 1050\,\text{mg}/\text{L}$) and nonremovable COD ($2500\,\text{mg}/\text{L}$). Therefore, we recommend that an aerobic treatment process (such as activated sludge) follow the anaerobic process to produce an effluent meeting the quality limits.
2. Another suggestion is to apply wet air oxidation (WAO) as a pretreatment step to remove biorecalcitrant compounds, which leads to the reduction of anaerobic reactor volume and also to the reduction of energy consumption. This combined WAO–anaerobic process achieves an overall performance to meet the national regulations of Mediterranean countries.

5.5.6   Combined Biological Treatment Processes

The following models are suggested for combined biological treatment processes of OMW. It has been referred to as the combined treatment in order to realize the following: partial treatment by high organic load in the first phase and full treatment by low organic load in the second phase.

Treatment on Site

Before discharge to a nearby water recourse, OMW could be subjected to either of the two subsequently proposed complete treatment systems.

Anaerobic–Aerobic Treatment. The combined model "anaerobic–aerobic treatment" (Fig. 5.10) may be considered quite practical, both environmentally and economically. This method can be applied without serious emissions into air, water, and soil, keeping to the key objectives of environmental policy adopted worldwide.

Anaerobic processes are especially suited for the treatment of high-load wastewater with a COD concentration of thousands (mg/L) in industry. Moreover, the climatic conditions in the olive-growing and production countries are optimal for anaerobic processes.

Combining anaerobic and aerobic processes lessens the disadvantages resulting from separate applications. The first step includes the advantages of the anaerobic process concerning degradation efficiency, energy self-sufficiency, and minimal excess sludge production. The

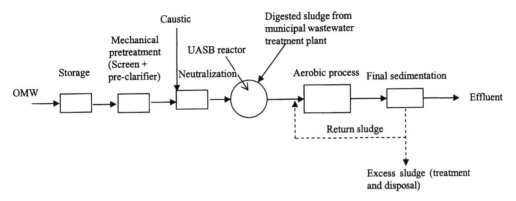

Figure 5.10　Combined anaerobic–aerobic treatment model (on site).

disadvantages of aerobic treatment are nearly compensated by the anaerobic preliminary stage. The high quantity of excess sludge that normally results is strongly reduced. At the same time, the aeration energy needed for the aerobic process is also considerably minimized. With regard to treatment efficiency, plant reliability, and costs, the anaerobic–aerobic model well suits the treatment of olive oil mill wastewater (alpechin) from both ecological and economical aspects [38].

Two-Stage Aerobic Treatment.　This is a combined treatment model of two-stage aerobic treatment based on an activated sludge process, as illustrated in Figure 5.11.

Treatment in Combination with Municipal Wastewater.　In the case where full treatment onsite is not possible, OMW after pretreatment should be drained to a municipal wastewater treatment plant in the vicinity. Figure 5.12 illustrates clearly the combined treatment of OMW with municipal wastewater, where two streams (a and b) are suggested.

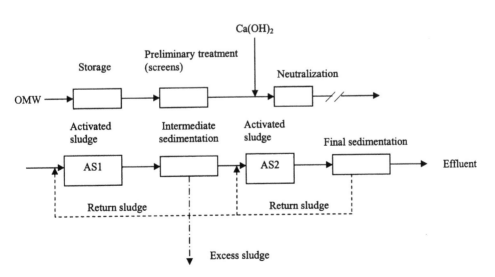

Figure 5.11　Combined treatment model of two-stage activated sludge process (on site). (*Note*: In dispensing with the primary sedimentation tank, it is recommended here to recirculate the return sludge from the final sedimentation to both the AS1 and AS2. Consequently, excess sludge will be discharged only from the intermediate sedimentation tank.)

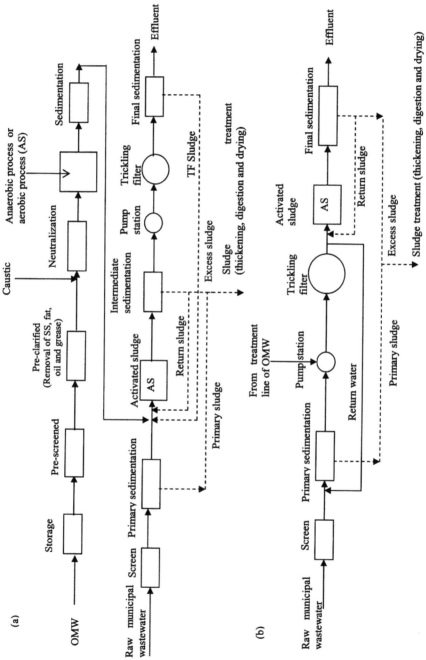

Figure 5.12 Combined treatment of OMW with municipal wastewater. (*Note:* Aerobic process may need addition of nutrients in order to maintain the ratio COD : N : P at 100 : 5 : 1, this ratio being commonly satisfactory for microorganism growth and activity.) (a) Where the activated sludge process is before a trickling filter process is preferable to line (b) in general, with the consideration that line (b) (trickling filter–activated sludge combined model) dispenses with the intermediate sedimentation basin.

The aforementioned combined models suggested for treatment of OMW realize different degrees of efficiency depending on the wastewater characteristics, discharge regulations, organic load in each phase, type and number of phases within the treatment line or plant. In this respect it is necessary that the treated wastewater meet the quality criteria of the water resources (drinking, irrigation, recreation, etc.), where it is supposed to be discharged. In the event the treated wastewater is intended to be used directly for irrigation, it should meet local criteria adopted in that country or those adopted by the Food and Agriculture Organization (FAO).

5.5.7 Design Example 4

To continue Example 3, assuming that an air-activated sludge plant follows the anaerobic process, a design for this plant is required under the following conditions to produce an effluent with a COD of 30 mg/L. The aerobic process parameters are: $T = 20°C$; $a = 0.5$; $F/M = 0.3\,\text{day}^{-1}$; $a' = 0.55$; $X_v = 2500\,\text{mg/L}$; $b = 0.15/\text{day}$ at $20°C$; power $= 1.5\,\text{lb}\,O_2/(\text{hp-hour})$ $(0.91\,\text{kg}\,O_2/\text{kW})$.

Solution

$$t = \frac{S_0}{X_v(F/M)} = \frac{1050}{2500 \times 0.3} = 1.4\,\text{day}$$

$$S_r = S_0 - S_e = 1050 - 30 = 1020\,\text{mg/L}$$

$$K = \frac{(S_0 S_r)}{t S_e X_v} = \frac{1050 \times 1020}{1.4 \times 30 \times 2500} = 10.2/\text{day}$$

The aeration tank volume is:

$$\forall = Q \cdot t = 180 \times 1.4 = 252\,\text{m}^3 \ (66{,}640\,\text{gal})$$

Calculate the degradable fraction X_d using the following equation:

$$X_d = \frac{aS_r + bX_v t - [(aS_r + bX_v t)^2 - 4bX_v t \times 0.8aS_r]^{1/2}}{2bX_v t}$$

$$= \frac{(0.5 \times 1020) + (0.15 \times 2500 \times 1.4) - [\ldots\ldots\ldots]^{1/2}}{2 \times 0.15 \times 2500 \times 1.4}$$

$$= \frac{(510 + 525) - [(510 + 525)^2 - (4 \times 525 \times 0.8 \times 510)]^{1/2}}{2 \times 525}$$

$$= \frac{1035 - 463}{1050} = 0.545$$

The oxygen required is:

$$O_2/\text{day} = (a'S_r + 1.4bX_d X_v t)Q$$

$$= [(0.55 \times 1020) + (1.4 \times 0.15 \times 0.545 \times 2500 \times 1.4)] \times 47{,}600\,\text{gal} \times 8.34 \times 10^{-6}$$

$$= 382\,\text{lb/day} = 16\,\text{lb/hour}\ (7.3\,\text{kg/hour})$$

The power required is:

$$hp = \frac{O_2/\text{hour}}{[1.5\,\text{lb}\ O_2/(\text{hp-hour})]} = \frac{16}{1.5}$$

$$= 10.7\,\text{hp}\ (8\,\text{kW})$$

Other olive oil mills wishing to economize their operations would like to join the abovementioned combined anaerobic–aerobic plant for the treatment of their wastewater ($45\,\text{m}^3/\text{day}$), without affecting the plant's efficiency.

- Compute the new effluent from the anaerobic process assuming (X_v) remains the same; what will the new gas production be?
- What modifications to the aerobic process must be made to maintain the same effluent quality? Assume the sludge settling characteristics are the same as originally and the volatile content of the sludge is 75%.

Solution

The load to the plant is increased to $225\,\text{m}^3/\text{day}$ (59,400 gal/day).

(a) *Anaerobic process*. New effluent concentration; from example 3: $SRT_{\min} = 15$ days; $T = 35°C$; $a = 0.14$; $b = 0.02$; $k' = 0.0005$ L/(mg-day); $X_v = 5000$ mg/L; $COD_R = 10,500$ mg/L; and volume $= 648\,\text{m}^3$ (0.1712 MG).

The new detention time is:

$$t' = \frac{\forall}{Q} = \frac{648}{225} = 2.9\,\text{day}$$

The COD effluent from the anaerobic process can be estimated by:

$$COD_E = \frac{COD_{\text{removed}}}{X_v K' t'} = \frac{(COD_R - COD_E)}{X_v K' t'}$$

$$= \frac{COD_R}{(1 + X_v K' t')}$$

$$= \frac{10,500}{(1 + 5000 \times 0.0005 \times 2.9)}$$

$$= 1273\,\text{mg/L}$$

The COD removed is:

$$COD_{\text{removed}} = COD_R - COD_E$$

$$= 10,500 - 1273$$

$$= 9227\,\text{mg/L}$$

Check SRT using the equation:

$$\text{SRT} = \frac{X_v t'}{\Delta X_v} = \frac{X_v t'}{a\text{COD}_{removed} - bX_v t'}$$

$$= \frac{5000 \times 2.9}{(0.14 \times 9227) - (0.02 \times 5000 \times 2.9)}$$

$$= 14.5 \, \text{day} \approx 15 \, \text{day} \qquad \text{OK}$$

New gas production. The sludge yield is:

$$\Delta X_v = (a\text{COD}_{removed} - bX_v t')Q$$

$$= (0.14 \times 9227 - 0.02 \times 5000 \times 2.9) \, \text{mg/L} \times 59{,}400 \, \text{gal/day}$$

$$\times 8.34 \times 10^{-6} \, (\text{lb/MG})/\text{mg/L}$$

$$= 496.4 \, \text{lb/day} \, (225.36 \, \text{kg/day})$$

The mass of COD removed per day is:

$$S_r = \text{COD}_{removed} \times Q$$

$$= 9227 \, \text{mg/L} \times 59{,}400 \, \text{gal/day} \times 8.34 \times 10^{-6}$$

$$= 4571 \, \text{lb/day} \, (2076 \, \text{kg/day})$$

or

$$S_r = 9227 \, \text{mg/L} \times 225 \, \text{m}^3/\text{day} \times 10^{-3} = 2076 \, \text{kg/day}$$

The methane production can be estimated from:

$$G = 5.62(S_r - 1.42\Delta X_v)$$

where G is given in ft^3 of CH_4/day

$$G = 5.62(4571 - 1.42 \times 496.4)$$

$$= 21{,}727 \, \text{ft}^3/\text{day} \, (615 \, \text{m}^3/\text{day})$$

(b) *Aerobic process.* The new detention time is:

$$t' = \frac{252 \, \text{m}^3}{225 \, \text{m}^3/\text{day}} = 1.12 \, \text{day}$$

The new COD removed:

$$S_r' = S_0' - S_e = 1273 - 30 = 1243 \, \text{mg/L}$$

From the equation:

$$\frac{S_0 - S_e}{X_v t} = K \frac{S_e}{S_0}$$

By rearrangement, the new MLVSS are obtained as

$$X_v' = (S_0' S_r')/(t' S_e K)$$

$$= (1273) \times (1243)/(1.12 \times 30 \times 10.2)$$

$$= 4617 \, \text{mg VSS/L}$$

and the MLSS are:

$$MLSS = 4617/0.75 = 6156\,mg/L$$

The new F/M is:

$$(F/M)' = S'_0/(X'_v t')$$
$$= 1273/(4617 \times 1.12) = 0.25/day$$

Power increase, the new degradable factor is:

$$X'_d = 0.50$$

The new oxygen required is:

$$O_2/day = (a'S'_r + 1.4bX'_d X'_v t')Q$$
$$= (0.55 \times 1243 + 1.4 \times 0.15 \times 0.5 \times 4617 \times 1.12) \times 59{,}400\,gal \times 8.34 \times 10^{-6}$$
$$= 608\,lb/day = 25.3\,lb/hour\,(11.5\,kg/hour)$$

The new power required is:

$$h'p = \frac{25.3\,lb/hour}{1.5} = 16.9\,hp\,(12.6\,kW)$$

The power increase is:

$$hp_{inc.} = 16.9 - 10.7 = 6.2\,hp\,(4.6\,kW)$$

5.5.8 Design Example 5

A 7500 m³/day (2.0 million gal/day) municipal activated sludge plant operates at an F/M of 0.3 day^{-1}. A group of olive oil mills needs to discharge 450 m³/day (0.12 million gal/day) of wastewater with a BOD of 8000 mg/L to the plant. What pretreatment is requested of the mills to reduce the BOD in their wastewater, in order to win the plant's approval?

Solution

(a) Municipal sewage: flow = 7500 m³/day (2.0 million gal/day); S_0 (BOD) = 300 mg/L; Soluble BOD = 100; $F/M = 0.3$; $X_v = 2500$ mg/L; S_e (soluble) = 10 mg/L; $K = 8$/day at 20°C. (b) Olive mill wastewater: flow = 450 m³/day (0.12 MG/day); S_0 (BOD) = 8000 mg/L; $K = 2.6$/day at 20°C; estimated MLVSS = 3500 mg/L.

Detention time is:

$$\frac{F}{M} = \frac{S_0}{X_v t}$$

$$t = \frac{300}{2500 \times 0.3} = 0.4\,day$$

Average reaction rate K will be:

$$\frac{7500(8) + 450(2.6)}{7950} = 7.7/day$$

The new detention time is $0.4 \times 7500/7950 = 0.38$. The influent to the plant to meet the permit can be calculated:

$$\frac{S_0 - S_e}{X_v t} = K \frac{S_e}{S_0}$$

$$S_0^2 - S_e S_0 - S_e K X_v t = 0$$

$$S_0 = \frac{S_e + \sqrt{S_e^2 + 4 S_e K X_v t}}{2} = \frac{10 + \sqrt{100 + (4 \times 10 \times 7.7 \times 3500 \times 0.38)}}{2}$$

$$= 325 \,\text{mg/L of soluble BOD}$$

The concentration of BOD in the pretreated mill wastewater can then be calculated by a material balance:

$$Q_s(S_{0,s}) + Q_l(S_{0,l}) = (Q_s + Q_l)S_{0,s+l}$$
$$7500(100) + 450(S_{0,l}) = 7950(325)$$

or

$$2.0(100) + 0.12(S_{0,l}) = 2.12(325)$$
$$S_{0,l} = 4075 \,\text{mg/L}$$

Pretreatment is required to reduce about 50% of the BOD in the mill wastewater.

(c) *Temperature effects*: Determine the change in MLVSS that will be required when the temperature coefficient θ increases from 1.015 to 1.04 due to an increase in soluble mill wastewater BOD:

$$\frac{K20}{K10} = (1.015)^{10} = 1.16 \,\text{sewage}$$

$$\frac{K20}{K10} = (1.04)^{10} = 1.48 \,\text{sewage–mill–wastewater}$$

The increase in MLVSS can be calculated as:

$$\frac{1.48}{1.16} \times 2500 \,\text{mg/L} = 3190 \,\text{mg/L}$$

Remarks

1. To achieve the BOD reduction of about 50% in the olive oil mill effluents, the anaerobic process should be recommended as pretreatment.
2. The municipal activated sludge plant could not achieve the quality limits or criteria of the water resources because of the high value of BOD in the mill wastewater (4075 mg/L). In such a case, an additional aerobic degradation stage is needed, such as activated sludge or trickling filter as illustrated in Figure 5.12.

5.5.9 Wet Air Oxidation and Ozonation

The clear advantages of the anaerobic process make it the process of choice for treating olive oil effluents [52]. However, many problems concerning the high toxicity and inhibition of biodegradation of these wastes have been encountered during anaerobic treatments, because some bacteria, such as methanogens, are particularly sensitive to the organic contaminants present in the OME. The biorecalcitrant and/or inhibiting substances, essentially phenolic compounds (aromatics), severely limit the possibility of using conventional wastewater anaerobic digestions [53] or lead to difficulties in the anaerobic treatment of OME [23].

Moreover, it was proved that the anaerobic sludge digestion of OME in UASB-like reactors was unstable after a relatively short period of activity [54]. Consequently, anaerobic biological treatment as a unique process showed limited efficiency in the removal of aromatics. Therefore, other treatments such as chemical oxidation have been investigated for olive oil mill wastewater and for table olive wastewater purification, with encouraging results.

This chemical oxidation proved to be very effective in treating wastewaters that contain large quantities of aromatics [55,56]. Recently, integrated physicochemical and biological technologies have been developed as efficient processes to achieve high purification levels in wastewaters characterized by difficult biotreatability [57].

The effectiveness of the combination of chemical oxidation and biological degradation relies on the transformation of nonbiodegradable substances into biogenic compounds readily assimilated by microorganisms [57].

Principle of Wet Air Oxidation (WAO)

The type of chemical preoxidation used in integrated processes is highly dependent on the characteristics and nature of the wastewater to be treated. Thus, in the case of effluents with a high content of phenol-type substances, oxidizing systems based on the use of oxygen or ozone at high temperatures and pressures have been shown to readily degrade phenolic structures [58]. Wet air oxidation (WAO) is an oxidation process, conducted in the liquid phase by means of elevated temperatures (400–600 K) and pressures (0.5–20 MPa). The oxidant source is an oxygen-containing gas (usually air).

As pressure increases, the temperature rises, which leads to an increasing degree of oxidation. With far-reaching material conversion, only the inorganic final stages of CO_2 and water (and possibly other oxides) are left. With incomplete degradation, the original components (which often are nondegradable) are decomposed to biodegradable fragments. Therefore, it is useful to install a biological treatment stage downstream of the wet oxidation stage (Fig. 5.13) (Case Study I).

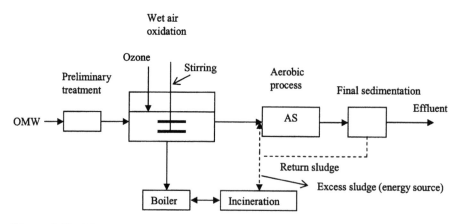

Figure 5.13   Wet air oxidation–aerobic process.

On other hand, Beltran-Heredia et al. [21] applied an opposite arrangement, that is, aerobic degradation followed by ozonation, in normal conditions where the temperature and the pH values were varied (Case Study II). Oxidizing chemicals are also used instead of oxygen so that even hardly degradable constituents of liquid waste from olive oil production can be destroyed or attacked. Possible oxidizing agents are ozone (O_3) or hydrogen peroxide (H_2O_2) [59].

The utilization of H_2O_2 has turned out to be environmentally friendly because this oxidizing agent has no negative effects. However, since H_2O_2 quickly undergoes decomposition, its ability to be stored is limited. The OH radicals formed during H_2O_2 decomposition have oxidative effects. Using suitable agents [e.g., titanium dioxide (TiO_2)] or UV radiation, the development of radicals can be considerably forced [38].

In oxidation systems, ozone in particular has many of the oxidizing properties desirable for use in water and wastewater treatment; it is a powerful oxidant capable of oxidative degradation of many organic compounds, is readily available, soluble in water, and leaves no byproducts that need to be removed. In addition, it may also be used to destroy bacteria, odors, taste, and coloring substances.

It has been reported in the literature that anions of phenolic compounds are more reactive towards oxidative processes than the noncharged species [58,60].

Case Studies

Case Study I. A considerable amount of work has been devoted to the integrated wet air oxidation–aerobic biodegradation process (Fig. 5.14) in treating olive-processing wastewater in the province of Badajoz, Spain [19]. The most representative parameters are the COD and BOD$_5$, with values of 24.45 and 14.8 g O_2/L respectively, and phenolic content 833 mg phenol/L. Chemical oxygen demand (COD) conversion in the range 30–60% (6 hours of treatment) was achieved by WAO using relatively mild conditions (443–483 K and 3.0–7.0 MPa of total pressure using air). Also noticed was a significant removal of phenolic content at the end of WAO process with conversion values 95%. Use of the homogeneous catalysts such as radical promoters (hydrogen peroxide) resulted in a higher efficiency of the process (between 16 and 33% COD removal improvement, depending on operating conditions). Biodegradability tests conducted after the oxidation pretreatment showed the positive effect of the WAO pretreatment on the aerobic biological oxidation of wastewater. Acclimation of microorganisms to oxygenated species formed in a chemical preoxidation step enhanced the efficiency of the biodegradation.

In conclusion, if WAO is used as a pretreatment step, the advantages associated with the use of the previous oxidation are based on the higher biodegradation rate and better properties of the activated sludge used in the biodegradation process to remove biorecalcitrant compounds.

As inferred and reported from this work [19], the following conclusions may be drawn:

- The WAO process may become thermally self-sustaining, because the COD of the influent is well above 15 g/L. In this case, the wastewater stream would not be diluted and more severe conditions should be applied.
- The seasonal character of these activities (fruit and vegetable related processes) may allow for the use of WAO mobile units, capable of processing up to a maximum of 400–500 L/hour of wastewater (more than needed for these types of industries). As a result, a permanent location is not needed, with subsequent savings in fixed capital costs.
- Use of *in situ* WAO shows additional advantages regarding necessary barreling and hauling to appropriate wastewater plants.

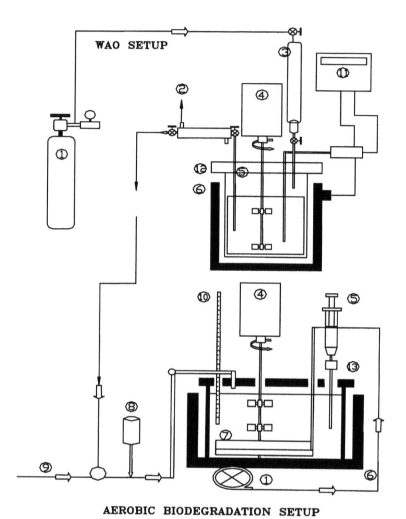

Figure 5.14 Experimental setup of WAO–aerobic processes (from Ref. 19). 1 = Air Cylinder; 2 = Cooling Water; 3 = Injection Port; 4 = Stirring System; 5 = Sampling Port; 6 = Thermostatic Bath; 7 = Porous Plate; 8 = pH Controller; 9 = Dilution Waterline; 10 = Thermometer; 11 = Temperature Controller; 12 = High-Pressure Reaction Vessel; 13 = Glass Bioreactor.

- The consequences of WAO pretreatment may also affect the operability of aerobic biological treatment itself. Thus the benefits are as follows. (a) The biodegradation rate was observed to increase from a nonpretreated effluent to a WAO pretreatment wastewater, which would imply a lower total volume of biological reactor and lower energy consumption (requirements for mixing and aeration) to achieve an overall performance to meet the limits of the environmental legislation. (b) The sludge volume index (SVI) decreased if the WAO pretreatment was applied. An average 20% decrease was observed for biological experiments using pretreated wastewater. This would help to prevent operational problems usually found in activated sludge plants, such as bulking sludge, rising sludge, and nocardia foam, and would allow a wider food-to-microorganisms (F/M) ratio for operation in the aeration tank and lower total volume of the secondary clarifier. (c) An excess of generated sludge as a result of biological oxidation could be recycled as an energy source by combustion or anaerobic treatment to use in the wastewater treatment plant, or it could even be treated by the same WAO system.

Case Study II. The original black-olive wastewater was obtained from a table olive processing plant in the Extremadura community (Spain). The treatment was carried out by ozonation, aerobic biological degradation, and the combination of two successive steps: an aerobic biological process followed by ozonation. For this purpose, the chemical oxygen demand (COD), the total aromatic content (A), and the total phenolic content (Tph), were selected as representative parameters to follow the overall purification process.

The experimental results [21] given for ozonation, where the temperature (10, 20, and 30°C) and the pH (7.9 and 13.6) were varied, are as follows: the COD conversions ranged between 42 and 55% depending on the operating conditions; the conversions of the total phenolic and aromatic compounds are around 75 and 67%, respectively.

A direct influence of temperature and pH on the COD and the phenolic compounds degradation was also observed. Thus, it may be concluded that ozone is an excellent oxidizing agent in the specific destruction of phenolic and aromatic compounds.

The experimental results from the aerobic biological treatment were as follows: the COD conversions ranged between 76 and 90%; the conversions of aromatic compounds ranged between 16 and 35%; and conversions ranged between 53 and 80% for total phenolics.

The combined process of an aerobic degradation followed by an ozonation produced a higher COD, phenolic and aromatic removal efficiency. This combined process reached a degradation level that cannot be obtained by any chemical or biological process individually under the same operating conditions.

There was a clear improvement in the second stage relative to ozonation, and biological pretreatment also led to an increase in the kinetic parameters. This implied that the aerobic pretreatment enhanced the later ozone oxidation by removing most of the biodegradable organic matter, while the ozonation step degraded some of the nonbiodegradable organic matter plus most of the phenolic compounds not removed previously.

Case Study III. This research focuses on the degradation of the pollutant organic matter present in wastewater obtained from an olive oil production plant located at the Extremadura Community (Spain), by combining two successive steps: (a) ozonation followed by aerobic degradation, and (b) aerobic degradation followed by ozonation. For this purpose, the chemical oxygen demand (COD), the total aromatic content (A) and the total phenolic content (Tph), were selected as criteria to monitor the overall degradation process [32]. The combined OMW degradation processes were studied with the goal of evaluating the influence of each respective pretreatment on the second stage. The first combined process (C-1) comprised ozone oxidation pretreatment followed by aerobic biodegradation. Table 5.12 summarizes the operating conditions, the initial and final COD concentrations, and the conversion values obtained (X_{COD}) in each stage individually considered, as well as the conversion achieved by the overall process. The total conversion obtained by the successive stage (C-1) was 84.6%, a higher value than achieved by either single process under the same operating conditions. This suggests that ozone pretreatment enhances the subsequent aerobic process, probably by removing some phenolic compounds capable of inhibiting biological oxidation. Similar to combination (C-1), the overall process achieved, by the second combined process (C-2), 81.8% degradation, which was greater than that obtained by the individual chemical or biological processes under the same operating conditions (Table 5.12). This suggests that aerobic pretreatment enhanced the subsequent ozone oxidation by removing most of the biodegradable organic matter. The ozonation step then

Table 5.12 Treatment of Olive Mill Wastewaters by Ozonation, Aerobic Degradation, and the Combination of Both Treatment Methods

C-1 Ozonation followed by aerobic degradation	C-2 Aerobic degradation followed by ozonation
C-1-A Ozonation stage Operating conditions: $T = 20°C$; $Po_3 = 1.73$ kPa; $pH = 7$; $CODo = 34.05$ g dm^{-3} Substrate removal obtained: $COD_1 = 29.9$ g dm^{-3}; $X_{COD} = 12.2\%$	C-2-A Aerobic degradation stage Operating conditions: $X = 0.53$ g dm^{-3}; $CODo = 41.95$ g dm^{-3} Substrate removal obtained: $COD_1 = 11.07$ g dm^{-3}; $X_{COD} = 73.6\%$
C-1-B Aerobic degradation stage Operating conditions: $X = 0.59$ g dm^{-3}; $CODo = 29.85$ g dm^{-3} Substrate removal obtained: $COD_1 = 5.22$ g dm^{-3}; $X_{COD} = 82.5\%$	C-2-B Ozonation stage Operating conditions: $T = 20°C$; $Po_3 = 1.69$ kPa; $pH = 7$; $CODo = 10.95$ g dm^{-3} Substrate removal obtained: $COD_1 = 7.63$ g dm^{-3}; $X_{COD} = 30.3\%$
Total removal in process C-1: $X_{COD} = 84.6\%$	Total removal in process C-2: $X_{COD} = 81.8\%$

Source: Ref. 32.

degraded some of the nonbiodegradable organic matter and much of the residual phenolic compounds.

In conclusion, the study shows that ozonation of OMW achieves a moderate reduction in the COD, and significant removal of aromatic and total phenolic compounds. The microbial aerobic treatment achieves significant removal of COD and phenolics but with less elimination of aromatic substances. The two processes combined, as presented in this case study, achieve higher COD removal efficiency than treatment by either stage separately under the same operating conditions. Together, the two processes may be used to treat OMW to meet discharge criteria or norms and reach treatment efficiencies required by national regulations, particularly in Mediterranean countries.

5.5.10 Fungal Treatment

Several types of industrial wastes contain phenols. Many of these compounds are extremely harmful as they are highly toxic both towards microorganisms and vertebrates [61]. Enzymatic approaches to removing phenols have been tried for some years as they have several advantages compared with the conventional methods (solvent extraction, chemical oxidation, absorbance on active carbons, etc.) [62].

Recently, results have been obtained for the removal of phenols using phenol oxidizers, which catalyze oxidative coupling reactions of phenol compounds and do not require hydrogen peroxide (H_2O_2) [63]. Olive oil mill wastewaters (OMW) contain large concentrations of phenol compounds, which are highly toxic. The structure of the aromatic compounds present in OMW can be assimilated to many of the components of lignin [64].

However, some microorganisms actively degrade lignin, among which the "white-rot" fungi are particularly efficient. These organisms utilize mainly peroxidases and phenol oxidizers [65]. Potential applications of white-rot fungi and their enzymes are gaining increasing importance in the detoxification of industrial wastewaters, reducing the toxicity of many aromatic compounds (pesticides, disinfectants, phenols) in several types of polluted environments.

Case Studies

Case Study I. This study investigates the application of "white-rot" basidiomycete *Pleurotus ostreatus* and the phenol oxidizers it produces, for reducing the phenol content and the toxicity of the olive wastewater at an olive oil factory in Abruzzo, Italy [61]. It was found that up to 90% of the phenols present in OMW could be removed by treatment with phenol oxidizer from a mixture containing aromatic compounds extracted from OMW, although no concomitant decrease of toxicity was observed.

Results show that *P. ostreatus* removed phenols and detoxified OMW diluted to 10% in the absence of any external added nutrient; the diluted wastewaters were also clarified from this treatment in a relatively short time (100 hours). The detoxifying activity of *P. ostreatus* was concomitant with a progressively increasing phenol oxidase expression. It was noticed that after 100 hours incubation with *P. ostreatus*, the concentration of phenol compounds decreased by 90% and the toxicity towards *Bacillus cereus* was reduced sevenfold compared with that of untreated waste.

Case Study II. This study focused on the ability of white-rot fungi isolated from Moroccan OMW (classified as *Phanerochaete chrysosporium* Burdsall M₁ to modify the polluting properties of diluted OMW in comparison with that of *P. ostreatus*. Olive oil mill wastewater (OMW) was collected from an olive oil factory in Marrakech, Morocco [20].

In order to study the effects of fungal treatment on OMW, two different white-rot fungi were tested in batch cultures of diluted OMW (20%). The maximum reduction of phenol content and COD was 62 and 52% for *P. ostreatus*, whilst it was 82 and 77% for *Ph. chrysosporium* after 15 days of treatment. The time course of absorbance decrease is similar to that of phenol content and COD reduction for both fungi, suggesting the existence of a correlation between these parameters and the colored components present in OMW. The results obtained indicate that *Ph. chrysosporium* is able to decolorize OMW and to degrade its phenolic component more efficiently than *P. ostreatus* can.

Toxicity tests performed on *B. cereus* revealed that fungal treatment of the waste (20 or 50%) causes the complete loss of OMW toxicity after 15 days of treatment. The optimal decolorization temperature for *Ph. chrysosporium* Busdsall M₁ was 28°C. Furthermore, the optimal pH for *Ph. chrysosporium* OMW treatment was in the 4.0–5.0 range. Since the pH of diluted OMW was between 4.0 and 5.0, the process did not require any pH alteration of the effluent.

Degradation of 20 or 50% OMW, expressed as color, phenol, and COD removal, was almost the same after 15 days of fungal growth. Hence, not only is this fungus able to grow in 50% OMW as the sole carbon source, but the degradation rate of the effluent increases in these cultural conditions. This proves that the isolated *Ph. chrysosporium* strain, which is able to grow using diluted OMW, and to notably reduce color, phenol content, and COD, would be a good candidate for the effective treatment of this wastewater.

5.5.11 Decolorization

Investigation of the effect of oxidative coloration on the methanogenetic toxicity and anaerobic biodegradability of aromatics showed that their oxidized solutions were less biodegradable in proportion to their color [66]. In contrast, the aerobic processes can have substantial aromatic removal efficiency, but these processes require sizeable energy expenditures in oxygen transfer and sludge handling [67].

An important step in the degradation of olive oil wastewater is the breakdown of colored polymeric phenolics (decolorization) to monomers, which may subsequently be mineralized.

A significant correlation has been demonstrated between sewage decolorization and reduction of total organic carbon and phenolic content. However, decolorization of wastewaters appears to be associated only with a partial depolymerization. A decrease in the content of the lower molecular mass components and an increase in the proportion of components of intermediary molecular mass have also been demonstrated.

Crude oil wastewater and solutions of its brownish pigment change in both color and solubility as the result of pH modification. It appears that sewage decolorization may be produced simply by a process of adsorption or by adsorption associated with subsequent chemical modification of chromophores.

The effluent is acidified as a consequence of fungal growth. A considerable decrease in pH and an elevated adsorption of lignin-derived products onto the biological matrix suggested that the decolorization process was an indirect effect of culture acidification. The sewage decolorization eventually stops with time, suggesting that the putative enzymes responsible for decolorization have a defined lifetime.

Many recalcitrant compounds from olive oil mill wastewater are present in the colored fraction. Optimum culture conditions will be identified for the decolorization of that sewage by *Phanerochaete flavido-alba* for subsequent use in bioremediation assays. Of several media tested, nitrogen-limited *P. flavido-alba* cultures containing 40 μg/mL Mn(II) were the most efficient at decolorizing oil wastewater. Decolorization was accompanied by a 90% decrease in the phenolic content of the wastewater. Concentrated extracellular fluids alone (showing manganese peroxidase, but not lignin peroxidase activity) did not decolorize the major olive oil wastewater, suggesting that mycelium binding forms part of the decolorization process [38].

In batch cultures, or when immobilized on polyurethane, *Ph. chrysosporium* is able to degrade the macromolecular chromophores of oil wastewater and decrease the amount of phenolic compounds with low molecular weight. *Pleurotus ostreatus* and *Lentinus edodes* also decrease the total phenolic content and reduce the color of cultures containing oil wastewater.

Decolorization of juices and wastewaters by Duolite XAD 761 resin is widely used on an industrial scale and is particularly useful for the removal of color, odor, and taste from various organic solutions in the food and pharmaceutical industries. It removes color, protein, iron complexes, tannins, hydroxymethyl furfural and other ingredients responsible for off-flavors, according to the Duolite Company. The degree of adsorption tends to increase with molecular weight in a given homologous series and has more affinity for aromatic than aliphatic compounds. Recovery of coloring compounds and pigments from agroindustrial products is a common practice [24].

The following case study offers detailed information about the efficiency of resin application in decolorization of olive mill effluents.

Case Study

Chemical and physical treatments of olive oil mill effluent (OME) were performed in this study [24]. The goal was to evaluate the efficiency of aromatic removal from undiluted OME through precipitation by iron sulfate and lime, adsorption on a specific resin, and chemical oxidation by hydrogen peroxide prior to anaerobic digestion as the final treatment method, in order to reduce the toxic effect of OME on bacterial growth and to reduce the coloring compounds in undiluted OME. Olive oil mill effluent was obtained from a local olive oil mill in Tunis and stored at $-20°C$. The main findings from this case study are as follows:

1. With regard to the decolorization of OME by iron as a complexing agent, it was noticed that many of the organic and inorganic OME components are susceptible to precipitation by iron. The decrease in the color of OME resulted in a decrease in COD.

The maximum amount of COD and OD removal that could be attained was close to 70% by using 30 g/L of ammonium iron(III) sulfate. Moreover, it seems that the removal of OME color corresponded to the same degree of COD removal. This means that COD is mostly due to the aromatic compounds that are responsible for the color. The complexing effect of iron was complete after 3 hours.

2. As for decolorization of OME by lime treatment and pure calcium hydroxide, the removal efficiency increased with increasing lime concentration. In total, 55% of COD and 70% of color (OD_{390nm}) removal were reached. However, for economic and biological considerations, treatment with 10 g/L calcium hydroxide was sufficient. The effect of lime was complete after 12 hours. It may be concluded that using only 10 g/L of iron and lime as complexing agents was sufficient to precipitate more than 50% of the initial COD and remove 50% of the initial color within a short contact time.

3. With regard to decolorization of OME by resin treatment, the Duolite XAD 761 resin as aromatic adsorbent was used in a column (28 cm long, 1.5 cm in diameter, and with a total volume of 50 cm^3). The results obtained after treating one, two, or three bed volumes of OME, were as follows: COD removal varied between 63 and 75%, and color decrease varied between 52 and 66% for OD_{280nm} and between 51 and 64% for OD_{390nm}. It was also shown that the coloring components in OME are the compounds most responsible for its pollution potential (COD). It may be concluded that the aromatic adsorbent resin retained more than 50% of the coloring compounds (chromophores) corresponding to removal of more than 60% of the initial COD after treating three bed volumes of crude OME. The efficiency depended on the volume treated.

4. As for oxidation of OME by hydrogen peroxide, it has already been shown before (Section 5.5.9) that chemical oxidation is very effective in treating wastewaters that contain large quantities of aromatics. The study was limited to the use of hydrogen peroxide (H_2O_2) concentrations of up to 3%. The effect of H_2O_2 on OME is clear: H_2O_2 removed the substituents of the aromatic rings, which resulted in a decrease in length of the coloring compounds in OME. However, they were not completely degraded, leading to shorter wavelength absorption. This chemical treatment was efficient in color removal but only 19% COD removal was possible. In all cases, simple aromatics were reduced, as determined by GPC analysis.

5. With regard to anaerobic digestion of pretreated OME, the anaerobic digestion of crude and treated OME was elucidated in order to evaluate the efficiency of the physical and chemical pretreatments of OME (Fig. 5.15). In general, it may be concluded that each pretreatment was efficient in removing the toxic effect in OME. The anaerobic digestibility of OME was improved, with iron and lime, and no inhibition was observed on methanogenic activity. Oxidation of coloring compounds in OME by H_2O_2 removed their toxic effect and did not generate new toxic chemicals to bacterial growth. Separation of aromatics by resin treatment seemed to be the most effective in removing the inhibitory effect of OME prior to anaerobic digestion. Nevertheless, the choice from these different alternatives must be based on economic considerations.

The following process was proposed for reducing environmental pollution by aromatic compounds: physicochemical reduction of most toxic compounds of OME, followed by anaerobic microbial decomposition of the main pollutants up to an insignificant amount (see Section 5.5.10 for case studies about the role of fungal treatment in decolorization of OME).

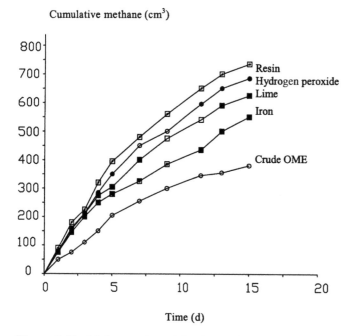

Cumulative methane (cm^3)

Figure 5.15 Methanogenic activity in relation to different treatments of OME (role of different treatments in decolorization of OME) (from Ref. 24).

5.5.12 Precipitation/Flocculation

Precipitation involves transforming a water-soluble substance into its insoluble particular form by means of a chemical reaction. Certain chemicals cause precipitation when they react with dissolved and suspended organic compounds. By adding flocculants and coagulation aids, the finest suspended compounds or those dissolved in colloidal form are then transformed into a separable form. This means that, in contrast to precipitation, flocculation is not a phase-transition process [38]. The wastewater may be further treated by activated carbon, ultrafiltration, or reverse osmosis. Figure 5.16 gives a general concept of the precipitation–flocculation process.

Iron sulfate and aluminum sulfate are commonly used as efficient chelating agents of complex organic compounds in certain wastewaters [68]. Their adsorption capacity is complex and depends on the composition of the precipitated molecule. Lime stabilization is a recognized means of treating municipal sludge prior to land application [69]. The addition of lime temporarily halts biological activity. Moreover, lime renders organic molecules more accessible to microorganisms [70].

In wastewater from olive oil mills (OMW), a purification efficiency of almost 70% of the organic and inorganic components could be removed or complexed by lime (calcium hydroxide) [24]. Disadvantages include the high consumption of chemicals and the large quantities of sludge formed in the process (about 20% of treated alpechin) [38]. For more information about the efficiency of lime and iron as complexing agents in removing COD and color from OMW, refer to the case study presented in Section 5.5.11 (Decolorization).

A proposed plant in Madrid for combined precipitation/flocculation treatment of OMW is presented as a good example of a complete treatment system [38]. This system consists of four phases. In the first phase, a flocculent is added, followed by discharge, filtration, or

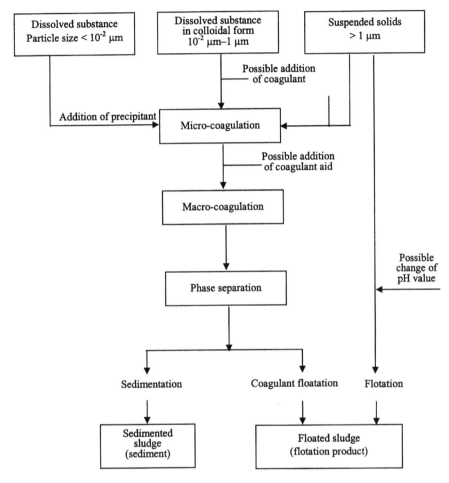

Figure 5.16 Precipitation–flocculation processes in general (from Ref. 38).

centrifugation. The resulting liquid has a dark red color, and its BOD_5 is about 10,000 mg/L. In the second phase, another flocculation occurs where the smaller size of the flocs are separated through filtration, and its BOD_5 reaches 8000 mg/L. The sludge from these two stages combined is 12% of the original alpechin. The third phase is biological and occurs in three or four stages in purification towers with a separation device for the solids (biomass) and biomass recirculation. The resulting wastewater has a BOD_5 of 2000 mg/L. The fourth phase consists of the filtration of the wastewater, ultrafiltration, and reverse osmosis. The concentrated and thickened sludge from the previous phase is then dried by means of band filters for further use as fertilizer.

5.5.13 Adsorption

Currently, the most commonly used methodologies for the treatment of aromatic-bearing wastewaters include solvent extraction, physical adsorption separation, and chemical oxidation [67]. The adsorption method, which refers to bonding of dissolved compounds (adsorbate) at the surface of solid matter (adsorbent), for example, activated carbon and bentonite, is used for adsorption of dissolved organic pollutants in water. In the field of olive oil wastewater, these are coloring substances (mainly tannic acid), hardly or nonbiodegradable pollutants, bactericidal or inhibiting compounds, which have to be removed. Adsorption not only takes place at the visible surface of the solid, but also in its pores. Activated carbon is especially suited because of its large inner surface (500–1500 m^2/g) and its high adsorptive capacity, but unfortunately it cannot be reused. However, the calorific value is very high so it can be incinerated without problems [38]. Activated carbons are the most common adsorbent, and they are made from different plants, animal residues, and bituminous coal [71,72]. Depending on the composition of the industrial wastewater, one type of carbon may be superior to another [73]. Between 60 and 80% of the organic constituents from alpechin can be adsorbed by activated carbon.

Strong contamination has negative effects on the workability of the plant; thus the alpechin should be pretreated, for example in an activated sludge tank (Fig. 5.17) [38].

The use of bentonite as an adsorbent for cleaning vegetable oils suggests its applications to reduce lipid inhibition on thermophilic anaerobic digestion [74]; bentonite was added to a synthetic substrate (glyceride trioleate, GTO) and turned out to stimulate methane production by binding the substrate on its surface and thus lowering GTO concentration in the liquid phase.

Laboratory-scale experiments were carried out on fresh OME obtained from an olive oil continuous centrifuge processing plant located in Itri, Lazio, Italy, in order to identify pretreatment type and condition capable of optimizing OME anaerobic digestion in terms of both kinetics and methane yield [75]. In this regard, a set of tests was carried out to evaluate the effect of adding bentonite to OME, both untreated and pretreated with $Ca(OH)_2$. Significant results were obtained by adding $Ca(OH)_2$ (up to pH 6.5) and 15 g/L of bentonite, and then feeding the mixture to the anaerobic biological treatment without providing an intermediate phase separation. Indeed, the biodegradable matter adsorbed on the surface of bentonite was gradually released during the biotreatability test, thus allowing the same methane yield (referred to the total COD contained in untreated OME) both in scarcely diluted (1 : 1.5) pretreated OMW and in very diluted (1 : 12) untreated OME.

These results suggest the application of a continuous process combining pretreatment [with $Ca(OH)_2$ and bentonite] and anaerobic digestion without intermediate phase separation [75]. Specific resin is an economic adsorbent alternative for separating complex organic compounds from wastewater. The Duolite XAD 761 resin is used industrially for the adsorption of mono- and polyaromatic compounds. A considerable number of experiments have focused on removal of coloring compounds in OME by resin treatment [24]. Crude OME was passed through a resin (Duolite XAD 761) column (28 cm long, 1.5 cm in diameter, and with a total volume of 50 cm^3) according to the suggested operating conditions reported by the Duolite

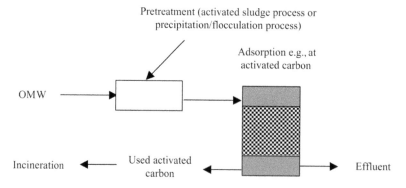

Figure 5.17 Adsorption process for treatment of olive oil mill wastewater (from Ref. 38).

Company. The pH of the resin was almost 4, and the pH of OME was corrected to 4 using 2 mol/ L HCl. The OME was passed through the resin bed at a rate of 50 cm^3/hour. Table 5.13 shows the results obtained after treating one, two, and three bed volumes of OME. With such treatment, it is clear that the removal of COD up to 75% and decrease in color (OD$_{280nm}$ and OD$_{390nm}$) up to 66.3 and 63.5%, respectively, could be achieved. Efficiency of the resin treatment decreased with OME volume, due to the saturation of the resin. Moreover, the ratio OD$_{280nm}$/OD$_{390nm}$ remained constant (almost 5) in crude and treated OME, which meant that adsorption of organic compounds on the resin occurred with the same degree of affinity. On the other hand, the decrease in OME color corresponded to the same degree of COD removal. (For more information about this process, refer to Section 5.5.11).

5.5.14 Biofiltration and Ultrafiltration

Physical processes including filtration, centrifugation, sedimentation, and ultrafiltration are highly efficient methods for phase separation. Filtration processes are used to remove solid material as far as possible from the wastewater. Particles and liquid are separated as a result of pressure difference between both sides of the filter, which enables the transport of water through the filter. During the filtering process, the solids accumulate in the filter and reduce the pore volume, resulting in a change of resistance to filtration and of the filtrate quality. As soon as the admissible resistance to filtration is reached, the filter must be backwashed by forcing clean water backwards through the filter bed. The washwater is a waste stream that must be treated [76].

Compounds that are already dissolved cannot be treated, except by biofiltration. In this case, the filter serves also as nutrient for bacteria so that dissolved organic substance can be aerobically degraded. The purification capacity of biofiltration plants is between 70 and 80%. Up to 100% of the solids can be reduced.

A prerequisite for biofiltration is sufficient oxygen supply. If the alpechin is insufficiently treated, the filter will be quickly clogged. The material kept back in the filter can be used in agriculture (Fig. 5.18).

A promising alternative method is based on a physicochemical pretreatment that removes lipids and polyphenols as selectively as possible before biological treatment. In this regard, the potential of filtration applied with other techniques for removal of COD, lipids, and polyphenols from OME has been studied in the following example [75].

A laboratory-scale experiment was carried out in order to choose the pretreatment operating conditions capable of optimizing the anaerobic digestion of OME in terms both of

Table 5.13 Treatment of OME Through Duolite XAD 761 Resin

OME	OD (280 nm)	OD (280 nm) removal (%)	OD (390 nm)	OD (390 nm) removal (%)	OD (280 nm)/OD (390 nm) ratio	COD (g/dm^3)	COD removal (%)
Crude OME	45.1	–	8.5	–	5.3	147	–
[V(o)/V(r)] = 1	15.2	66.3	3.1	63.5	4.9	37	75
[V(o)/V(r)] = 2	18.7	58.5	3.6	57.6	5.2	43.4	70.1
[V(o)/V(r)] = 3	21.7	51.8	4.2	50.6	5.1	54	63.2

Note: OD: optical density measures qualitatively the color darkness of OME. The OD values were measured at 390 nm and 280 nm.
Source: Ref. 24

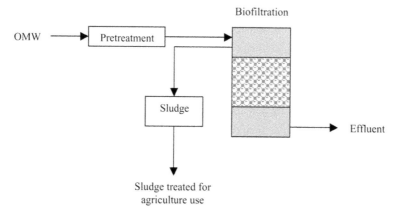

Figure 5.18 Biofiltration process for treatment of olive oil mill wastewater (from Ref. 38).

kinetics and biomethane yield. Fresh OME was obtained from an olive oil continuous centrifuge processing plant located in Itri, Italy. The OME (pH 4.4, total COD = 92.6 g/L) contained 5.1 g/L of polyphenols, 3.1 g/L of oleic acid, and 11.1 g/L of lipids. The first set of pretreatment tests was carried out by using only physical methods of phase separation: sedimentation, centrifugation, filtration, and ultrafiltration. In the sedimentation phase, after two hours of magnetic stirring, 50 mL of OME were left undisturbed for 24 hours. Afterwards, the OME were centrifuged at 4600 rpm for 15 minutes. The resulting intermediate phase was filtered under vacuum on filter at several pore sizes (25, 11, 6, and 0.45 μm). After filtration on 0.45 μm filters, 20 mL of OME were ultrafiltrated on membranes at 1000 and 10,000 D cutoff threshold (a micron ultrafiltration cell; operating pressure, 4 bar by nitrogen gas).

Table 5.14 shows the results obtained. The highest removals of oleic acid (99.9%) and polyphenols (60.2%) were obtained through ultrafiltration (at 1000 D). However, COD removed by this technique (65.1%) was much higher than COD associated to lipids and polyphenols removal. While very efficient as a separation technique, ultrafiltration subtracts too much biodegradable COD from the pretreated OME, thus lowering the potential for methane production.

Table 5.14 Removal of COD, Oleic Acid, and Polyphenols from OME by Means of Physical Methods of Separation

Method of separation	Removal of COD (%)	Removal of oleic acid (%)	Removal of polyphenols (%)
Sedimentation	38.4	96.1	0
Cenrifugation	38.6	95.4	10.2
Filtration [pore size (μm)]			
25	36.7	96.6	12.2
11	37.6	97.6	13.4
6	38.9	98.1	13.4
0.45	40.3	99.0	13.1
Ultrafiltration [cutoff (D)]			
10,000	51.5	99.8	37.2
1000	65.1	99.9	60.2

Source: Ref. 75.

Therefore, ultrafiltration is considered here as a separation technique with poor selectivity. Moreover, the application of ultrafiltration to OME pretreatment might encounter serious problems of membrane fouling as well as of treatment of the concentrated stream. Among the other separation techniques, centrifugation demonstrated the important advantages of producing smaller volumes of separated phases. Further details about this and other sets of pretreatment tests in connection with anaerobic biotreatability may be found in Ref. 75.

5.5.15 Evaporation/Drying

Evaporation is a method used to concentrate non-steam-volatile wastewater components. The evaporation plant contains a vapor condenser by which vapor and steam-volatile compounds are separated from the concentrate. While the concentrate is then recycled into the evaporator, the exhaust steam can be used for indirect heating of other evaporator stages (Fig. 5.19).

The degree of concentration of the wastewater components depends on different factors, for example [38]:

- reuse of the concentrate (e.g., reuse in production, use as fodder, recovery of re-cyclable material);
- type of disposal of the concentrate (e.g., incineration, landfill)
- properties of the concentrate (e.g., viscosity, propensity to form incrustation, chemical stability).

Advantages of this method include:

- the residue (dried oil wastes) can be reused as fodder and fertilizer;
- only a small area is needed;
- exhaust steam can be reused as energy;
- considered state of the art in the food industry [38].

Disadvantages are:

- the exhaust steam from evaporation is organically polluted and needs treatment;
- rather high operation and maintenance costs;

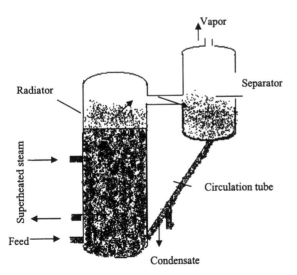

Figure 5.19 Evaporation/drying processes for treatment of olive oil mill wastewater (from Ref. 38).

- requires high energy;
- requires trained personnel.

Details about drying processes, including case studies for the treatment of olive oil mill wastes, are discussed in Section 5.6.2.

5.5.16 Electrolysis

There are methods still in the experimental stage for treatment of olive oil mill wastewater, one of which is electrolysis. This method is based on electrolytic oxidation of margine constituents, using titanium/platinum for the anode and stainless steel for the cathode. The following data are drawn from laboratory experience (Fig. 5.20) [38].

The process has the following components:

- electrolytical cell;
- recirculation reactor;
- margine input;
- pH control;
- cooling system.

The performance of the electrolytic cell was tested with a 4% NaCl density in the margine (alpechin) at 42°C, with the temperature remaining constant during the course of the experiment. Four tests lasting 10 hours each were carried out under the same conditions. After 10 hours of electrolysis, the organic load was reduced by 93% in COD and by 80.4% in TOC (total organic carbon). The greatest disadvantage of this method is its high energy consumption (12.5 kW per kg of margine). Therefore, it should be applied only as part of the biological pretreatment of the wastewater. Energy consumption then reaches 4.73 kW/kg within the first three hours [38].

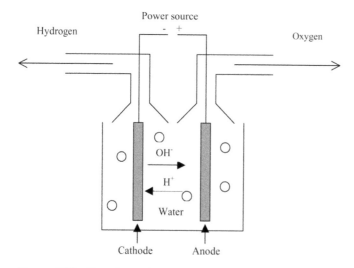

Figure 5.20 Experimental setup of electrolysis for olive oil wastewater treatment (from Ref. 38).

5.5.17 Bioremediation and Composting

The aim of bioremediation is to repurpose the liquid waste (alpechin) or the liquid fraction of alpeorujo (aqueous fraction that can be separated from fresh alpeorujo by percolation or soft pressing) by diverse aerobic fermentation. The composting of the solid waste (orujo) or the solid fraction of alpeorujo produces a useful material for plant growth.

Several years' research work at the Laboratory for Microbiology of the Athens University has shown that margine (alpechin) is a good substrate for certain microbial fauna. It is especially useful for producing fertilizer for agricultural purposes. Under aerobic conditions, the margine content aids the qualitative breeding of nitrogen-consuming bacteria, especially of acetobacter. This feature was taken into consideration when developing a treatment method for the wastewaters from olive oil production with high organic load. Using this method, a substrate for soil improvement with high nutrient content is obtained from the wastewaters.

Case Studies

Case Study I. A pilot plant was put into operation in an oil mill of the Romano-Pylias region. The first big treatment unit was built in 1997 within the framework of the LIFE program for a total of six oil mills in the region of Kalamata (Peloponnes). In addition, a second plant with lower performance was built to treat wastewater from the oil mill in the Arta district. The method consists of two phases [38]. In Phase 1, the margine is neutralized by adding CaO at a pH between 7 and 8. The substrate is mixed in a reactor, which is equipped with a mechanical stirring device. The undiluted residues from the decanter are fed into the stirring reactor. In Phase 2, the contents of the stirring reactor are fed into the bioreactor where sessile microorganisms (especially *Acetobacter vinelandii*) degrade the substances with phytotoxic effect. These bacteria consume nitrogen and take in oxygen from atmospheric air, which is provided by a turntable air distribution system. This leads to increased nitrogen consumption of the bacteria, degradation of the phytotoxic substances, formation of polymers, and secretion of reproduction factors like auxines, cytocynines, which support plant growth.

Retention time in the reactor is 3 days (repeated fed batch culture). The advantages of this method lie in the possibility of applying it directly to olive oil mill wastewater without oil separation, and the high removal efficiency of COD and decolorization.

We propose the possibility of replacing the bioreactor (Phase 2) with the process of natural composting, where the content of Phase 1 is to be mixed, in a well studied way, with municipal solid waste. On the other hand, the main disadvantages here are the long duration (one month or more) needed for aerobic degradation and the need for a large area to conduct the aerobic process.

The final product from the bioreactor or from the natural waste composting plant has a pH of 7.5–8, and, mixed with any quantity of water, can be used to improve soil. Moreover, it has the following characteristics:

- It shows a high content of organic nitrogen (by consumption of atmospheric nitrogen), and substances like auxines support plant growth.
- All nutrients and trace elements present in the olive can be found again in the substrate improved soils.
- The product is able to improve the soil structure and to increase its water retention capacity, due to the biopolymers contained therein.

Case Study II. A study was carried out on isolating bacteria from the alpeorujo composting system at Kalamata, Greece [77]. The main results were:

- Identifying bacterial diversity using biochemical techniques of lipid analysis and the molecular biological techniques.
- Demonstration of detoxification of compost by indigenous bacteria.
- Possibility of using a combination of traditional microbiological and modern molecular biological approaches, to follow the changes in microbial flora within the composting material in a qualitative manner.

Strain A of *Azotobacter vinelandii* was used as an agent in the bioremediation process, which was studied in an aerobic, biowheel-type bioreactor, under nonsterile conditions. Before inoculation, the pH of the liquid function of alpeorujo was adjusted to 8.5 by adding CaO. The inoculation was then added at a rate of 10^5 cells/cm^3. The main experimental findings were:

- The alpeorujo liquid fraction (ALF) is very phytotoxic, and inhibitory to the growth of pleurotus and other fungi and many bacteria.
- When ALF is diluted with water (tenfold or more) it can be used as substrate for Azotobacter, Fusarium, Pleurostus and some yeasts (Candida).
- *A. vinelandii* (strain A), while it can degrade and utilize phenolic compounds, grows slowly during the first 3 days because of the antimicrobial properties of OMW.

Standard bioremediation conditions are of major importance, since (a) the OMW quality is largely dependent on the olive mill machinery and storage facilities and on the quality of the raw material (olives); and (b) bioremediation cycles are performed during wintertime in plants that are exposed to variable environmental conditions.

A continuous composting process was followed. It was observed that alpeorujo, unlike the extracted press cake of the three-phase decanters, is highly unsuitable and cannot be used as a pleurotus substrate. This is due to its high concentration of phenolics. This toxicity is more acute in the pulp fraction of alpeorujo. The wet olive pulp represents 60% of alpeorujo. It is acidic (pH 4.6–4.8), almost black in color mass with moisture content of 65–67% (wet basis), having a smooth doughlike structure. It is also rich in organic and inorganic constituents, especially potassium. Nevertheless, its chemical composition is not compatible with the composting process, and so the olive pulp poses quite a serious obstacle to waste treatment and hinders alpeorujo recycling efforts.

In the course of this case study, the possibility of composting both alpeorujo and pulp was also investigated. The major experimental findings were:

- Composting of alpeorujo is feasible when it is mixed with bulky material at a proportion of 3 : 1.
- The mature alpeorujo compost or compost taken from the end of the thermophilic phase offers an ideal microbial consortium to act as starter.
- For alpeorujo and deoiled alpeorujo a self-sustainable composting process was elaborated. Bulky material is only required for the initiation of the process.

In addition, a novel thermophilic process of composting based on the use of hydrogen peroxide (H_2O_2) was developed, due to the fact that hydrogen peroxide exerts a triggering effect on the composting process. The key points include:

- The long-term rise of temperature reflects intensification of microbiological activity in the catabolic processes.

- The formation of glucose from cellulose yields hydrogen peroxide, hydroxyl, and superoxide radicals that are needed to initiate in a snowball reaction the breakdown of the lignin skeleton.
- Similar evidence has bean reported in the case of the brown rot fungus *Gloephyllum trabeum.*

These findings have led to the establishment of a new method for assessing compost stability [77].

With regard to positive effect on plant growth and control of soil fungal pathogens, it was noticed that *A. vinelandii* possesses the ability to induce soil suppressiveness against some notorious soil-borne root pathogens, such as *Pythium*, *Phytophthora*, and *Rhizoctonia* species through its intrinsic ability to produce siderophores.

At the end of this project, the compost produced satisfied farmers, who expressed commercial interest in its use. The compost extract gave similar or even better control against potato blight when compared with commercial organic preparations. Therefore, composting and subsequent utilization in agriculture appears to be the most suitable procedure for treatment of (solid–liquid) waste (alpeorujo). However, large-scale application and more intensive investigation must follow before these procedures may be introduced to the market.

5.5.18 Livestock Feeding

Several methods may be used to enrich OMW with fungi and yeasts so that it becomes suitable for animal feed. The following is a summary of successful experiments performed in Greece as part of the Improlive project, an "International Project to Improve Environmental Compatibility in Olive Oil Production" (during the period 1997–1999) within the European FAIR Programme "Quality of Life and Management of Living Resources."

Case Study

Research [78] was conducted by the University of Athens (1997–1999) with the objective of enriching the two-phase system waste "alpeorujo" with fungal or yeast protein through microbial fermentation and subsequent amino acid production. To give a clear picture of the microorganisms (such as fungi, yeasts, and bacteria) present in the alpeorujo, various techniques and methodologies were applied: serial dilution and selective culture media, application of different inoculation techniques and enrichment of cultures and subcultures, as well as variation in growth temperature and anaerobic conditions. The isolated microorganisms were analyzed for their morphological and biochemical features, then classified into 27 bacteria strains, nine yeasts and three more fungal strains. In order to study the fermentation of bacteria and yeasts, the microcosm system was selected, while a solid-state fermentation bioreactor was used for the fungal strain of *Paecilomyces variotii*. In the microcosm system, and as for as the bacteria concerned, their population declined immediately after inoculation and showed no survival after 72 hours. Total sugars and tannins of the fermented products decreased shortly after each growth cycle of the inoculums. Total lipid content increased after fermentation in all cases.

The microcosm system was followed by solid-state fermentation experiments, which were used to study the growth and activity of selected strains of yeasts and fungi and relevant control conditions, leading to findings such as (a) protein content increased after fermenting the substrate (alpeorujo) with *P. variotii*; (b) the best growth temperature is 35°C for *P. variotii*; (c) long-term experiments are suitable for the best fermentation of alpeorujo substrate. Another step performed was the enrichment of alpeorujo with molasses, which is an inexpensive, renewable industrial byproduct with a very high sugar concentration.

The following conclusions may be drawn from the case study:

- The main constituents of alpeorujo are tannins, lipids, proteins, sugars, and lignocellulosic materials. The chemical profile of alpeorujo makes it adequate for supporting microbial growth by providing plenty of carbon, nitrogen, and energy sources. The results confirm this assumption: alpeorujo is a suitable substrate for the growth of fungi and yeasts and metabolite production.
- Apart from the aerobic bacteria growing at 30°C, several thermophilic bacteria have been isolated and identified, in addition to yeasts (for example, *Candida* genus) and fungi such as *Rhizopus* and *Penicillium*.
- The enrichment of alpeorujo with molasses produced satisfactory results. The increase in the final protein content is around 45%. This increase is a very positive result for the use of the waste material.
- The industrial application of *P. variotii* as a means of increasing the protein content seem feasible, giving the excellent ability to grow in a variety of high-polluted industrial effluents, such as molasses, wood hydrolysates, and spent sulfite liquor. This fungus has an optimum growth at 35°C, while the optimum pH was 4.

The enrichment of alpeorujo with molasses could be a good solution to increase the final protein content and for the optimization of waste materials to be used as animal feed or food additives.

The final conclusion is that *P. variotii* is a fungus that can better utilize the substrate and grows well in it. The resulting increase in the final protein content allows for the possibility of using it as an animal feed or as a feed additive. In addition, not only the fresh but also dried (solid/liquid) waste can be used for fermentation experiments. It is more convenient, however, to use the latter since it is easily handled as a substrate. Further experiments are needed to test the nutrition value of the derived products and their safety for animal consumption.

5.5.19 Ocean Outfalls

The authors proposed for study and application the following method for disposal of olive oil mill wastewaters through submarine outfalls. This section will introduce this method and present its advantages, defects, success conditions, quality limits of sea water, design criteria of marine outfalls, and the required specific pretreatment.

Significance of Submarine Outfalls

Discharge of sewage to the sea through sea outfalls was introduced more than 50 years ago. Outfalls can range in length from a few hundred meters up to more than 15 km; diameters typically vary from 0.5 m up to 8 m and the number of diffuser ports can range from one to several hundred. Sea conditions vary significantly from protected estuaries to open coasts with strong currents and breaking waves [79].

The discharge of industrial and domestic wastewater through submarine outfalls and diffuser systems is one of the most economic solutions for the final disposal process in coastal areas. This disposal system represents a viable alternative for the many population and industrial centers of the world located on sea coasts, particularly for developing countries where financial resources are limited. The capital costs of constructing inland treatment works are often similar to those for an equivalent marine treatment scheme. However, the operational cost of inland treatment is much greater.

Diffusion of industrial and domestic wastewater into marine receiving water, after the degree of treatment deemed necessary for a location, from a properly designed and sited marine outfall system is one of the most environmentally safe options for populations near open coast areas. Such systems can make maximum utilization of the natural assimilating capacity of the sea water environment, which serves as a treatment and disposal facility, and when properly planned, will not produce an undesirable impact upon marine water.

Specific Pretreatments and Quality Limits

Marine treatment via a sea outfall must be considered as a part of the wastewater treatment in conjunction with land treatment, and is one of the most efficient processes to treat effluents with high contaminations. However, since wastewater discharged from inadequately designed or poorly maintained sea outfalls can be a major source of pollution in many coastal areas, the EPA and the EEC have developed some restrictive legislation regarding this issue [80].

In some cases, sea outfalls are used to discharge toxic effluents without proper pretreatment and, consequently, are responsible for some ecological damage. However, it is widely accepted by scientists and engineers that the use of long sea outfall with an adequate control of the discharged effluent quality is an environmentally safe, waste disposal option.

Materials diffused through marine outfalls may or may not affect the ecology of the receiving water area. Consequently, the oceanography, biology, and ecology of receiving water areas were studied to determine sensitivities to contaminants and design allowing diffusion below sensitivity levels. By satisfying these requirements, marine outfalls could have a positive impact on the coastal water, including the presence of fertilizers, such as nitrogen, phosphorus, and carbon in wastewater that maintain life productivity [81].

Sea discharge of industrial and municipal effluents should meet the quality limits of coastal waters used for fisheries, swimming, and recreational purposes, and meet the design criteria given at national level. If a coastal country has no such limits or standards, it may benefit from other countries' experience in this respect. Turkey is a good example in the Mediterranean area (Tables 5.15 and 5.16).

Table 5.15 Required Characteristics of Industrial Wastewater for Sea Discharge in Turkey

Parameter	Value	Remarks
pH	6–9	
Temperature (°C)	35	
SS (mg/L)	350	
Oil and grease (mg/L)	10	
Floating matter	None	
BOD_5 (mg/L)	250	
COD (mg/L)	400	
Total N (mg/L)	40	
Total P (mg/L)	10	
Surface active agents (mg/L)	10	
Other parameters	–	Special care for hazardous wastes

Source: Ref. 82.

Table 5.16 Design Criteria for Marine Outfalls Systems in Turkey

Parameter	Limits
Temperature	2°C (max) increase after initial dilution
Total coliform (fecal coliform bacteria/100 mL)	1000 in 90% of samples
Initial dilution (D_x)	40 (min)
Discharge depth (m)	20 (min)
Discharge length (m)	1300 m (min) for discharge depth less than 20 m

Source: Ref. 82.

If the receiving water body and/or wastewater characteristics are not deemed acceptable, then marine outfall is not permitted [82]. Table 5.17 shows the necessity for pretreating some polluting constituents such as particle, oil, grease, and floatables prior to sea discharge through submarine outfalls, with special concentration on refractory substances and heavy metals that require specific treatment at source in conformity with the quality limits of the sea water.

Table 5.18 presents the removal of significant constituents by pretreatments (milliscreens or rotary screens and by primary sedimentation) [83]. It is noted that the main differences in effluent characteristics relate to the removal of settleable solids and suspended solids and, to a lesser extent, to removal of grease. However, milliscreens remove floatables and particulate fat, which is the material of significance regarding aesthetic impact on the marine environment. The only adverse impact of the discharge of grease relates to slick formation, but when initial dilution is sufficient, the concentration of such material in the mixed effluent/sea water plume is very low and this problem is eliminated [84].

In addition, the data show that screens with openings of less than 1.0 mm require extensive maintenance for cleaning whereas those of 1.0 mm do not.

Disposal of OME Through Submarine Outfalls

With regard to olive industry wastewater, which is mainly characterized by a high content of polyphenols, fats, COD, and solid matters, Table 5.17 shows that sea water can play a role in the treatment and disposal of biodegradable organics. Refractory organics should be subjected to proper treatment at the source (mill). Fats, floatables, settleable and suspended solids should be pretreated by rotary screens or milliscreens and primary sedimentation. It is possible to treat polyphenols by the decolorization process, which has demonstrated significant correlation

Table 5.17 The Role of Sea Water in Removal of Wastewater Constituents and the Required Pretreatment Process Prior to Sea Discharge Through Submarine Outfalls

Constituent	Pretreatment	The required process
Particle	Partly needed	Mechanical pretreatment (preliminary treatment + primary sedimentation)
Oil, fats, and floatables	Needed	
Biodegradable organics	Not needed	–
Nutrients	Not needed	–
Pathogenic bacteria	Not needed	–
Refractory organics	Needed	Proper treatment at source
Heavy metals	Needed	Proper treatment at source

Table 5.18 Removal of Wastewater Constituents by Milliscreens and Primary Treatment

	Percentage removal		
	Milliscreens		
Constituent	0.5 mm apertures	1.0 mm apertures	Primary treatment
Settleable solids	43	23	95–100
Suspended solids	15	10	50
Oil and grease	43	30	50–55
Floatable solids	99	96	95–100

Source: Ref. 82.

between the sewage decolorization and reduction of total organic carbon and phenolic content. It is also advisable to conduct intensive research about sea water's role in reducing these compounds. In cases where pH is less than or equal to 5, it is necessary to apply neutralization within the pretreatment. The criteria given in Table 5.16 can be referred to for planning and designing the submarine outfalls. Other references provide further details about design criteria and modeling [85]. For economic reasons, it is recommended that several neighboring mills associate in one submarine outfall.

The possible impact of effluents on public health and the environment (aesthetic) should be assessed through monitoring stations for effluent discharge and bathing water (Fig. 5.21) to achieve national or international standards (fats, COD, and polyphenols).

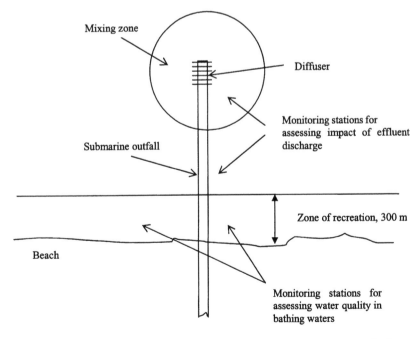

Figure 5.21 Monitoring stations location for olive oil mill wastewater discharge through submarine outfalls (from Ref. 48).

5.6 SOLID WASTE TREATMENT METHODS

Many of the abovementioned treatment methods for liquid waste are suitable for the treatment of solid/liquid waste arising from the two-phase decanter (alpeorujo). Some of these methods are also appropriate for the treatment of solid waste (orujo), such as recycling methods (composting and livestock feeding). In this respect, a distinction should be made between aerobic treatment systems for liquid waste (such as activated sludge, trickling filter, bioremediation) and aerobic treatment systems for solid waste (such as composting).

Based on the various experiments and published research for waste, especially solid waste and liquid–solid waste, we can propose suitable methods for treating waste from olive oil mills (Table 5.19). These treatments are classified into three groups: physical, biotechnological, and chemical processes [1]. At the same time, it should be realized that no specific treatment or solution can be generalized. Each case must be studied and evaluated according to local circumstances.

5.6.1 Biotechnological Processes

Biotechnological processes mainly include aerobic (composting), anaerobic (mixed fermentation), solid fermentation, and fungal treatments. A detailed description of methodologies, results, and case studies related to these processes was discussed in Section 5.5.

Other points of considerable importance can be added in this respect [1]:

- Because olive oil mills are operated over limited periods, that is, about 3 months only per year, an ideal treatment method would be one that could be shifted for treating other types of waste after the end of the olive oil production season.
- The composting method for solid waste treatment is preferable to other methods. This process takes place without serious emissions into air, water, or soil and therefore conforms to the key objectives of Mediterranean environmental policy. Since operational and personnel costs are rather low, this process might also be accepted by plant operators.

Table 5.19 Treatment Methods for the Solid Waste from Olive Oil Production

Treatment method of orujo and alpeorujo
Physical processes
• Drying
• Evaporation
• Thermal treatment
Biotechnological treatment
• Aerobic (composting)
• Solid fermentation
• Anaerobic/mixed fermentation
• Fungal treatment
Chemical processes
• Incineration
• Combustion
• Pyrolysis
• Gasification

Adapted from Ref. 1.

- The costs of a composting plant strongly depend on the sales potential for the final product in the individual countries. In Greece, for example, higher receipts from compost selling are possible than in Spain. As a result the total costs of a plant also change [2].
- The start-up time of the compost process is only 2 weeks. It runs in a cycle, which means that additional structuring material is required only in the beginning, and the compost itself is used later as structuring material. The final product is of a high quality and well suited to be used as fertilizer in agriculture.
- Anaerobic treatment by itself is not suitable for solid waste because of its low water content. Problems with mixing and clogging may arise during treatment. Moreover, anaerobic treatment requires further treatment measures, causing additional costs. Another problem is the long start-up time of the process after a longer shutdown period. These problems were behind the breakdown of anaerobic plants in Greece. In the meantime, these plants have been shut down. An economically reasonable solution is to combine this treatment for existing fermentation plants. For this purpose, however, the local situation must be suitable, that is, the fermentation plant should have free capacity and be situated near the olive oil production to avoid high transportation costs and start of digestion of the solid waste.

5.6.2 Physical Processes

Evaporation/drying processes and their advantages and disadvantages in liquid waste treatment have already been discussed. In solid waste treatment, these processes can be discussed in detail as follows. Two of the most important problems related to the treatment of solid waste or solid/liquid waste (alpeorujo) are the optimization of drying and oil recovery by physical means (to get, as much as possible, olive oil instead of orujo oil).

The following case studies discuss new driers based on the combination of fluidized and moving beds, in addition to different pilot-plant treatments of pit separation, drying in a ring drier, and deoiling solid waste in oil mills.

Principle of Fluidized/Moving Beds (Flumov)

The fluidized/moving bed (flumov) combines a fluidized bed with a top section in the form of a fixed/moving bed. The main problem that must be dealt to is the control of the circulation of solids to obtain almost-perfect mixing flow of the solids through the fluidized bed and a plug-flow of the solids in the moving bed (Fig. 5.22).

The drying of solid waste (or alpeorujo) is required before this waste may be used to recover orujo oil by extraction with hexane and for other processes such as the production of compost, activated coal, biopolymers, and so on. The classical driers, for example, rotary kilns (trommels) and trays, have a low thermal efficiency due to the poor air–solid contact and can present several problems because of the high moisture and sugar contents of the alpeorujo. The presence of the moving zone in flumov allows the fresh product feed to have a higher degree of moisture. Moreover, it favors the solid transport to the fluidized bed contactor, since part of the water is eliminated in the moving zone and the solid enters into the fluidized zone with a relatively low level of moisture [86].

We were particularly interested in confirming the filtering action of the moving bed zone. The filter effectiveness would improve the performance of conventional filtering units usually required for eliminating the suspended solids in the outgoing gas, and even eliminate the necessity of using these units. The stability of the vault, which forms between both beds, requires the input of secondary air into the conical zone to regulate the flow rate of solids from moving

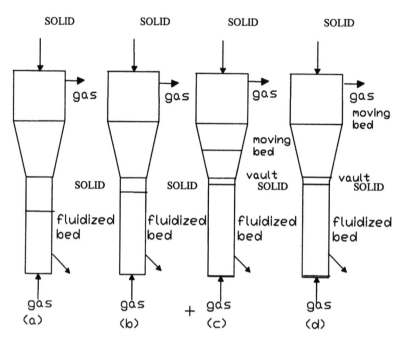

Figure 5.22 Concept of the flumov state: (a) fluidized bed; (b) expanded fluidized bed; (c) formation of vault; (d) fluidized moving bed (from Ref. 1).

bed to the fluidized bed. The experimental results of residence time distribution of the solid agree with combined models of flow and illustrate the almost plug-flow in the moving bed and the perfect mixing in the fluidized zone of the flumov. The filtering effectiveness of the moving zone is very high and the fines in the output air are mostly eliminated.

Case Studies

Case Study I: Flumov Drier. A fluidized/moving bed drier was constructed and operated [87,88]. It consisted of a cylinder 5.4 cm (inner diameter) and 40 cm height (fluidized bed zone) jointed by a conical device to an upper cylinder 19.2 cm (inner diameter) and 30 cm height (moving bed zone). The feed and removal of solids is made with the aid of J-valves especially designed for this work [89]. The system is a small pilot plant capable of treating up to 5 kg/hour of solid or solid–liquid waste (alpeorujo) (Fig. 5.23). The drying of waste was studied in batch, semibatch, and continuous operation. Several runs were made in both a conventional fluidized bed drier and a flumov drier with input air between 70 and 200°C and temperature inside the beds between 50 and 150°C. Fresh alpeorujo contained 50–60% moisture (wet basis) and the dried alpeorujo obtained was rather homogeneous. The extracted oil had the same quality as the oil obtained from dried alpeorujo obtained by other drying methods. The filtering effectiveness of the moving bed was very high. In order to solve the operative problems derived from the high moisture content of alpeorujo and the high viscosity of the semidried one, two solutions were found: mixing dry and wet alpeorujo and using pulses of a secondary air injection into the conical zone. Using these two conditions, the dry/wet mixture circulated more effectively along the whole system than the fresh wet alpeorujo. The feeding from the moving bed to the fluidized zone was also well controlled, the air–solid contact improved and the flumov drier was able to operate at a low temperature, about 60°C, inside the fluidized zone (implying a better thermal efficiency balance and allowed for improvement in the dry solid characteristics).

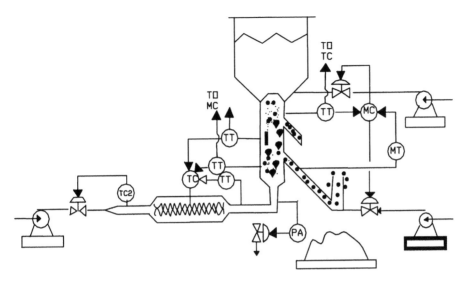

Figure 5.23 Drier with the implemented advanced control prototype (from Ref. 1). Temperature (TC) and Moisture (MC) control system; Pressure alarm (PA) and temperature transmitters (TT). Control prototype designed by Cognito Quam Electrotechnologies Ltd.

The energy consumption of the flumov drier was between 0.71 and 1.11 kWh/kg water. The mechanical power consumption was similar to other industrial driers, 0.05 kWh/kg water. From the results obtained in the small pilot plant, the flumov drier is a feasible and competitive solution for drying waste. The possibility of drying at low temperatures resulted in a better thermal efficiency balance, lower operating and energetic costs, and improved solid characteristics in use of subsequent solid treatments (high quality of the orujo oil extracted). The main advantages of the system are: reduced total volume, filtering capacity, and ability of using low temperature sources to recover heat from several systems, for example, combined cycle systems and exhaust gases. The details about the control system and prototype, and moisture sensor are in the reference materials [1,89].

Case Study II: Ring Drier. (a) Deoiling of the waste. In southern Spain, Westfalio Separator A.G. installed a batch pilot plant with a capacity of approximately 1 m^3 per batch (Fig. 5.24) [1]. This plant allowed for an efficient pretreatment of solid/liquid waste (alpeorujo), the separation of the phases as well as a subsequent drying. Owing to product variation, the actual daily quality of the waste was determined as a basis for the planning of the tests. Thus, for each sample a standard test was carried out and several runs were carried out under different process combinations in order to reach a better deoiling of the fresh waste. For this aim, the pits were partially separated, different malaxing times were tested, enzymes or talcum were added to the malaxing process, small quantities of water were added, or other measures were tested for an improvement of the oil yield.

All these measures changed the characteristics of the raw material and, consequently, contributed in improving the drying process of the deoiled waste. After the deoiling, different intermediate products were generated, that is, partially deoiled orujo and partially depitted orujo. The following parameters were adjusted or the following aids were used [1].

- Enzymes: combination of pectinase and cellulase;
- Talcum: type "talco" 2%;

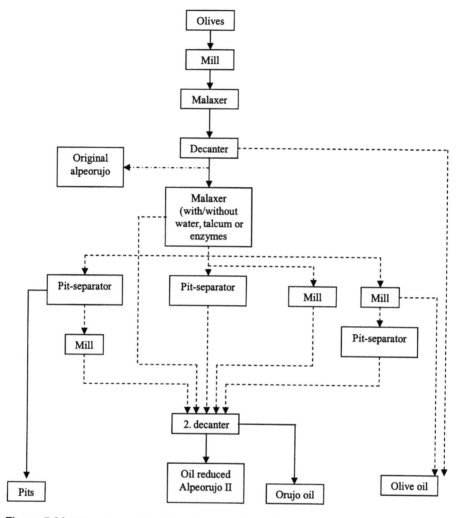

Figure 5.24 Flow sheet of deoiling pilot plants (from Ref. 1).

- Mill: 2.5, 3, and 4 mm screen;
- Pit separator: 3 mm and 4 mm screen;
- Storage tank: 2 m^3;
- Malaxer: 1 m^3;
- Feed rate: up to 500 kg/h (waste).

(b) Drying of the waste. For the drying, a ring drier was installed to dry different alpeorujos. The intermediate products generated by the deoiling pilot plant were stored and dried. This drier was fueled by propane gas, and hot air was produced with this gas heater. The temperature of this hot air can be varied between 160 and 400°C. With the help of the horizontal screw, one part of the dried waste was mixed with the raw stuff. Both pit-reduced waste and simple deoiled waste were dried as a result. By using the ring drier, the humidity of the waste (alpeorujo) was reduced to approximately 10–15%. The dried material is a powder, the fractions of which are: pit fragments, skin, fruit flesh particles, or agglomerates. The thermal energy requirement for the drier is 1.13 kWh/kg evaporated water. After drying, the oil content vs. drying substance (DS) is sometimes higher than the original material. Another conclusion can be drawn here from deoiling and drying of waste in ring drier. Pit separation before processing is a good solution, in order to produce pit that can be used as a fuel directly in the oil mill, and can raise the throughput. On the other hand, the oil yield is a little bit lower than in the basic version. It is worth noting that drying of solid or solid/liquid waste (alpeorujo) is supposed to precede composting or combustion, and is even indispensable for the latter.

5.6.3 Chemical Processes

Incineration plants are widely known as the conventional means for municipal solid waste treatment for many decades up to the present day. This method, which consists of oxidation of organic substances in high temperatures, has its advantages and disadvantages. Pyrolysis, in-contrast to incineration, is a thermic-decomposed reaction of materials containing a high percentage of carbon (without oxygen) in high temperatures. Thus, pyrolysis is a reduction process that might trigger degasification. It is possible to introduce gasification when there is a partial reaction of coke and water with oxygen. These substances react to carbon oxide and hydrogen. The heat obtained in this process helps to crack heavy molecules. Although the pyrolysis can be used to recycle solid residuals and produce heat, it has not become widespread for technical and/or economical reasons [90]. Additionally, there are no known successful applications, even at pilot plant scale of either incineration or pyrolysis in treatment of olive oil mill waste.

We will discuss a new technique that applies combustion and gasification together in a pilot plant, and has tested successfully in treatment of olive oil mill solid waste concentration [1]. This technique depends on a fluidized moving system, which is a good concept of the gasifier because of the special configuration of the reactor zones. In the bottom part of the gasifier, the fluidized bed permits the required combustion, which represents exothermic reactions, necessary to maintain the thermal balance inside the whole reactor. In the upper part, the moving bed zone does not allow the combustion process to occur but only the endothermic gasification processes. This is due to the fact that the rising gas that reaches the moving bed contains a very low concentration of oxygen and has a high temperature (800–850°C). So only the gasification process can be performed in the moving bed.

Case Study

A fluidized/moving bed reactor was designed to serve as combustor and gasifier. The pilot plant was capable of processing 1–5 kg/hour of solids. The control system in the reactor could regulate the mass flow of air, temperature, and level in the fluidized bed and solid feed. The gasifier is a flumov system, a rather new concept of reactor, and was based on a combination of (a) fluidized bed in the bottom part, where mainly combustion processes take part, and (b) moving bed in the upper part, where the solids are preheated and gasified (Fig. 5.25) [86]. A special characteristic of the flumov system is that the moving bed filters the flue gases.

The solid used for gasification was orijullo (deoiled orujo and deoiled alpeorujo) of mean particle size 1.4 mm, and pits (ground stone) of mean particle size 2.57 mm. The fluidized bed was filled with sand of mean particle size 0.21 mm, or in some runs, with dolomite with a mean particle size 0.35 mm.

The ultimate analysis of orujillo and stone showed that both have the same composition (dry ash free analysis: 47% C, 6% H, 1% N, 46% O, and <0.01% S). The content of ash is about 3.2% by weight. One of the main elements in ash is potassium (8–30% in K_2O), the ingredient

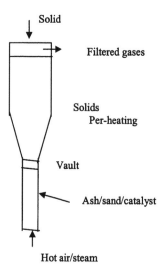

Solid

Filtered gases

Solids
   Per-heating

Vault

Ash/sand/catalyst

Hot air/steam

Figure 5.25   Flumov gasifier (from Ref. 86).

that makes the ashes useful as fertilizer additives. The main process operation variables were temperature, air/water ratio, and equivalent ratio (ER). The presence of sand and dolomite in the fluidized bed had no positive effect on the tar production in the moving bed nor on the flue gas composition (10% H_2, 2% CH_4, 8% CO). Many runs were carried out to find out the best operating conditions, both in combustion and in gasification to obtain the best thermal efficiency. The optimal operating conditions for obtaining the best flue gas were:

- Equivalent ratio (ER = actual air/stoichiometric): 0.20–0.30;
- Temperature in the moving bed: 750–800°C;
- Temperature in the fluidized bed: 800–824°C;
- Throughput: 400–500 kg solid/hm^2 fluidized bed;
- Airflow rate: 1.3 Nm^3/hour;
- Water/air ratio: 0.2 kg water/kg air.

An assessment of the energetic validation by combustion and gasification of orujillo and pits was made. The gas produced in the fluidized/moving bed gasifier supported the expected composition of gasification flue gases and could be suitable for applications in the electrical power production by means of classical explosion motors.

5.6.4   Examples of Technologies and Treatments

After reviewing various case studies applied in different regions, we can conclude that the most appropriate treatment depends not only on intrinsic factors but also on the capacity and system of production of the plants (olive mills and extraction plants and other industries or activities) [1].

As an example, the present practice in Greece and Italy is decanting in three-phase conditions (Fig. 5.26) with generation of alpechin and treatment of orujo in extraction plants that use hexane to extract the orujo oil. Part of the deoiled orujo (orujillo) is used to dry wet orujo in its own extraction plant. The excess orujillo is sold as solid fuel (ceramic manufacture furnaces, cement kilns, domestic heating), or used as raw material for composting and as additive for animal feed.

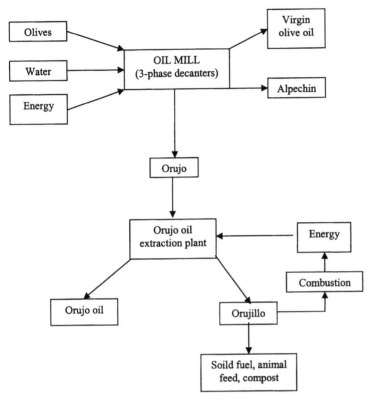

Figure 5.26 Common integration of treatments for orujo from three-phase decanting method (from Ref. 1).

Spain is a different case, especially in the southern regions, where production is carried out almost exclusively by medium and big cooperatives, and where the two-phase decanting method has been adopted by more than 95% of producers (Fig. 5.27). The main waste is alpeorujo. Nowadays the "repaso" or second decanting of alpeorujo in the same oil mill is producing a new kind of wastewater, not equal to alpachin but nevertheless representing a growing environmental problem. The orujo oil can still be extracted by extraction plants, but the oil content decreases over time due to the deoiling of alpeorujo made in the oil mills. This means that some producers have decided to burn deoiled alpeorujo to produce electricity. Recent normative, with assured advantages for producers of energy from biomass, has also contributed to the use of orujillo as fuel in small electrical power plants (15 MW). Other new applications such as the production of active coal are also emerging [1].

Currently there is a tendency in some countries to move from the traditional pressing system to the three-phase system and from three-phase to the two-phase system, so the use of different models is constantly changing. Since there are no general unified solutions, every case should be studied according to the local conditions.

As we have seen in the previous section, in the case of waste resulting from the two-phase decanting process, separation into pulp, alpeorujo liquid fraction (ALF), and pits allows for the application of selective treatments and techniques such as composting, bioremediation, and gasification. Another valuable point is worth mentioning here: mixing alpeorujo with other wastes such as molasses improves the production of animal feed with a high protein content.

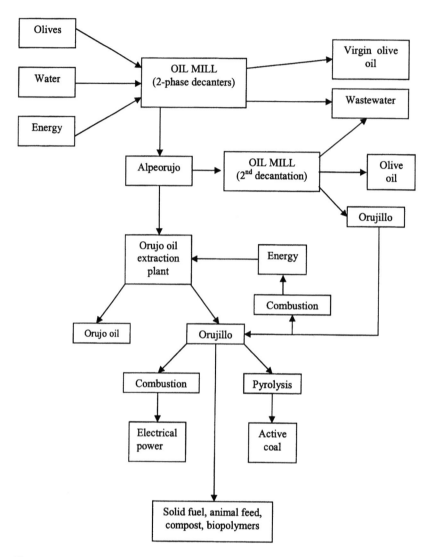

Figure 5.27   Common integration of treatments for orujo from two-phase decanting method (from Ref. 1).

With regard to the energy value of wastes, it is important to consider that the integration of energy cycles will optimize costs and environmental impacts, for example, by burning pits to dry, or predry the waste or alpeorujo, and combustion/gasification of it to recover energy and combustible gases to obtain and use electrical energy [1].

Furthermore, there should be always specific training programs for operators and supervisors in oil mills and related waste treatment units.

5.7   ECONOMY OF TREATMENT PROCESSES

Many food-processing-related industries, including the manufacture of olive oil and table olives, are of a seasonal nature, and consequently waste is not generated throughout the entire year.

Capital and operating costs of an *in situ* complete treatment (physical-chemical and biological processes) of these waste streams are inevitably high [91]. Thus, if a factory is located in an urban area, the most common practice for dealing with these kinds of effluents is to deliver the industrial effluent to the nearest municipal wastewater treatment plant and to pay the appropriate fee. However, the presence of inhibitory or toxic substances may have a serious effect on the overall treatment system, particularly the biological treatment process, from an operational and economical viewpoint. Thus, in the activated sludge process, phenol-type compounds in concentrations of >200 mg/L and >10 mg/L are known to inhibit carbonaceous removal and nitrification, respectively [92]. As a result, some action must be taken before discharging these industrial effluents into municipal sewers and treatment facilities.

As discussed before, several anaerobic processes or techniques have been applied only to the treatments of diluted OME, such as an upflow anaerobic sludge blanket (UASB) reactor, a combined sludge blanket reactor with fixed-bed filter, anaerobic contact reactors, and anaerobic filters. In these biological treatments, OME has to be diluted prior to biological digestion, otherwise the bioreactors need high volumes due to the relatively low loading rates that could be applied and the high pollution potential of OME. At the same time, physical and chemical methods are widespread and applied for treatment of OME. These methods, as discussed before (treatment sections), are considered partial treatments, for example, precipitation by iron and lime, adsorption on a specific resin, and chemical oxidations by hydrogen peroxide and ozone. It was noticed that each pretreatment was efficient in removing the toxic effect of OME. Furthermore, the aerobic pretreatment of OME with different microorganisms (such as *Azotobacter* and *Aspergillus*) reduces considerably the COD and the total phenolic compounds concentration of the waste, which is responsible for its biotoxicity.

It is important to consider that any of these alternatives (physical, chemical, biological) must depend on economic factors, taking into account the possible combination of two or more alternatives. The physical or chemical pretreatment of OME could resolve the problems of time-variable composition and of pollution potential [24]. As a result, dilution for further biological treatment could be reduced, which is an important factor in the evaluation of its economy. The precise evaluation of the cost and feasibility of each of these treatment alternatives depends on several factors, such as capacity of production, waste amount, waste state (liquid or solid), site requirements, specific training of the workers, noise and odor emissions, industrial and agriculture–ecological surroundings, local laws [93].

As reported in the literature, wet air oxidation (WAO) is an economically acceptable technology used to treat aqueous wastes containing oxidizable pollutants at concentrations too high or too toxic for aerobic biological treatment [94]. An exhaustive economic evaluation of WAO is a rather difficult task, given the high number of parameters involved in the process. Thus, for a continuous process, there are several operating variables (influent flow rate, temperature, pressure, contamination level, cooling and steam process water temperature, effluent temperature, final contamination level, biodegrability, etc.). Obviously, kinetic and thermodynamic data of the wastewater to be processed must also be considered (specific heat, heat of reaction, rate constants, etc.). These parameters will determine the residence time of the wastewater in the reactor and the energy needed and released in the process [19,95].

An economic assessment compared WAO and incineration processes for treatment of industrial liquid waste with a high content of phenol-type substances. The outcome was that incineration resulted in roughly four times the expense of WAO [96,97].

Another example focuses on solid waste treatment by gasifier/combustion flumov system to produce the optimal flue gas. Economic and industrial estimations were made of the gasifier's industrial design. The size and cost of a gasifier for treating 15 T/hour of solids capacity was estimated at 3.6 million euro (fluidized bed 2.6 × 8 m, moving bed 8 × 8 m) [1].

As previously discussed, it is important from an economic perspective to develop profitable uses for the final waste product, such as organic fertilizer, soil conditioner, and livestock feed. In this regard, it is worth pointing out that an opportunity exists to obtain a new type of renewable and low-cost activated carbon (J-carbon) from the processed solid residue of olive mill products. This is due to the fact that olive mills generate a huge amount of waste, which can be suitable as a raw material with economic value, and as a supportive means for pollutant removal from wastewater [98]. A study was performed to compare the capability of J-carbon with commercial activated carbon to remove ammonia (NH_3), total organic carbon (TOC), and some special organics from Flexsy's (Rubber) wastewater treatment plant as tertiary treatment [99].

In this regard the final result was that the J-carbon has almost similar behavior and efficiency as the commercial activated carbons (powder activated carbon and granular activated carbon). Therefore, it was concluded that the J-carbon, as well as other commercial activated carbons, could be used in the treatment of industrial wastewater to improve efficiency of the treatment plant. The exhausted carbon would be settled by gravity and disposed of with the sludge as a carbon–sludge mixture. Thus, there would be no need for regeneration since the J-carbon is a renewable and very low-cost adsorbent.

5.8 SUMMARY

This chapter is based around the fact that the olive oil industry is in continuous growth due to its nutritious and economic importance, particularly for Mediterranean countries. This is accompanied by vast waste generation from different olive oil technologies (traditional and pressing decanting processes). The wastewater is mainly characterized by a high degree of organic pollution, polyphones, and aromatics forming inhibitor or toxic substances, which constitute a serious environmental problem for soil, rivers, and groundwater.

The great variety of components found in liquid waste and solid waste requires different appropriate technologies to eliminate those that have harmful effects on the environment. From an economic perspective it is important to develop profitable uses for the final waste product, such as organic fertilizer, soil conditioner, and livestock feed.

The optimal disposal and management of olive oil mill waste should be viewed within a multidisciplinary integrated frame that comprises specific procedures, such as extraction by decanter centrifuge, liquid/solid waste treatments, aerobic bioremediation and composting, enrichment of waste with fungal/yeast protein, drying and gasification in fluidized moving beds, recovery of orujo oil, and recovery of energy and combustible gases.

Prospective research should take into consideration the new advances in biotechnology, treatment reactors, control, new products and processes, composting from different wastes mixtures, all for the service of minimizing the impact on the environment, and reducing the use of valuable natural and living resources within the course of sustainable development.

REFERENCES

1. Aragón, J.M. *Improvement of Treatments and Validation of the Liquid–Solid Waste from the Two-Phase Olive Oil Extraction*, Final Report, (Annex A2) Project IMPROLIVE (FAIR. CT96-1420); Department of Chemical Engineering, Univesidad Complutense de Madrid: Madrid, Spain, 2000.

2. Stölting, B.; Bolle, W.F. Treatment processes for liquid and solid waste from olive oil production. In *Proceedings of Workshop "IMPROLIVE-2000"*, (Annex A1), Seville, Spain, April 13–14, 2000; 29–35.

3. FAOSTAT data base. http://appsfao.org/default.htm (2000).

4. IMPROLIVE Project. http://fiw.rwth-aachen.de/info/improlive (2002).

5. Fedeli, E. Olive oil technology. SSOG, Milano, *Toronto Special, OLIVE*, **1993**, *45*, 20–23.

6. Hruschka, S.; Geissen, K. Global contemplation of the olive oil production with regard to the economy, quality and ecology. In *Proceedings of Symposium "Olive Waste*," Culture Center of Kalamata, Greece, November 5–8, 1997.

7. Kiritsakis, A.K. Olive oil – from the tree to the table. In *Food and Nutrition*; Press Inc.: Trumbull, CT, USA, 1998.

8. Baccioni, L. Producing first-rate extra virgin oil. *Oil and Fats Intern.*, **1999**, *15* (*1*), 30–31.

9. Gasparrini, R. Treatment of olive oil residues. *Oil and Fats Intern.*, **1999**, *15* (*1*), 32–33.

10. Boskou, D. Olive oil. In *Chemistry and Technology*; AOCS-Press: New York, 1996.

11. Kiritsakis, A.K. The olive oil quality. In *8th Congress Seccion Latinoamericana AOCS*, Santiago de Chile, October 24–27, 1999.

12. Durán, R.M. Relationship between the composition and ripening of olive and quality of the oil. In *Proceedings of I Intl, Sympos, on Olive Growing*, Rallo, L. Universidad Autonoma de Cordoba, Acta Horticulturae 286, 1989.

13. Pompei, C.; Codovilli, F. Risultari preliminari sul trattamento di deparazione delle acque di vegetazione delle olive per osmosi inversa. Scienza e Tecnologia Degli Alimenti **1974**, 363–364.

14. Fiestas, R. et al. The anaerobic digestion of wastewater from olive oil extraction. In *Anaerobic Digestion*; Traue-muende: Poster, Germany, 1981.

15. Steegmans, S.R. *Optimierung der anaeroben Verfahrenstechnik zur Reinigung von organischen hochverschmutzten Abwässern aus der Oliven Ölgewinnung*; Forschungsinstitut fuer Wassertechnologie an der RWTH Aachen (Hrsg.), Forschungsbericht AZ 101/81 der Oswald-Schulze-Stiftung: Aachen, 1992.

16. Hamdi, M. Toxicity and biogradability of olive mill wastewaters in batch anaerobic digestion. Bioprocess Engineering, Heft **1993**, *8*, 79.

17. Andreozzi, R. et al. Integrated treatment of olive oil mill effluents (OME): Study of ozonation coupled with anaerobic digestion. J. Wat. Res. **1998**, *32* (*8*), 2357–2364.

18. Garcia, G.P.; Garrido, A.; Chakman, A.; Lemonier, J.P.; Overrend, R.P.; Chornet E. Purifction de aguas residuales ricas en poliven-oles: Aplication de la oxidaction húmeda a los effluentes acuosos derivados de las industrias olivarerns. Grasos y Aceites **1989**, *40* (*4*–*5*), 291–295.

19. Rivas, F.J.; Beltran, F.J., Gimeno, O.; Alvares, P. Chemical-biological treatment of table olive manufacturing wastewater. J. Environ. Engng ASCE, **2001**, *127* (*7*), 611–619.

20. Kissi, M.; Mountador, M.; Assobhei, O.; Gargivlo, E.; Palmieri, G.; Giardira, P.; Sannia, G. Roles of two white-rot basidiomycete fungi in decolorisation and ditoxification of olive mill wastewater. J. Appl. Microbiol. Biotech. **2001**, *57*, 221–226.

21. Beltran-Heredia, J.; Torregrosa, J.; Dominguez, J.R.; Garcia, J. Treatment of black-olive wastewaters by ozonation and aerobic biological degradation. J. Wat. Res. **2000**, *34* (*14*), 3515–3522.

22. Ubay, G.; Özturk, I. Anaerobic treatment of olive mill effluents. J. Wat. Sci. and Technol. **1997**, *36* (*2*), 287–294.

23. Beccari, M.; Bonemazzi, M.; Majone, M.; Riccardi, C. Interaction between acidogenesis and methanogenesis in the anaerobic treatment of olive oil mill effluents. J. Wat. Res. **1996**, *30* (*1*), 183–189.

24. Zouari, N. Decolorization of olive oil mill effluent by physical and chemical treatment prior to anaerobic digestion. J. Chem. Technol. Biotechnol. **1998**, *73*, 297–303.

25. Borja, R.; Martin, A.; Alonso, V.; Garcia, I.; Banks, C.J. Influence of different aerobic pretreatments on the kinetics of anaerobic digestion of olive mill wastewater. J. Wat. Res. **1995**, *29* (2), 489–495.

26. IOOC. *Olive Oil Quality Improvement*, Technical Report. International Olive Oil Council: Madrid.

27. Lopez, R. et al. Land treatment of liquid wastes from the olive oil industry (Alpechin). Fresenivs Envir. Bull. *1*, **1992**, 129–134.

28. Borja, R.; Martin, A.; Maestro, R.; Alba, J.; Fiestas, J.A. Enhancement of the anaerobic disgestion of olive mill wastewaters by removal of phenolic inhibitors. J. Processes Biochem. **1992**, *27*, 231–237.

29. Annesini, M.C. et al. Treatment of olive oil wastes by distillation. J. Effluent and Water Treatment **1983**, *June*.

30. Geissen, K. *Aufbereitung von Alpeorujo aus der Olivenoelproduction in Spanien*; Westfalia Separator AG (HRSG), Interne Firmenschrift: Oelda, Germany, 1995.

31. Özturk, I.; Ubay, G.; Sakar, S. *Anaerobic Treatability of Olive Mill Effluents and Bioenergy Recovery*, Research Report; The Scientific and Technical Research Council of Turkey, Project No. DEBCAG-56, Turkey, 1990.

32. Benitez, F.J.; Beltran-Heredia, J.; Torregrosa, J.; Acero, J.L. Treatment of olive mill wastewaters by ozonation, aerobic degradation and the combination of both treatments. J. Chem. Technol. Biotechnol. **1999**, *74*, 639–646.

33. Wu, J.; Taylor, K.E.; Biswas, N. Optimization of the reaction conditions for enzymatic removal of phenol from wastewaters in the presence of polyethlene glycerol. J. Wat. Res. **1993**, *27*, 1701–1706.

34. Perez, D.J.; Esteban, E.; Gomez, M.; Dolaraf, G. Effect of wastewater from olive processing on seed germination and early plant growth of different vegetable species. J. Environ. Sci. Health. **1986**, *49*, 349–357.

35. Yesilada, O.; Sam, M. Toxic effects of biodegraded and detoxified olive oil mill wastewaters on the growth of Pseudomonas aeruginosa. Toxic. Environ. Chem. **1998**, *65*, 87–94.

36. Bahlo, K.; Wach, G. *Naturnahe Abwasserreinigung*, 3. Aufl.; Oekobuch Verlag: Staufen bei Freiburg, Germany, 1995.

37. Kumaran, P.; Paruchuri, L. Kinetics of phenol biotransformation. J. Wat. Res. **1997**, *31* (*1*), 11–22.

38. FIW (Forshungsinstitut für Wasser and Abfallwirtschaft, RWTH Aachen, Germany)–IMPROLIVE web site. www.fiw.rwth-aachen.de/improlive/improlive.htm.

39. J.G. Press Inc., Seville. Recycling olive mill residuals in Andalusia. Bio-cycle **1995**, *36* (6), 24.

40. Eckenfelder, W.W. *Industrial Water Pollution Control*, 2nd Ed.; McGraw-Hill International, 1989.

41. Davis, M.; Cornwell, D. *Introduction to Environmental Engineering*, 2nd Ed.; McGraw-Hill International, 1991.

42. US-EPA. *Environmental Pollution Control Alternatives, Municipal Wastewater*; Environmental Protection Agency: Washington DC, 9–12.

43. Chmielowski, J. et al. Zesz. Nauk. Politech Slaska Inz. (Polish) **1965**, *8*, 97.

44. Aveni, A. Biogas recovery from olive oil mill wastewater by anaerobic digestion. In *Anaerobic Digestion and Carbohydrate Hydrolysis of Waste*; Ferrero, G.L., Ferranti, M.P., Neveau, H. Eds.; Elsevier, 1984; 489–490.

45. Carrieri, C.; Balice, V.; Rozzi, R. Comparison of three anaerobic treatment processes on olive mills effluents. In *Proceeding. Int. Conf. on Environment Protection*, Ischia, Italy, October 5–7, 1988.

46. Boari, G.; Brunetti, A.; Passin, R.; Rozzi, A. Anaerobic digestion of olive oil mill wastewaters. J. Agric. Wastes **1984**, *10*, 161–175.

47. Özturk, I.; Sakar, S.; Ubay, G.; Eroglu, V. Anaerobic treatment of olive mill effluents. In *Proceedings of the 46th Industrial Waste Conf.*, Purdue University, West Lafayette, IN, May 14–16, 1991.

48. Rozzi, A.; Boari, G.; Liberti, L.; Santori, M.; Limoni, N.; Menegatti, S.; Longobardi, C. Anaerobic-aerobic combined treatment of urban and olive oil mill effluents – pilot scale experimentation. J. Ing. Sanit. **1989**, *37* (4), 44–54.

49. Wang, Y.T.; Gabbard, H.D.; Pai, P.C. Inhibition of acetate methanogenesis by phenols. Environ. Engng **1991**, *117*, 487–500.

50. Koster, I.W.; Gramer, A. Inhibition of methanogenesis from acetate in granular sludge by long-chain fatty acids. J. Appl. Environ. Microbiol. **1980**, *53*, 403–409.

51. Craveiro, A.M.; Rocha, B.M. Anaerobic digestion of vinasse in high-rate reactors. In *proceedings of NWA-EWPCA Conference (Aquatech 86) on Anaerobic Treatment*, Amsterdam, September 15–19, 1986; 307–320.

52. Borja, R.; Martin, A.; Garrido, A. Anaerobic digestion of black-olive wastewater. J. Biores. Technol. **1993**, *45*, 27–32.

53. Capasso, R.; Cristinzio, G.; Evidente, A.; Scognamiglio, F. Isolation, spectroscopy and selective phyto-toxic effects of polyphenols from vegetable wastewaters. J. Phytochemistry **1992**, *31*, 4125–4128.

54. Zouari, N.; Ellouz, R. Toxic effect of colored olive compounds on the anaerobic digestion of olive oil mill effluent in UASB like reactors. J. Chem. Technol. Biotechnol. **1996**, *16*, 414–420.

55. Eisenhauser, H.R. Ozonation of phenolic wastes. J. Wat. Pollut. Control Fed. **1986**, *40*, 188–193.

56. Gould, J.P.; Weber, W.J. Jr. Oxidation of phenols by ozone. J. Wat. Pollut. Control Fed. **1976**, *48*, 47–52.

57. Scott, J.P.; Ollis, D.F. Integration of chemical and biological oxidation processes for water treatment: review and recommendation. Environ. Prog. **1995**, *14* (2), 88–103.

58. Kolaczkowski, S.T.; Beltran, F.J.; McLurgh, D.B.; Rivas, F.J. Wet air oxidation of phenol: Factors that may influence global kinetics. *Trans. Instn. Chem. Engs, Part B, Process, Safety and Environ. Protection* **1997**, *75*, 257–265.

59. Debellefontaine, H.; Chakchouk, M.; Foussard, J.N.; Tissot, A.; Striolo, P. Treatment of organic aqueous wastes: Wet air oxidation and wet peroxide oxidation. J. Environ. Pollut. **1996**, *92* (2), 155–164.

60. Beltran, F.J.; Kolaczkowski, S.T.; Crittenden, B.D.; Rivas, F.J. Degradation of o-chlorophenol with ozone in water. *Trans. Instn Chem. Eng, Part B, Process, Safety and Environ. Protection* **1993**, *71*, 57–65.

61. Martirani, L.; Giardina, P.; Marzullo, L.; Sannia, J. Reduction of phenol content and toxicity in olive oil mill wastewaters with the ligninolytic fungus Pleurotus ostreatus. J. Wat. Res. **1996**, *30* (8), 1914–1918.

62. Nannipieri, P.; Ballag, J.M. Use of enzymes to detoxify pesticide-contaminated soils and waters. J. Environ. Qual. **1991**, *20*, 510–517.

63. Atlow, S.C.; Bonadonna, M.; Apero, L.C.; Klibanov, A.M. Dephenolization of industrial wastewaters catalyzed by polyphenol oxidase. J. Biotechnol. Bioeng. **1984**, *26*, 599–603.

64. Sanjust, E.; Pompei, R.; Rescigno, A.; Rinaldi, A.; Ballero, M. Olive milling wastewaters as a medium for growth of four pleurotus species. J. Appl. Biochem. Biotech. **1991**, *31*, 223–235.

65. Eriksson, K.E.L.; Blanchette, R.A.; Ander, P. Biodegradation of lignin. In *Microbial and Enzymatic Degradation of Wood and Wood Components*; Timmel, T.E., Eds.; Springer-Verlag: Berlin, 1990; 215–232.

66. Field, J.A.; Lettinga, G. The effect of oxidative coloration on the methanogenic toxicity and anaerobic biodegradability of phenols. J. Biol. Wastes **1989**, *29*, 161–179.

67. Wang, Y.T.; Suidan, M.T.; Rittman, B.E. Anaerobic treatment of phenol by an expanded-bed reactor. J. Wat. Pollut. Control Fed. **1986**, *58*, 227–232.

68. Jackson-Moss, C.A.; Duncan, J.R. The effect of aluminum on anaerobic digestion. Biotech. Lett. **1991**, *13*, 143–148.

69. Christy, R.W. Sludge disposal using lime. J. Wat. Environ. Technol. **1990**, *4*, 56–61.

70. Bevins, R.E.; Longmaid, F.M. Stabilization of sewage-sludge cake by addition of lime and other materials. J. Wat. Pollut. Control **1984**, 9–12.

71. Pollard, S.J.T., Thompson, F.E.; McConnachie, G.L. Microporous carbons from moringa oleifera husks for water purification in less developed countries. J. Wat. Res. **1995**, *29*, 337–374.

72. Wigmans, T. Industrial aspects of production and use of activated carbons. J. Carbon. **1989**, *27*, 13–22.

73. Lankford, P.W.; Eckenfeldes, W.W. *Toxicity Reduction in Industrial Effluents*; VAN Nostrand Reinhold: New York, 1990.

74. Angelidaki, I.; Peterson, S.P.; Ahring, B.K. Effect of lipids on thermophilic anaerobic digestion and reduction of lipid inhibition upon addition of bentonite. J. Appl. Microbiol. Biotechnol. **1990**, *33*, 469–472.

75. Beccari, M.; Majone, M.; Riccardi, C.; Savarese, F.; Torrisi, L. Integrated treatment of olive oil mill effluents: effect of chemical and physical pretreatment on anaerobic treatability. Wat. Sci. & Technol. **1999**, *40* (1), 347–355.

76. Davis, M.L.; and Cornwell, D.A. *Introduction to Environmental Engineering*, 2nd Ed.; McGraw-Hill Int., 1991.

77. Balis, C.; Antonokou, M. Composting and bioremedation. In *Proceedings of Workshop "IMPROLIVE-2000"*, (Annex A1), Seville, Spain, April 13–14, 2000; 13–18.

78. Giannoutsou, E.P.; Karagouni, A.D. Olive oil waste: could microbial fermentation be the solution. In *Proceedings of Workshop "IMPROLIVE-2000"*, (Annex A1), Seville, Spain, April 13–14, 2000; 23–28.

79. Larsen, T.; Burrows, R.; Engedahl, L. Unsteady flow and saline intrusion in long sea outfalls. J. Wat. Sci. & Technol. **1992**, *25* (*9*), 225.

80. Monteiro, A.J.; Neves, J.R.; Sousa, R.E. Modeling transport and dispersion of effluent outfalls. J. Wat. Sci. & Technol. **1992**, *25* (*9*), 143.

81. Garber, F.G.; Neves, J.R.; Roberts, P. Marine disposal systems. J. Wat. Sci. & Technol. **1992**, *25* (*9*), IX.

82. Özturk, I.; Eroglu, V.; Akkoyunlu, A. Marine outfall applications on the Turkish coast of the Black Sea. J. Wat. Sci. Technol. **1992**, *25* (*9*), 204.

83. Bannatyne, A.N.; Speir, J. Milli-screening – a pretreatment option for marine disposal. In *Proceedings of the IAWPRC Marine Disposal Seminar*, Rio de Janeiro, 1987, *18*, 11.

84. Ludwig, G.R. *Environmental Impact Assessment – Sitting and Design of Submarine Outfalls*; The Monitoring and Assessment Research Center (MARC) and WHO, 1988, Report 43, 5–6.

85. Awad, A. *Submarine Outfalls of Wastewater*, 1st Ed.; Kuwait Foundation for the Advancement of Science, Department of Authorship, Translation and Publications: Kuwait, 1998.

86. Aragón, J.M.; Palancar, M.C.; Serrano, M.; Torrecilla, J.S. Fluidised/moving beds: applications to driers and gasifiers. In *Proceedings of Workshop "IMPROLIVE-2000"*, (Annex A1), Seville, Spain, April 13–14, 2000; 19–22.

87. Aragón, J.M.; Palancar, M.C.; Torrecilla, J.S.; Aparicio, J.T. Modeling fluidized bed dryers by artificial neural network. In *Chemical and Engineering Congress*, Chisa 98, 1998.

88. Aragón, J.M.; Palancar, M.C.; Torrecilla, J.S.; Serrano, M. Drying a high viscosity solid–liquid waste in fluidized-moving bed. In *Chemical and Engineering Congress*, Chisa 98, 1998.

89. Aragón, J.M.; Palancar, M.C.; Serrano, M.; Torrecilla, J.S. Design and fluid-dynamic behaviour of J-valve systems for feeding and discharging solid in fluidized beds. In *Chemical and Engineering Congress*, Chisa 98, 1998.

90. Awad, A. Pyrolysis for disposing municipal wastes. In *Proceedings of the Atmospheric Pollution Symposium*, Damascus, August 1985; 219–225.

91. Kim, J.S.; Kim, B.G.; Lee, C.H.; Kim, S.W.; Jee Koh, J.H.; Fan, A.G. Development of clean technology in alcohol fermentation industry. J. Cleaner Prod. **1997**, *5* (*4*), 263–267.

92. Metcalf & Eddy (Eds.) *Wastewater Engineering*, 3rd Ed.; McGraw-Hill, Int., 1991.

93. European Commission. *Improlive 2000*, Final Report (Accompanying Measures), Program (Quality of Living Sources), No. QLK1-1995-30011, Madrid, June 2000.

94. Baillod, C.; Faith, B. *Wet Oxidation and Ozonation of Specific Organic Pollutants*, EPA-600-2-83-060; US. Environmental Protection Agency: Washington DC, 1983.

95. Baillod, C.R.; Lamparter, R.A.; Barra, B.A. Wet oxidation for industrial waste treatment. J. Chem. Engng Prog. **1985**, *March*, 52–56.

96. Zimpro Environmental Inc. *Commercial-Scale Wet Air Oxidation Process*; Rothschild, WI, 1990.

97. Zimpro Environmental Inc. Wet air oxidation cleans up black wastewater. J. Chem. Eng **1993**, *Sept.*, 175–176.

98. Gharaibeh, S.H.; Abu El Sha'r, W.Y.; Al Kofahi, M.M. Removal of selected heavy metals from aqueous solutions using processed solid residue of olive mill products. J. Wat. Res. **1998**, *32* (*2*), 498–502.

99. Gharaibeh, S.H.; Moore, S.V.; Buck, A. Effluent treatment of industrial wastewater using processed solid residue of olive mill products and commercial activated carbon. J. Chem. Technol. Biotechnol. **1998**, *71*, 291–298.

6
Potato Wastewater Treatment

Yung-Tse Hung and Howard H. Lo
Cleveland State University, Cleveland, Ohio, U.S.A.

Adel Awad and Hana Salman
Tishreen University, Lattakia, Syria

6.1   INTRODUCTION

In the past two decades, the potato industry has experienced rapid growth worldwide, accompanied by a staggering increase in the amount of water produced. It is estimated that the US potato industry alone generates about 1.3×10^9 kg of wastes each year [1]. Large volumes of wastewater and organic wastes are generated in potato processing as result of the water used in washing, peeling, and additional processing operations.

The potato industry is well known for the vast quantities of organic wastes it generates. Treatment of industrial effluents to remove organic materials, however, often changes many other harmful waste characteristics. Proper treatment of potato processing wastewaters is necessary to minimize their undesirable impact on the environment.

Currently, there is an increasing demand for quality improvement of water resources in parallel with the demand for better finished products. These requirements have obliged the potato industry to develop methods for providing effective removal of settleable and dissolved solids from potato processing wastewater, in order to meet national water quality limits. In addition, improvement and research have been devoted to the reduction of wastes and utilization of recovered wastes as byproducts.

This chapter discusses (a) the various potato processing types and steps including their sources of wastewaters; (b) characteristics of these wastewaters; (c) treatment methods in detail with relevant case studies and some design examples; and (d) byproduct usage.

6.2   POTATO PROCESSING AND SOURCES OF WASTEWATER

High-quality raw potatoes are important to potato processing. Potato quality affects the final product and the amount of waste produced. Generally, potatoes with high solid content, low reducing sugar content, thin peel, and of uniform shape and size are desirable for processing. Potatoes contain approximately 18% starch, 1% cellulose, and 81% water, which contains dissolved organic compounds such as protein and carbohydrate [2]. Harvesting is an important operation for maintaining a low level of injury to the tubers. Improved harvesting machinery reduces losses and waste load.

The type of processing unit depends upon the product selection, for example, potato chips, frozen French fries and other frozen food, dehydrated mashed potatoes, dehydrated diced potatoes, potato flake, potato starch, potato flour, canned white potatoes, prepeeled potatoes, and so on. The major processes in all products are storage, washing, peeling, trimming, slicing, blanching, cooking, drying, etc.

6.2.1 Major Processing Steps

Storage

Storage is needed to provide a constant supply of tubers to the processing lines during the operating season. Potato quality may deteriorate in storage, unless adequate conditions are maintained. The major problems associated with storage are sprout growth, reducing sugar accumulation, and rotting. Reduction in starch content, specific gravity, and weight may also occur. Handling and storage of the raw potatoes prior to processing are major factors in maintaining high-quality potatoes and reducing losses and waste loads during processing.

Washing

Raw potatoes must be washed thoroughly to remove sand and dirt prior to processing. Sand and dirt carried over into the peeling operation can damage or greatly reduce the service life of the peeling equipment. Water consumption for fluming and washing varies considerably from plant to plant. Flow rates vary from 1300 to 2100 gallons per ton of potatoes. Depending upon the amount of dirt on the incoming potatoes, wastewater may contain 100–400 lb of solids per ton of potatoes. For the most part, organic degradable substances are in dissolved or finely dispersed form, and amount to 2–6 lb of BOD_5 (biological oxygen demand) per ton of potatoes [3].

Peeling

Peeling of potatoes contributes the major portion of the organic load in potato processing waste. Three different peeling methods are used: abrasion peeling, steam peeling, and lye peeling. Small plants generally favor batch-type operation due to its greater flexibility. Large plants use continuous peelers, which are more efficient than batch-type peelers, but have high capital costs [4].

Abrasion peeling is used in particular in potato chip plants where complete removal of the skin is not essential. High peeling losses, possibly as high as 25–30% may be necessary to produce a satisfactory product.

Steam peeling yields thoroughly clean potatoes. The entire surface of the tuber is treated, and size and shape are not important factors as in abrasion peeling. The potatoes are subjected to high-pressure steam for a short period of time in a pressure vessel. Pressure generally varies from 3 to 8 atmospheres and the exposure time is between 30 and 90 sec. While the potatoes are under pressure, the surface tissue is hydrated and cooked so that the peel is softened and loosened from the underlying tissue. After the tubers are discharged from the pressure vessel, the softened tissue is removed by brushers and water sprays [4]. Screens usually remove the peelings and solids before the wastewater is treated.

Lye peeling appears to be the most popular peeling method used today. The combined effect of chemical attack and thermal shock softens and loosens the skin, blemishes, and eyes so that they can be removed by brushes and water sprays. Lye peeling wastewater, however, is the most troublesome potato waste. Because of the lye, the wastewater pH is very high, usually between 11 and 12. Most of the solids are colloidal, and the organic content is generally higher than for the other methods. The temperature, usually from 50 to 55°C, results in a high dissolved starch content, and the wastewater has a tendency to foam.

The quality of the peeling waste varies according to the kind of potato processing product, peeling requirements, and methods. Table 6.1 represents the difference in waste quality among the peeling methods in potato processing plants.

6.2.2 Types of Processed Potatoes

Potato Chips

The processing of potatoes to potato chips essentially involves the slicing of peeled potatoes, washing the slices in cool water, rinsing, partially drying, and frying them in fat or oil. White-skinned potatoes with high specific gravity and low reducing sugar content are desirable for high-quality chips. A flow sheet of the process is shown in Figure 6.1 [3].

Frozen French Fries

For frozen French fries and other frozen potato production, large potatoes of high specific gravity and low reducing sugar content are most desirable. After washing, the potatoes are peeled by the steam or lye method. Peeling and trimming losses vary with potato quality and are in the range 15–40%. After cutting and sorting, the strips are usually water blanched. Because the blanching water is relatively warm, its leaching effect may result in high dissolved starch content in the wastewater. Surface moisture from the blanching step is removed by hot air

Table 6.1 Wastewater Quality in the Different Applied Peeling Methods in Potato Processing Plants

Parameters	Potato peeling method		
	Abrasion[a]	Steam[b]	Lye[c]
Flow (gal/ton, raw potato)	600	625	715
BOD	20 lb/ton (4000 ppm)	32.6 lb/ton (6260 ppm)	40 lb/ton (6730 ppm)
COD	–	52.2 lb/ton (10,000 ppm)	65.7 lb/ton (11,000 ppm)
Total solids	–	53.2 lb/ton (10,200 ppm)	118.7 lb/ton (20,000 ppm)
Volatile solids	–	46.8 lb/ton (9000 ppm)	56.4 lb/ton (9500 ppm)
Suspended solids	90 lb/ton (18,000 ppm)	26.8 lb/ton (5150 ppm)	49.7 lb/ton (8350 ppm)
pH	–	5.3	12.6

[a] Waste quality in a dehydration plant [5].
[b] Waste quality in a potato flour plant [6].
[c] Waste quality in a potato flake plant [6].
Source: Refs 5 and 6.

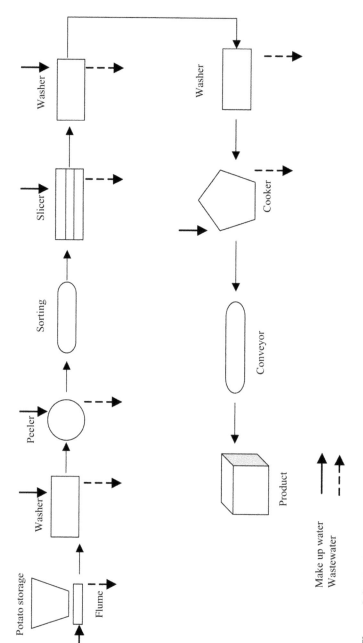

Figure 6.1 Typical potato chip plant (from Ref. 3).

prior to frying. After frying, the free fat is removed on a shaker screen and by hot air stream. The fries are then frozen and packed. Figure 6.2 is a flow diagram of the French fry process [3].

Dehydrated Diced Potato

Potatoes with white flesh color and low reducing sugar content are desirable for dice production. After washing and preliminary sorting, the potatoes are peeled by the steam or lye method. Minimum losses amount to 10%. One important factor during trimming is minimizing the exposure time. The tubers are cut into different sized pieces. After cutting and washing, the dice are blanched with water or steamed at 200–212°F. Following blanching, a carefully applied rinsing spray removes surface gelatinized starch to prevent sticking during dehydration. Sulfite is usually applied at this point as a spray solution of sodium sulfite, sodium bisulfite, or sodium metabisulfite. Calcium chloride is often added concurrently with sodium bisulfite or sodium metabisulfite. Following drying, the diced potatoes are screened to remove small pieces and bring the product within size specification limits. Finally, the potatoes are packed in cans or bags [3].

Dehydrated Mashed Potatoes: Potato Granules

Potato granules are dehydrated single cells or aggregated cells of the potato tuber that are dried to about 6–7% moisture content. A flow diagram of the potato granules is shown in Figure 6.3. After peeling and trimming, the potatoes are sliced to obtain more uniform cooking. The slices are cooked in steam at atmospheric pressure for about 30–40 minutes. After cooking is completed, the slices are mixed with the dry add-back granules and mashed to produce a moist mix. This mix is cooled and conditioned by holding for about 1 hour before further mixing and then dried to about 12–13% moisture content [3,4].

Potato Flakes

Potato flakes are a form of dehydrated mashed potatoes that have been dried on a steam-heated roll as a thin sheet and then broken into small pieces for packaging. Potatoes for flake processing have the same characteristics as those for potato granule processing. A flow diagram of the process is shown in Figure 6.4. After prewashing, the potatoes are lye or steam peeled. Following trimming, the tubers are sliced into 0.25–0.50 in. slices and washed prior to precooking in water at 160–170°F for about 20 minutes [6]. After cooking, the potatoes are mashed and then dried on a single drum drier in the form of a sheet. The sheet is broken into flakes of a convenient size for packaging.

Potato Starch

Potato starch is a superior product for most of the applications for which starch is used. Figure 6.5 shows a flow diagram of a typical starch plant. After fluming and washing, the potatoes are fed to a grinder or hammer mill and disintegrated to slurry, which is passed over a screen to separate the freed starch from the pulp. The pulp is passed to a second grinder and screened for further recovery of starch. The starch slurry, which is passed through the screen, is fed to a continuous centrifuge to remove protein water, which contains soluble parts extracted from the potato. Process water is added to the starch, and the slurry is passed over another screen for further removal of pulp. Settling vats in series are used to remove remaining fine fibers. The pure starch settles to the bottom while a layer of impurities (brown starch) forms at the top. The latter is removed to the starch table consisting of a number of settling troughs for final removal of white starch. The white starch from the settling tanks and the starch table is dried by filtration or centrifugation to a moisture content of about 40%. Drying is completed in a series of cyclone driers using hot air [3].

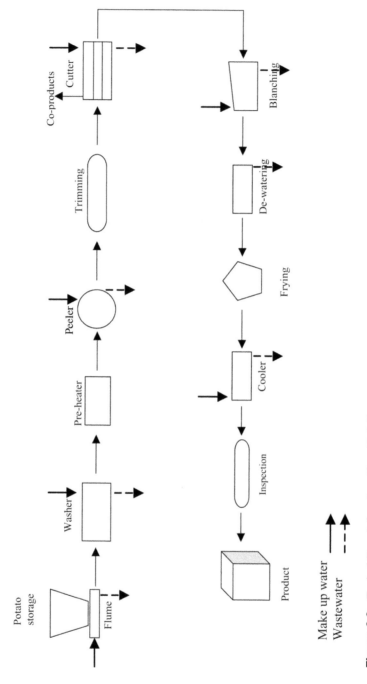

Figure 6.2 Typical French fry plant (from Ref. 3).

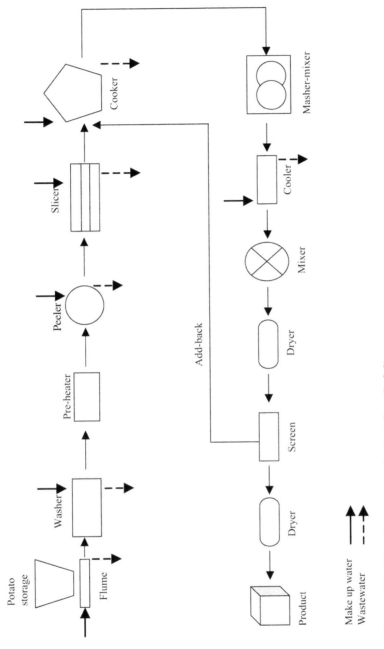

Figure 6.3 Typical potato granule plant (from Ref. 3).

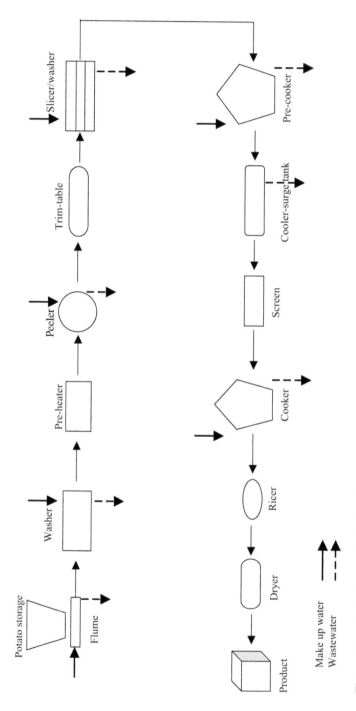

Figure 6.4 Typical potato flake plant (from Ref. 3).

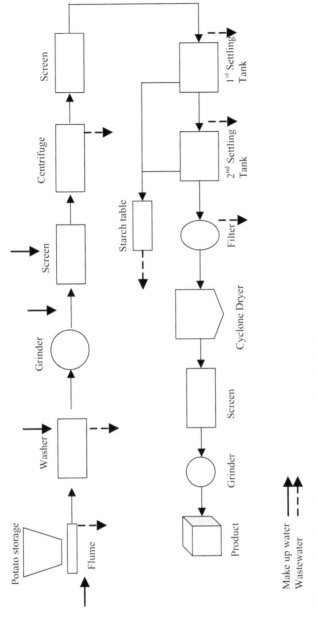

Figure 6.5 Typical potato starch plant (from Ref. 3).

Potato Flour

Potato flour is the oldest commercial processed potato product. Although widely used in the baking industry, production growth rates have not kept pace with most other potato products. A flow diagram of the process is shown in Figure 6.6. After the prewash, the potatoes are peeled, usually with steam. Trimming requirements are not as high as for most potato products. The flaking operation requires well-cooked potatoes; the tubers are conveyed directly from the cooker to the dryer, where 4–5 applicator rolls along one side of the drum contribute a thin layer of potato mesh. The mesh is rapidly dried and scraped off the drum at the opposite side by a doctor knife. The dried sheets are passed to the milling system where they are crushed by a beater or hammer mill and then screened to separate granular and fine flour [3].

Besides the above products, other types include canned potatoes, prepeeled potatoes, and even alcohol. The quantities and qualities of the wastewaters resulting from the mentioned potato processing plants are discussed in the next section.

6.3 CHARACTERISTICS OF POTATO PROCESSING WASTEWATER

6.3.1 Overview

Because potato processing wastewater contains high concentrations of biodegradable components such as starch and proteins [7,8], in addition to high concentrations of chemical oxygen demand (COD), total suspended solids (TSS) and total kjeldahl nitrogen (TKN) [9], the potato processing industry presents potentially serious water pollution problems. An average-sized potato processing plant producing French fries and dehydrated potatoes can create a waste load equivalent to that of a city of 200,000 people. About 230 million liters of water are required to process 13,600 tons of potatoes. This equals about 17 L of waste for every kilogram of potatoes produced. Raw potato processing wastewaters can contain up to 10,000 mg/L COD. Total suspended solids and volatile suspended solids can also reach 9700 and 9500 mg/L, respectively [10]. Wastewater composition from potato processing plant depends on the processing method, to a large extent. In general, the following steps are applied in potato processing: washing the raw potatoes; peeling, which includes washing to remove softened tissue; trimming to remove defective portions; shaping, washing, and separation; heat treatment (optional); final processing or preservation; and packaging.

The potato composition used in potato processing operations determines the components of the resultant waste stream. Foreign components that may accompany the potato include dirt, caustic, fat, cleaning and preserving chemicals. A typical analysis of potato waste solids from a plant employing steam or abrasive peeling is shown in Table 6.2. Generally, the various waste streams are discharged from the potato plant after being combined as effluent. It is difficult to generalize the quantities of wastewater produced by specific operations, due to the variation in process methods. Many references and studies in this respect show wide variations in water usage, peeling losses, and methods of reporting the waste flow. Several publications on the characteristics of wastewaters resulting from various types of potato processing are summarized in Table 6.3 for French fries [11,12], Table 6.4 for starch plants [12], and Table 6.5 for the other types of potato processing plants (chips, flakes, flour, mashed) [13–18].

Processing involving several heat treatment steps such as blanching, cooking, caustic, and steam peeling, produces an effluent containing gelatinized starch and coagulated proteins. In contrast, potato chip processing and starch processing produce effluents that have unheated components [11].

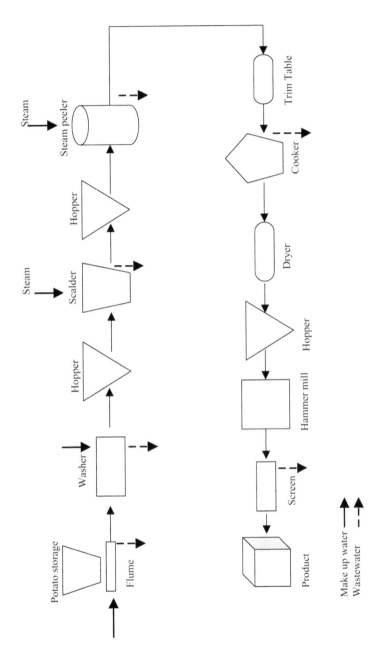

Figure 6.6 Typical potato flour plant (from Ref. 3).

Table 6.2 Composition Percentage of Potato Waste
Solids

Component	Amount (%)
Total organic nitrogen as N	1.002
Carbon as C	42.200
Total phosphorus as P	0.038
Total sulfur as S	0.082
Volatile solid	95.2

Source: Ref. 11.

As for the starch plant effluent, the resulting protein water and pulp form about 95% of the total organic load in the effluent. Table 6.4 represents the composition of waste streams of starch plants and summarizes a survey of five starch plants in Idaho/United States, with and without pulp.

It is evident that if the pulp is kept and not wasted, the organic load is significantly reduced. Potato pulp has been proven to be a valuable feed for livestock when mixed with other ingredients and thus represents a valuable by-product [19]. Protein water is difficult to treat because of the high content of soluble organic water [3].

In plants of joint production of starch and alcohol found in some countries, the pulp and protein water from the starch production is used for alcohol fermentation. As for the wastewater streams in French fries plants, it can be noted from Table 6.3 that the spray washer forms the main organic load (BOD and COD) in comparison to other waste streams. The large variations in wastewater composition can be observed in the potato processing plants as presented in Table 6.5, particularly in COD and TSS concentrations and pH values.

Depending on the abovementioned characteristics of potato processing wastewater, the following should be highly considered:

- Potential methods for reducing the load of waste production including in-plant measures for water conservation, byproduct recovery, and water recycling.
- Choosing the wastewater treatment systems that take into account the wide variations of wastewater compositions, due to wide variation in potato processing steps and methods, in order to reduce the wastewater contaminants for meeting in-plant reuse or the more stringent effluent quality standards required in the potato processing industry.

6.3.2 Case Study [20]

J.R. Simplot Company, an international agribusiness company, operated a potato processing plant in Grand Forks, North Dakota, United States. The company's frozen potato product line, which was produced locally in Grand Forks, consists of more than 120 varieties of French fries and formed products. In all, J.R. Simplot produced more than 2 billion pounds of French fries annually, making it one of the largest processors of frozen potatoes. Its local plant in Grand Forks employed nearly 500 people.

Sources of Wastewater [20]

The main sources of wastewaters consist of silt water and process wastewater. The silt waste resulted from raw potato washing and fluming operations. It contained a large amount of soil removed from the raw potatoes. Process wastewater results from potato processing operations including peeling, cutting, blanching, and packing. The process wastewater included caustic

Table 6.3 Characteristics of Wastewater from French Fry Plants

Parameters	French fries						Caustic peel	French fries and starch plant				
	Spray washer	Trimming	Cutting	Inspection	Blanch	Plant composite		Wash water	Peel waste	Trim table	Blanch waste	Plant effluent
COD (mg/L)	2830	45	150	32	1470	1790	–	100–250	10,000–12,000	150–200	600–700	6450
BOD (mg/L)	1950	30	77	5	1020	1150	4300	–	10,000–15,000	–	–	4100
Total solids (mg/L)	14,900	270	880	260	2283	8100	11,550	700	–	600	1600	7794
Suspended solids (mg/L)	2470	7	16	15	60	1310	–	–	–	–	–	4050
Settleable solids (mg/L)	–	–	–	–	–	–	–	2.0–5.5	200–400	0.6	2–3	–
Total nitrogen (mg/L)	60	–	–	–	–	20	–	–	–	–	–	224
Total phosphorus (mg/L)	81	27	29	14	160	80	–	–	–	–	–	23
pH	11.5	6.9	7.2	6.9	4.7	11.1	–	7.0	–	6.2	5.1	10.7

Source: Refs. 11 and 12.

Table 6.4 Characteristics of Wastewater from Starch Plants

Type of waste	Plant capacity (tons/day)	Flow rate (gal/ton)	BOD mg/L	BOD lb/ton	COD mg/L	COD lb/ton	Solid content (%wt)	Protein in solid (%wt)
Waste stream								
Flume water	–	1740[a]	100	0.4	260	1.5	–	–
Protein water	–	670	5400	30.1	7090	40.3	1.7	38.5
First starch washwater	–	155	1680	2.2	2920	3.3	0.46	31.1
Second starch washwater	–	135	360	0.4	670	0.8	–	–
Brown starch water	–	30	640	0.2	1520	0.4	0.81	–
Starch water	–	25	150	0.0	290	0.0	–	–
Pulp (dry basis)[b]	–	–	–	24.8	–	56.8	–	–
Total organic load without pulp								
Plant I	200	–	–	45.3	–	–	–	–
Plant II	250	–	–	27.7	–	–	–	–
Plant III	150	–	–	26.2	–	–	–	–
Plant IV	62.5	–	–	31.7	–	–	–	–
Plant V	180	–	–	35.0	–	–	–	–
Average				33.3				
Total organic load with pulp								
Plant I	200	–	–	70.1	–	–	–	–
Plant II	250	–	–	52.5	–	–	–	–
Plant III	150	–	–	51.0	–	–	–	–
Plant IV	62.5	–	–	56.5	–	–	–	–
Plant V	180	–	–	59.8	–	–	–	–
Average				58.1				

[a]No recirculation.
[b]An average of 55.5 lb of pulp (on dry basis) were produced per ton of potatoes processed.
Source: Ref. 12.

Table 6.5 Characteristics of Wastewater from Different Potato Processing Plants

| Parameters | Wastewater after settling (Austerman-Haun, et al. 1999)[13] | Wastewater after screening and presettlement (Zoutberg and Eker, 1999)[14] | | | Wastewater from potato chips plant (Hadjivassilis, et al. 1997)[8] | (Kadlec, et al. 1997)[15] | Wastewater influent (Hung, 1989)[16] | |
		Smith food	Peka Kroef	Uzay Gida			Wastewater from potato juice	Wastewater from mashed potato
Total daily flow (m³/day)	1700	912	1600	890	115	–	–	–
Hourly peak flow (m³/hour)	–	(38 av.)	90 (67 av.)	(37 av.)	15	–	–	–
COD (mg/L)	4000	5000	7500	4500	7293	1100–3100	2546	1626
BOD (mg/L)	–	–	–	–	5450	–	–	–
Total suspended solids (mg/L)	–	–	–	–	1300	280–420	18,107	33,930
VSS	–	–	–	–	–	–	–	–
Total TKN (mg/L)	120	286 (max. 400)	50–200	20–70	–	95–145	–	–
Total P (mg/L)	60	–	10–50 (PO$_4$-P)	2–10 (PO$_4$-P)	–	10–15	–	–
pH	6.6 (adjusted)	4.5–7.5	4.5 (after buffering)	5–9	4–10	–	7.6	7.3

(continues)

Table 6.5 Continued

Parameters	Wastewater from potato starch	Primary settling tank effluents (Hung, 1984)[17]	Potato chips (slicing and washing) (Cooley et al. 1964)	Potato flakes (slicing, washing, precooking and cooling) (Cooley et al. 1964)[6]	Potato flour (raw screened waste)	
					(Cooley et al., 1964)[6]	(Olson et al., 1965)[18]
Total daily flow (m³/day)	–	–	1140 gal/ton (4.3 m³/t)	1540 gal/ton (5.8 m³/t)	–	–
Hourly peak flow (m³/hour)	–	–	–	–	–	–
COD (mg/L)	1270	2500	7953	4373	12,582	8314
BOD (mg/L)	–	–	2307	2988	7420	3314
Total suspended solids (mg/L)	62,444	500	5655	1276	6862	4398
VSS	–	450	6685	4147	6480	3019
Total TKN (mg/L)	–	–	–	–	–	–
Total P (mg/L)	–	–	–	–	–	–
pH	7.8	6.7	7.4	5.2	4.2	6.9

Source: Refs. 6, 8, 13–18.

potato peeler and barrel washer discharges, as well as all other liquid wastes from the processing operations, including cleanup water.

Characteristics of Wastewater [20]

The characteristics of the potato processing wastewater were influenced by potato processing operations. Potato peeling was the first stage of potato processing. Caustic soda was used to soften the potato skin so that it can be removed by the scrubbing and spraying action of the polisher. The liquid effluent from the polisher, which contained a majority of the contaminants of wastewater, accounted for about 75% of the alkalinity of the wastewater from the plant. It was also high in COD and BOD, with values of about 2000 and 1000 mg/L, respectively. The TDS (total dissolved solids) and TSS (total suspended solids) were about 29,000 and 4100 mg/L, respectively.

Polished potatoes were then conveyed to the cutter. The degree of size reduction depended upon the requirements of the final product. Here the surface of the potato and the amount of water used for washing determine the quantity of soluble constituent in the waste stream. The pH of the stream was about 7. The COD and BOD values were about 50% of those of the effluent from the polisher. The TDS and TSS were approximately 1390 and 460 mg/L, respectively. The blanching process removed reducing sugar, inorganic salts, gelatinized starch, and smaller amounts of protein and amino acids. The effluent stream from this operation had pH 6.2, total dissolved solids 1500 mg/L, phenols 8.2 mg/L, COD 1000 mg/L, and BOD 800 mg/L, respectively.

The wastewater treatment processes used in the plant included shaker, primary settling tank, aerated lagoon, and final settling tank. The effluent from the final settling tank was discharged to the municipal sewer and was transported to Grand Forks Municipal Wastewater Treatment Plant, Grand Forks, North Dakota, for treatment. A portion of the final settling tank effluent was treated by tertiary sand filter. The filtered water was reused inside the plant.

During the period of September 1978 to March 1979, primary effluent had an average concentration of 4250 mg/L COD and 3000 mg/L TSS. After primary settling tank treatment, the effluent had an average concentration of 2500 mg/L COD and 500 mg/L TSS. After the aerated lagoon and final settling tank treatment, the effluent had an average concentration of 410 mg/L COD and 350 mg/L TSS and pH 7.55. The aerated lagoon had 4900 mg/L MLSS (mixed liquor suspended solids) and 4100 mg/L MLVSS (mixed liquor volatile suspended solids). The onsite treatment plant removed 90.35% COD and 88.33% TSS.

6.4 TREATMENT METHODS

Wastewater from fruit and vegetable processing plants contains mainly carbohydrates such as starches, sugars, pectin, as well as vitamins and other components of the cell wall. About 75% of the total organic matter is soluble; therefore, it cannot be removed by mechanical or physical means. Thus, biological and chemical oxidations are the preferred means for wastewater treatment [21,22].

In the United States, there are three geographical areas of major potato processing activity: (a) Idaho, eastern Oregon, and eastern Washington; (b) North Dakota and Minnesota; and (c) Maine. Most plants are located in sparsely populated areas where the waste load from the plants is extremely large compared to the domestic sewage load [11]. By contrast, potato chips and prepeeled potato plants, while expanding in number and size, are largely located near metropolitan areas, where the waste effluent is more easily handled by municipal facilities. In general, these plants are much smaller than French fry or dehydrated potato plants and produce less waste load.

6.4.1 Waste Treatment Processes

An integrated waste treatment system usually consists of three phases: primary treatment, secondary treatment, and advanced treatment. Primary treatment involves the removal of suspended and settleable solids by screening, flotation, and sedimentation. Secondary treatment involves the biological decomposition of the organic matter, largely dissolved, that remains in the flow stream after treatment by primary processes. Biological treatment can be accomplished by mechanical processes or by natural processes.

The flow from the biological units is then passed through secondary sedimentation units so that the biological solids formed in the oxidation unit may be removed prior to the final discharge of the treated effluent to a stream. When irrigation is used as the secondary treatment system, bacteria in the topsoil stabilize the organic compounds. In addition, the soil may accomplish removal of some ions by adsorption or ion exchange, although ion exchange in some soils may fail. In all cases, great importance should be given to the steps that contribute to reducing the waste load in the plant itself. As for the industrial wastewaters, most of them require equalization (buffering) and neutralization prior to biological treatment, according to the characteristics of the resultant effluents.

In many parts of the world, potato processing wastewater treatment systems employed primary treatment from 1950 until 1970 to 1980. Thereafter, potato processing plants involved either secondary treatment or spray irrigation systems. Currently the most commonly used treatment methods, particularly in the United States, depend on screening, primary treatment, and settling of silt water in earthen ponds before discharging to municipal sewers or separate secondary treatment systems.

Many countries that have potato processing industries have determined current national minimum discharge limits following secondary treatment or in-land disposal. For example, the US Environmental Protection Agency (EPA) has proposed nationwide such limits for potato processing effluents [12].

To meet national effluent limits or standards, advanced waste treatment is needed in many cases to remove pollutants that are not removed by conventional secondary treatment. Advanced treatment can include removal of nutrients, suspended solids, and organic and inorganic materials. The unit processes for treating potato processing effluent are shown in sequence in Table 6.6.

Figure 6.7 illustrates a general treatment concept typical for the treatment of potato processing effluent: advanced treatment is added as a result of the growing environmental requirements. Currently, different treatment units are combined as a highly effective system for the secondary (biological) treatment that covers both anaerobic and aerobic processes. Note that it is quite acceptable and applicable that wastewater after preclarification (screening and primary treatment) can be discharged into the public sewer system to be treated together with sewage water in the municipal treatment plants.

The following describes in detail the current wastewater treatment units and subsystems.

In-Plant Treatment

Minimizing waste disposal problems requires reduction of solids discharged into the waste stream and reduction of water used in processing and clean-up. To reduce the solids carried to waste streams, the following steps should be undertaken [11]:

- improvement of peeling operation to produce cleaner potatoes with less solids loss;
- reduction of floor spillage;

Table 6.6 Treatment Units, Unit Operation, Unit Processes, and Systems for Potato Processing Wastewater

Treatment unit or subsystem	Unit operation/unit process/ treatment system	Remarks
In-plant	• Conservation and reuse of water • Process revisions • Process control • New products	• Reduction of waste flow and load
Pretreatment	• Screening (mesh size: 20 to 40 per inch)	• 10–25% BOD_5 removal
Primary treatment	• Sedimentation • Flotation • Earthen ponds	• 30–60% BOD_5 removal • 20–60% COD removal
Equalization Neutralization Secondary treatment	• Balancing tank/buffer tank • Conditioning tank	• Constant flow and concentration • pH and temperature corrections • 80–90% BOD_5 removal
1. Aerobic processes	• Natural systems – Irrigation land treatment – Stabilization ponds and aerated lagoons – Wetland systems • Activated sludge • Rotating biological contactors • Trickling filters	• 70–80% COD removal
2. Anaerobic processes	• Upflow anaerobic sludge blanket (UASB) reactors • Expended granular sludge bed (EGSB) reactors • Anaerobic contact reactors • Anaerobic filters and fluidized-bed reactors	• 80–90% BOD_5 removal • 70–80% COD removal
Advanced treatment	• Microstraining • Granular media filtration • Chemical coagulation/ sedimentation • Nitrification–denitrification • Air stripping and ion exchanging • Membrane technology (reverse osmosis, ultrafiltration)	• 90–95% BOD_5 removal • 90–95% COD removal (Sometimes >95%)

Notes: BOD_5 and COD removal percentage depended on experience of the German and other developed countries. There are other advanced treatment methods (not mentioned in this table) used for various industrial wastewater such as activated carbon adsorption, deep well injection, and chlorination that are not expected to be highly used in potato processing wastewater treatment.

- collection of floor waste in receptacles instead of washing them down the drains;
- removal of potato solids in wastewater to prevent solubilization of solids.

Water volume can be reduced by reusing process water, with several advantages. First, the size of wastewater treatment facilities can be decreased accordingly. Secondly, with

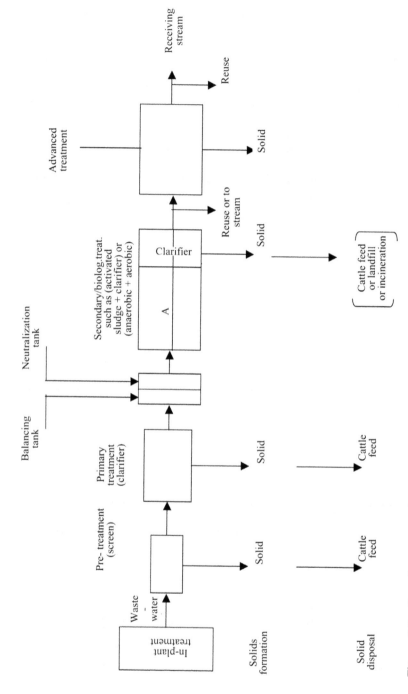

Figure 6.7 General treatment scheme for potato processing effluent.

concentration of the waste, the efficiency of a primary settling tank is increased. In the final processing stages, chlorinated water should be utilized to prevent bacterial contamination of the product. Other steps to reduce wastewater volume include alternate conveying methods of transporting potatoes other than water fluming, improved cleaning facilities for equipment and floors (high-pressure nozzles, shut-off nozzles for hoses), collecting clean waste streams, and discharge to natural drainage or storm water systems.

Pretreatment (Screening)

Typically, the screen is the first device encountered by wastewater entering the treatment plant. Screening is often used to remove large pieces of waste so that the water can be reused within the processing plant. Three types of screens are commonly used: stationary gravity screens, rotary screens, and vibratory screens. These units are similar to screens used in dewatering products during processing. Coarse solids are normally removed in a fine screen with a mesh size of 1 mm. The simplest type of stationary screen consists of a number of bars eventually spaced across the wastewater channel (bar rack). In modern wastewater treatment plants, the racks are cleaned mechanically. Rotary screens are used to a large extent and a variety of types are available. The most common type is the drum screen, which consists of a revolving mesh where wastewater is fed into the middle of the drum, and solids are retained on the peripheral mesh as the water flows outward. Another type of rotary screen is the disc screen, which is a perforated plate of wire mesh disc set at right angles to the waste stream. The retained solids are removed at the top of the disc by brushes or water jets. Vibratory screens may have reciprocating orbital or rocking motion, or a combination of both. The wastewater is fed into the horizontal surface of the screen, and the water passing through the retained solids is bounced across the screen to a discharge point.

The waste screen should be carefully located and elevated. Plant wastewaters can be collected in a sump pit below the floor level of the plant, from which they are pumped to the screen. The screen is elevated so that the solid wastes may fall by gravity into a suitable hopper. Then, the water flows down into the primary treatment equipment or to the sewer. With suitable elevations, the screen can be located below the level of the plant drains. After screening, the solid waste is conveyed up to the waste hopper and the water pumped into the clarifier, or other disposal system.

Primary Treatment

Sedimentation. Sedimentation is employed for the removal of suspended solids from wastewater. After screening, wastewater still carries light organic suspended solids, some of which can be removed from the wastewater by gravity in sedimentation tanks called clarifiers. These tanks/clarifiers can be round or rectangular, are usually about 3.5 m deep, and hold the wastewater for periods of 2 to 3 hours [23]. The required geometry, inlet conditions, and outlet conditions for successful operation of such units are already known. The mass of settled solids is called raw sludge, which is removed from the clarifiers by mechanical scrapers and pumps. Floating materials such as oil and grease rise to the surface of the clarifier, where they are collected by a surface skimming system and removed from the tank for further processing.

Figures 6.8 and 6.9 show cross-sections of typical rectangular and circular clarifiers. Construction materials and methods vary according to local conditions and costs.

In the primary treatment of potato wastes (Fig. 6.10), the clarifier is typically designed for an overflow rate of 800–1000 gal/(ft^2/day) (33–41 m^3/m^2/day) and a depth of 10–12 ft (3–3.6 m). Most of the settleable solids are removed from the effluent in the clarifier. The COD

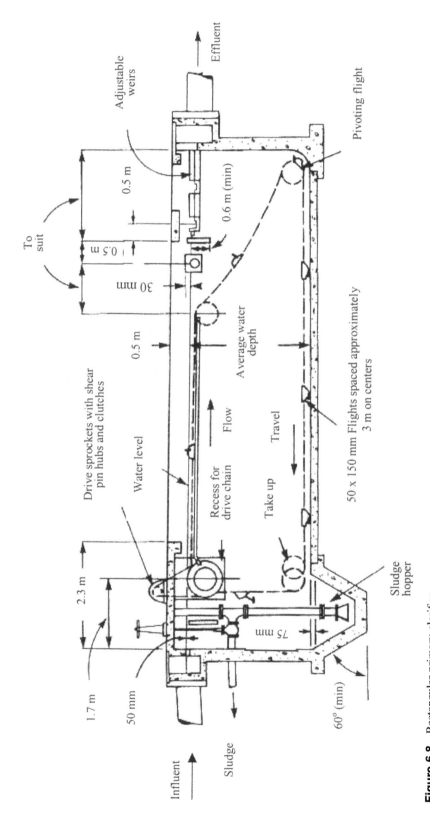

Figure 6.8 Rectangular primary clarifier.

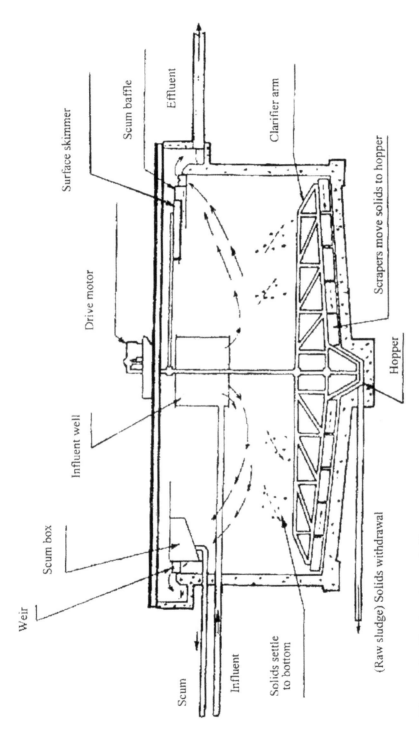

Figure 6.9 Circular primary clarifier.

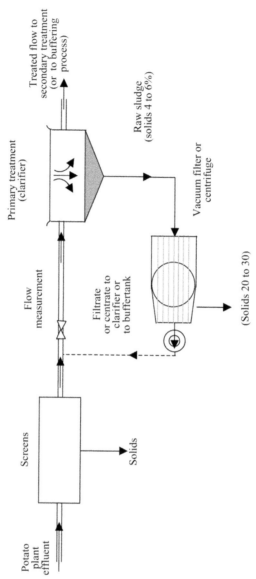

Figure 6.10   Schematic representation of primary treatment for potato wastes (from Ref. 11).

removal in this primary treatment is generally between 40–70% [11]. In comparison with cornstarch wastes, it was reported that BOD removals of 86.9% were obtained from settling this kind of waste [24].

To reduce the volume of the settled waste, which contains 4–6% solids, vacuum filters or centrifuges are used.

Withdrawal of the underflow from the bottom of the clarifier is accomplished by pumping. The resulting solids from caustic peeling have a high pH. The optimum pH level for best vacuum filtration of solids differs from plant to plant. However, when the underflow withdrawal is adjusted to hold the solids in the clarifier for several hours, biological decomposition begins and the pH of the solids falls greatly. At a pH of between 5 and 7, these solids will dewater on a vacuum filter without the addition of coagulating chemicals.

As for the solids resulting from steam or abrasive peeling operations, these will also undergo biological degradation in a few hours. With a longer duration, however, dewatering of solids becomes more difficult.

Flotation. Flotation is another method used for the removal of suspended solids and oil and grease from wastewater. The pretreated waste flow is pressurized to 50–70 lb/in^2 (345–483 kPa or 3.4–4.8 atm) in the presence of sufficient air to approach saturation [24]. When this pressurized air–liquid mixture is released to atmospheric pressure in the flotation unit, minute air bubbles are released from the solution. The suspended solids or oil globules are floated by these minute air bubbles, which become enmeshed in the floc particles. The air–solids mixture rises to the surface, where it is skimmed off by mechanical collectors. The clarified liquid is removed from the bottom of the flotation unit. A portion of the effluent may be recycled back to the pressure chamber.

The performance of a flotation system depends upon having sufficient air bubbles present to float substantially all of the suspended solids. This performance in terms of effluent quality and solids concentration in the float, is related to an air/solids ratio that is usually defined as mass of air released per mass of solids in the influent waste.

Pressure, recycle ratio, feed solid concentration, and retention period are the basic variables for flotation design. The effluent's suspended solids decrease and the concentration of solids in the float increase with increasing retention period. When the flotation process is used for primary clarification, a detention period of 20–30 min is adequate for separation and concentration. Rise velocity rates of 1.5–4.0 gal/(min/ft^2) [0.061–0.163 m^3/(min/m^2)] are commonly applied [24].

Major components of a flotation system include a pressurizing pump, air-injection facilities, a retention tank, a backpressure regulating device, and a flotation unit, as shown in Figure 6.11. The pressurizing pump creates an elevated pressure to increase the solubility of air. Air is usually added through an injector on the suction side of the pump or directly to the retention tank. The air and liquid are mixed under pressure in a retention tank with a detention time of 1 to 3 min. A backpressure regulating device maintains a constant head on the pressurizing pump.

Equalization

Equalization is aimed at minimizing or controlling fluctuations in wastewater characteristics for the purpose of providing optimum conditions for subsequent treatment processes. The size and type of the equalization basin/tank used varies with the quantity of waste and the variability of the wastewater stream. In the case of potato processing wastewater, the mechanically pretreated or preclarified wastewater flows into a balancing tank (buffer tank). Equalization serves two purposes: physical homogenization (flow, temperature) and chemical homogenization (pH,

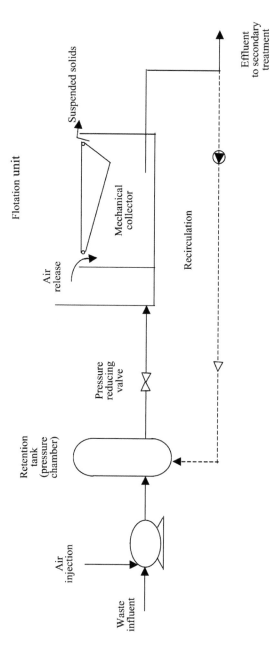

Figure 6.11   Schematic diagram of flotation system (from Ref. 24).

nutrients, organic matter, toxicant dilution). For proper homogenization and insurance of adequate equalization of the tank content, mixing is usually provided, such as turbine mixing, mechanical aeration, and diffused air aeration. The most common method is to use submerged mixers.

Neutralization

Industrial wastewaters that contain acidic or alkaline materials should be subjected to neutralization prior to biological treatment or prior to discharge to receiving wastes. For biological treatment, a pH in the biological system should be maintained between 6.5 and 8.5 to ensure optimum biological activity. The biological process itself provides neutralization and a buffer capacity as a result of the production of CO_2, which reacts with caustic and acidic materials. Therefore, the degree of the required preneutralization depends on the ratio of BOD removed and the causticity or acidity present in the waste [24].

As for potato processing wastewater in general, the water from the balancing tank (buffer tank) is pumped into a conditioning tank where the pH and temperature of the wastewater are controlled or corrected. Continuous monitoring of the pH of the influent is required by dosing a caustic or acidic reagent, according to the nature of resulting wastewater. The required caustic or acidic reagent for dosing in the neutralization process is strongly related to the different peeling methods used in the potato processing plant, since peeling of potatoes forms the major portion of the organic load in potato processing waste. Three different peeling methods are used extensively today: abrasion peeling, steam peeling, and lye peeling. Between lye and steam peeling wastes, the biggest difference is the pH of the two wastes. While steam peeling wastes are usually almost neutral (pH values vary between 5.3 and 7.1), lye peeling wastes have pH values from 11 to 12 and higher [3].

Secondary Treatment

Secondary treatment is the biological degradation of soluble organic compounds from input levels of 50–1000 mg/L BOD or more to effluent levels typically under 15–20 mg/L. In all cases, the secondary treatment units must provide an environment suitable for the growth of biological organisms that carry out waste treatment. This is usually done aerobically, in an open aerated tank or lagoon. Also, wastewaters may be pretreated anaerobically, in a pond or a closed tank. After biotreatment, the microorganisms and other carried-over solids are allowed to settle. A fraction of this sludge is recycled in certain processes. However, the excess sludge, along with the sedimented solids, must be disposed of after treatment.

As for potato waste, the most full-scale secondary treatment systems have been applied since 1968, although considerable research works of a pilot-plant scale have been conducted prior to that date. The description or characteristic data of these pilot-scale secondary treatment designs have been presented in detail [11]. Among the different known aerobic processes for secondary treatment of wastewater, we concentrate here on the most common treatment processes for potato processing wastewater with relevant case studies.

Natural Treatment Systems: Irrigation Land Treatment. Land treatment of food-processing wastewater resulting from meat, poultry, dairy, brewery, and winery processes has proved successful mainly through spray irrigation, applied as various types and methods in many areas. By 1979, there were an estimated 1200 private industrial land-treatment systems [24]. Potato processing wastewater can be utilized as irrigation water to increase the crop yield, because they are not polluted biologically. Irrigation systems include ones in which loading rates are about 2–4 in./week (5–10 cm/week).

Factors such as the crops grown, soil type, groundwater, and weather determine the required land area for irrigation. Some potato processors choose land disposal systems (spray or flood irrigation) because other treatment systems, while they give a higher efficiency rate, are exposed to operational problems.

Loamy, well-drained soil is most suitable for irrigation systems. However, soil types from clays to sands are acceptable. A minimum depth to groundwater of 5 ft (1.5 m) is preferred to prevent saturation of the root zone [24]. If a 5 ft depth is not available due to higher groundwater, underdrained systems can be applied without problems. As for potential odors issued from spray irrigation, they can be controlled by maintaining the wastewater in a fresh condition in order not to become anaerobic.

Water-tolerant grasses have proved to be the most common and successful crops for irrigation disposal, due to their role in maintaining porosity in the upper soil layers. The popular cover crop is reed canary grass (*Phalaris arundinacea*), which develops extensive roots that are tolerant to adverse conditions. In addition, water-tolerant perennial grasses have been widely used because they are able to absorb large quantities of nitrogen, require little maintenance, and maintain high soil filtration rates.

In some cases, wastewaters have been sprayed into woodland areas. Trees develop a high-porosity soil cover and yield high transpiration rates. Irrigation systems normally consist of an in-plant collection system, screens, low-head pump station, pressure line, pumping reservoir, high-head irrigation pumps, distribution piping, spray nozzles, and irrigation land. It is preferable in this respect to preclarify the potato processing wastewater by using a primary settling tank with a minimum 1.5 hours detention time to decrease the suspended solids content, in order to prevent closing of spray nozzles and soil. If the effluent has excess acid or alkali, it should be neutralized prior to discharging to land so that cover crops may be protected. Groundwater contamination from irrigation can be a serious problem and must be addressed during the predesign phase of a project, with the consideration that continuous monitoring of groundwater is necessary at all times in the irrigated area.

Design Example 1.   A potato processing industry plans to treat its resultant wastewater by a land irrigation system. Determine the area required under the specific conditions: flow = 0.2 MG/day (756 m^3/day), BOD concentration = 2600 mg BOD/L, N concentration = 100 mg N/L. The regulation limits are: loading rates are 2 in./week (5 cm/week) and 535 lb BOD/acre/day (0.06 kg/m^2/day), nitrogen loading rate for crop's need of grass is 250 lb N/acre (0.028 kg/m^2) (the spraying period for the grass is 16 weeks).

Solution: Prescreened wastewater: assuming that 20% BOD is removed by using fine screen with mesh size 1 mm. Residual BOD: 2600 × 0.8 = 2080 mg/L.

$$\frac{Qm}{A} = \frac{r}{258}$$

where Qm is in million gallons per day, A is in acres, and r is the average wastewater application rate (inches per week).

$$\frac{0.2}{A} = \frac{2}{258}$$

and $A = 26$ acres (10.5 ha $= 105,000$ m^2).

$$\text{Daily loading of BOD} = \frac{2080 \text{ mg BOD}}{L} \times 0.2 \text{ MG/day} \times 8.34 \frac{\text{lb/MG}}{\text{mg/L}}$$

$$= 3469.4 \text{ lb/day} \ (1575 \text{ kg/day})$$

$$A = \frac{3469.4 \text{ lb/day}}{535 \text{ lb/acre.day}} = 6.5 \text{ acres} \ (2.6 \text{ ha} = 26,000 \text{ m}^2)$$

$$\text{Total loading of N} = \frac{100 \text{ mg N}}{L} \times 0.2 \text{ MG/day} \times 8.34 \frac{\text{lb/MG}}{\text{mg/L}}$$

$$\times 16 \text{ weeks} \times 7 \text{ days/week}$$

$$= 18,682 \text{ lb N} \ (8482 \text{ kg})$$

$$A = \frac{18,682 \text{ lb N}}{250 \text{ lb N/acre}} = 75 \text{ acres} \ (30.4 \text{ ha} = 304,000 \text{ m}^2)$$

or

$$\frac{Qm}{A} = \frac{NC}{58.4 \, nT}$$

where NC is nitrogen removal by the growing crop (lb/acre), n is nitrogen concentration of the wastewater (mg/L), and T is the number of weeks of the irrigation season.

$$\frac{0.2}{A} = \frac{250}{58.4 \times 100 \times 16}$$

and $A = 75$ acres (30.4 ha $= 304,000$ m^2) or, in metric units:

$$\frac{Qm}{A} = \frac{143 NC}{nT}$$

$$\frac{756 \text{ m}^3/\text{day}}{A} = \frac{143 \ (0.028 \text{ kg/m}^2)}{100 \times 16}$$

where $A = 304,000$ m^2 $= 30.4$ ha (75 acres).

The area required is 75 acres (30.4 ha).

Natural Treatment Systems: Stabilization Ponds and Aerated Lagoons. A wastewater pond, sometimes called a stabilization pond, oxidation pond, or sewage lagoon, consists of a large, shallow earthen basin in which wastewater is retained long enough for natural processes of treatment to occur. Oxygen necessary for biological action is obtained mainly from photosynthetic algae, although some is provided by diffusion from the air. Lagoons differ from ponds in that oxygen for lagoons is provided by artificial aeration.

Depending on the degree of treatment desired, waste stabilization ponds may be designed to operate in various ways, including series and parallel operations. In some cases such as industrial wastewater treatment, they are referred to as tertiary ponds (polishing or maturation ponds), in order to remove residual pollutants and algae prior to effluent discharges.

The majority of ponds and lagoons serving municipalities and industries are of the facultative type, where the wastewater is discharged to large ponds or lagoons. Usually the ponds vary from 3 to 6 ft (0.9 to 1.8 m) deep, for a period of 3 weeks and longer, while lagoons vary from 6 to 15 ft (1.8 to 4.6 m), for a period of 2 weeks and longer.

Climatic conditions play an important role in the design and operation of both ponds and lagoons. Air temperature has a great effect on the success of this type of treatment. Within naturally occurring temperature ranges, biological reactions roughly double for each 10°C increment in water temperature. This fact encourages countries with warmer climates to utilize ponds and lagoons for wastewater treatment, particularly where land is abundant, thus providing considerable savings in both capital and operating costs.

The use of a stabilization pond in treating combined wastewaters of potato processing wastewaters and domestic wastewaters has been examined [25]. Extensive treatment loading rates for stabilization ponds were recommended in the range 5.6–6.7 kg BOD/1000 m^3/day.

High-strength wastewaters require long detention times, increasing heat loss, and decreasing efficiency in cold climates. Additionally, highly colored wastewaters cannot be treated effectively by facultative ponds, where oxygen generation is supplied mainly by photosynthesis, which depends on light penetration. Therefore, it is necessary to use aerated lagoons in which the required oxygen is supplied by diffused or mechanical aeration units. The biological life in such lagoons contains a limited number of algae and is similar to that found in an activated sludge system. In addition, aerated lagoons prevent the completion of anaerobic conditions with their attendant odor problems.

There are two types of aerated lagoons: aerobic and facultative lagoons. They are primarily differentiated by the power level employed. In aerobic lagoons, the power level is sufficiently high to maintain all solids in suspension and may vary from 14 to 20 hp/MG (2.8–3.9 W/m^3) of lagoon volume, depending on the nature of the suspended solids in the influent wastewater [24].

In facultative lagoons or aerobic–anaerobic lagoons, the power level employed is only sufficient to maintain a portion of the suspended solids in suspension, where the oxygen is maintained in the upper liquid layers of the lagoon. The employed power level in such lagoons for treating industrial wastewater is normally lower than 1 W/m^3.

As for the design of facultative ponds and aerated lagoons, several concepts and equations have been employed, and they can be found in many publications. The following is a design example for the treatment plant of potato processing wastewater.

Design Example 2. A potato processing wastewater flow of 1150 gal/ton of raw potatoes (4.35 m^3/ton) has a BOD of 2400 mg/L and a VSS content of 450 mg/L (nondegradable). It is to be pretreated in an aerobic lagoon with a retention period of one day. The k is 36/day; the raw potatoes processed are 150 tons/day. Estimate the following: the effluent soluble BOD concentration; the effluent VSS concentration; the oxygen required in mass/day; where $a = 0.5$, $a' = 0.55$, $b = 0.15$/day.

Solution: Effluent soluble BOD (S_e), by rearranging the equation:

$$\frac{S_e}{S_o} = \frac{1 + bt}{akt}$$

$$S_e = \frac{S_o(1 + bt)}{akt} = \frac{2400 \, \text{mg/L}(1 + 0.15/\text{day} \times 1 \, \text{day})}{0.5 \times 36 \times 1 \, \text{day}}$$

$$S_e = 153 \, \text{mg/L}$$

Effluent volatile suspended solids ($VSS_{effl.}$): the mixed liquor volatile suspended solids can be predicted from the equation:

$$X_v = \frac{aS_r}{1 + bt} + X_i$$

where X_i = influent volatile suspended solids not degraded in the lagoon.

$$X_v = \frac{0.5(2400 - 153)\,mg/L}{1 + 0.15/day \times 1.0\,day} + 450\,mg/L$$

$$= 977 + 450$$

$$= 1427\,mg/L$$

Oxygen required, using equation:

$$O_R = [a'(S_o - S_e) + 1.4bX_vt]Q$$

$$= [0.55(2400 - 153)\,mg/L + 1.4 \times 0.15/day \times 977\,mg/L \times 1\,day]$$

$$\times 4.35\,m^3/ton \times 150\,ton/day$$

$$= (1235.85 + 205.17)\,652.5 \times 10^{-3}$$

$$= 940.27\,kg/day\ (2069\,lb/day)$$

Remark: The pretreated wastewater in an aerobic lagoon can be discharged to a municipal treatment system, or to facultative ponds followed the aerobic lagoon.

Natural Treatment System: Wetland Systems. Wetland treatment technology of wastewater dates back to 1952 in Germany, starting with the work of Seidel on the use of bulrushes to treat industrial wastewaters. In 1956, Seidel tested the treatment of dairy wastewater with bulrushes, which may be regarded as the first reported application of wetland plants in food processing industries [26].

Throughout the last five decades, thousands of wetland treatment systems have been placed in operation worldwide. Most of these systems treat municipal wastewater, but a growing number of them involve industrial wastewaters. Frequently targeted pollutants are BOD, COD, TSS, nitrogen, phosphorus, and metals.

The design and description of treatment wetlands involves two principal features, hydraulics and pollutant removal [9], while the operational principles include biodegradation, gasification, and storage. Food-processing wastes are prime candidates for biodegradation. The attractive features of wetland systems are moderate capital cost, very low operating cost, and environmental friendliness. The disadvantage is the need for large amounts of land.

Reed beds in both horizontal and vertical flows have been successfully used in treating wastewater of the potato starch industry [27]. Several types of meat processing waters have been successfully treated using wetland systems [28–30]. The vertical flow of the integrated system has been used with favorable results in several domestic wastewater treatment applications [31–33].

Engineered natural systems have been used successfully to treat high-strength water from potato processing. Such integrated natural systems consist in general of free water surface and vertical flow wetlands, and a facultative storage lagoon (Fig. 6.12) [34]. (For a detailed description of wetland components with regard to their operational results and performance refer to case studies.)

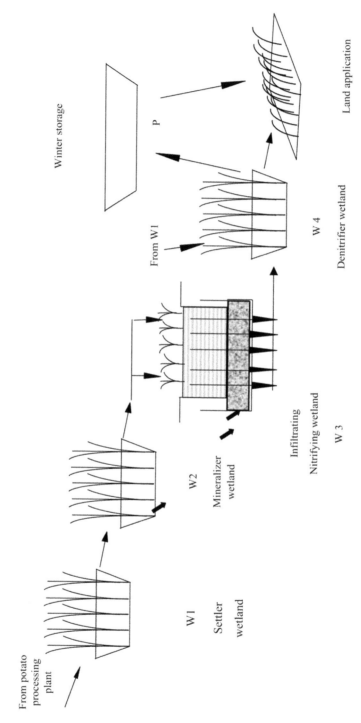

Figure 6.12 Schematic layout of an integrated natural system (wetland) for treatment of potato processing wastewater (from Ref. 15).

Case Studies

Case Study I. A full-scale integrated natural system has been used to treat high-strength potato processing water for 2 years [34]. The integrated natural system consists of free water surface and vertical flow wetland, and a facultative storage lagoon. Wetland components were designed for sequential treatment of the wastewater. Wastewater is pumped from a primary clarifier to ten hectares of free water surface wetlands constructed for sedimentation and mineralization of wastewater (W1/W2). The process water from the W1/W2 wetlands is sprayed onto 4 hectares of vertical flow wetland (W3) for oxidation of carbon and nitrogen. These wetlands were filled with 0.9 m of a local sand (D50 = 2.6 mm) excavated on site. These vertical flow wetlands were operated as intermittent sand filters with duty cycles of 6–72 hours. They were not planted with *Phragmites australis* due to poor growth when sprayed with the wastewater [15]. Water flows by gravity from the W3 into 2 hectares of denitrifying free water surface wetlands (W4). Raw process water is supplemented to augment denitrification in the wetlands. Treated process water flows into a 0.48 million m^3 lagoon (126 million gallon), which provides facultative treatment and storage prior to land application (Fig. 6.12).

The wetlands were constructed in stages throughout 1994 and 1995 in Connell, Washington. Connell is located centrally in the Colombia Basin, which is an arid agricultural area sustained by irrigation water from the Colombia River. All wetlands were lined with 1.0 mm (40 mil) HDPF liner impregnated with carbon black for UV resistance. All free water surface wetlands had 20–30 cm (8–12 in.) of native soil placed on the liners as soil for *Typha* sp. and 2 spaces of *Scirpus* sp.

The wetlands system is designed to treat an annual average flow of 1.4 mgd (approx. 5300 m^3/day) of wastewater with an annual average concentration of 3150 mg/L COD, 575 mg/L TSS, 149 mg/L TKN, and 30 ml/L NH$_4$-N. The winter design temperature was 1°C, with the consideration that the flow to the engineered natural system was lower in the winter season, due to operational difficulties in the water supply system.

Regarding the operational results of the integrated natural system, there were excellent reductions of TSS and COD, while organic nitrogen was effectively mineralized. TKN was reduced by about two-thirds, which is the requisite amount for balancing irrigation and nitrogen supply to the crop [15].

The net COD removal through the system was greater than 90% all year round. The W1/W2 wetlands removed about 85–90% of the COD, and 80–90% of the TSS. The average COD loading to the W1/W2 was 0.5 kg/m^3/day (31 lb/1000 ft^3/day) and 0.3 kg/m^3/day (18 lb/1000 ft^3/day) for the summer and winter, respectively. This loading rate is similar to the low rate covered anaerobic lagoons used for COD reduction in food processing. The effluent concentrations from the wetlands are lower in COD and TSS than from equivalently loaded covered anaerobic lagoons [35,36].

The effluent TSS from W1/W2 wetlands is consistently less than 75 mg/L. The W1/W2 wetland plants have proven to be very effective in solids removal. The TSS concentration increases in the lagoon due to algae growth.

In terms of nitrogen removal, the treatment objective of the system is a 53% reduction in total nitrogen (TN). The wastewater application permit requires an annual nitrogen load of 500 kg/ha/year on 213 hectares of land used to grow alfalfa and other fodder crops. The results related to TN removal indicate that the wetlands operate better than design expectation.

With regard to organic carbon, the potato water mineralizes very rapidly so that >60% of the organic carbon was mineralized to NH$_4$-N prior to entering the wetlands. This mineralization continued in the W1/W2 wetlands so that <15 mg/L organic nitrogen remained.

More than 60% of the TN entering the W1/W2 wetlands was in the form of NH_4-N, and 10–20% of the NH_4 was removed from the W1/W2. The pH in the W1/W2 was always >7.0 and may have contributed to volatilization of NH_4-N. The NH_4-N removal through the vertical flow wetlands averages 85% during the summer and 30–50% during the winter.

Removal of nitrate and nitrite is critical for compliance with TN removal goals in order to minimize the amount of oxidized N applied in land. Reduction of COD or BOD is often viewed as a prerequisite to establishment of nitrifying conditions [37]. Dissolved oxygen is slightly higher in the winter, but most of the system is anoxic except for the vertical flow component. Alkalinity is sufficient to support nitrification (ca. 1000 mg/L) [15]. The majority of the denitrification occurred in the W4 wetlands. Endogenous carbon in the W4 wetland was inadequate to support significant denitrification. Addition of raw potato water allows >90% denitrification, but also resulted in increased effluent NH_4-N concentrations. Approximately 5–7 NO_3-N were removed for each NH_4-N added.

Regarding the problem of odor, which generates from the decomposition of potato products, the strongest odors arose from the death of a large population of purple sulfur bacteria in the W1/W2 wetlands and the resulting sulfides >40 mg/L.

The integrated natural system is effective in reducing sulfate concentrations, from about 40 mg/L to 10 mg/L, in wetland W1. Because W1 is devoid of oxygen, sulfate has been reduced to sulfides or sulfur, including the possibility of hydrogen sulfide formation. The effluent of the treatment system has no serious odors. The final product is high-quality water with available nutrients and no odor problem during land application.

In comparing this integrated natural system with other treatment wetlands for treating food processing wastewaters, such as meat processing waters, it may be concluded that potato processing water is comparable to meat processing effluents in treatability [15]. Furthermore, it has been demonstrated that the use of this full-scale engineered natural system is a cost-effective treatment alternative for high-strength industrial wastewater. Continued research and development in operations and design of the full-scale system have resulted in better performance than that of the original design.

Activated Sludge Processes.   In these processes, the preclarified wastewater is discharged into aeration basins/tanks, where atmospheric oxygen is diffused by releasing compressed air into the wastewater or by mechanical surface aerators. Soluble and insoluble organics are then removed from the wastewater stream and converted into a flocculent microbial suspension, which is readily settleable in sedimentation basins, thus providing highly treated effluent.

There is a number of different variants of activated sludge processes such as plug-flow, complete mixing, step aeration, extended aeration, contact stabilization, and aerobic sequential reactors. However, all operate essentially in the same way. These variants are the result of unit arrangement and methods of introducing air and waste into the aeration basin and they have, to a large extent, been modified or developed according to particular circumstances.

For the treatment of food and vegetable industrial wastewater, the common activated sludge methods are shown in Figure 6.13.

With regard to potato wastewater treatment, the first full-scale activated sludge system was applied in the United States toward the end of the 1970s, by the R.T. French Company for treating their potato division wastewaters in Shelley, Idaho. Thereafter, many other potato processors installed biological treatment systems, most of which were activated sludge processes (Table 6.7).

Hung and his collaborators have conducted extensive research in various treatment processes for potato wastewater [10,16,17,20,38–41]. These included activated sludge processes with and without addition of powdered activated carbon, a two-stage treatment system of an activated sludge process followed by biological activated carbon columns, a two-stage

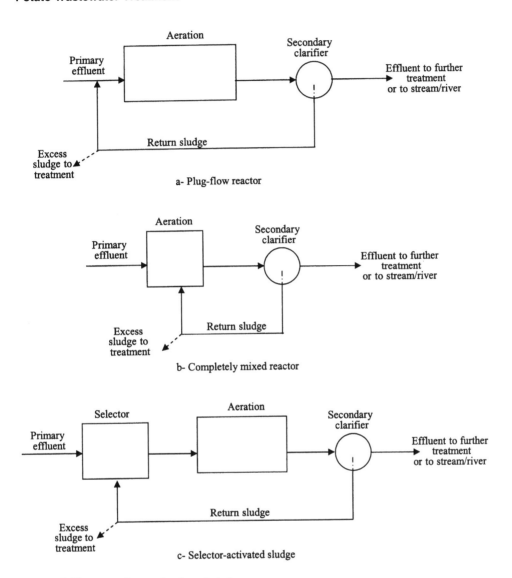

Figure 6.13 Flow sheets of activated sludge processes.

treatment system of an anaerobic filter followed by an activated sludge process, anaerobic digestion, and bioaugmentation process in which bacterial culture products were added to the activated sludge and anaerobic filter processes, and activated carbon adsorption process. In a laboratory study activated sludge treatment removed 86–96% of COD from primary settled potato wastewaters with 2500 mg/L COD and 500 mg/L TSS. Activated sludge followed by activated carbon adsorption removed 97% COD from primary settled wastewaters with a final effluent COD of 24 mg/L [17]. The hydraulic detention time in the aeration tank was 6.34 hours and in the sludge was 20 days.

A comparison study for potato wastewater treatment was conducted for a single-stage treatment system activated sludge reactor with and without addition of powdered activated carbon (PAC) and a two-stage treatment system using activated sludge followed by the

Table 6.7 Data of Various Full-Scale Secondary Treatment Designs (*Source*: Refs. 11 and 12)

Treatment process and process modification	Type of process water	Volumetric organic loading	Detention time	BOD removal (%)	Remarks
Complete mixing activated sludge	Dry caustic peel	32–39 lb/ (1000 ft³.day)	2 days	73	Sludge bulking
Complete mixing activated sludge	Lye peel	28–84 lb/ (1000 ft³.day)	1–2 days	70–90	Removal varies with sludge bulking
Complete mixing activated sludge	Lye peel	60–180 lb/ (1000 ft³.day)	14 hours	87	Sludge bulking will reduce removal
Multiple aerated lagoons	Lye peel	3–6 lb/ (1000 ft³.day) in aerated lagoons	16–20 days in aerated lagoons 105 days in aerobic lagoons	98	Algal blooms will reduce removal
Anaerobic pond and lye peel activated sludge	Lye peel	25–80 lb/ (1000 ft³.day) to activated sludge	1 day	95	Sludge bulking will reduce removal
Activated sludge and lye peel aerated lagoons	Lye peel	60–150 lb/ (1000 ft³.day) in aeration basin	14 hours in aeration basin	99	Sludge bulking and algal blooms will reduce removal
		55 lb/ac in aerated lagoons 8.5 lb/ac in aerobic lagoons	52 days in aerated lagoon 60 days in aerobic lagoon		

Note: lb/(1000 ft³/day) = 0.016 kg/(m³/day). Excess sludge: 0.2–0.5 lb/lb COD removed at about 2.0% solid concentration.

biological activated carbon (BAC) column [10,41]. The primary settled wastewater contained 2668–3309 mg/L COD. Results indicated that 92% of COD was removed in the non-PAC activated sludge reactors, while 96% COD was removed in the PAC activated sludge reactors. For the non-PAC activated sludge process, increasing hydraulic detention time in the aeration tank from 8–32 hours reduced effluent COD from 304 to 132 mg/L. With the addition of powdered activated carbon in the activated sludge tank, effluent COD was further improved to 78 mg/L at a hydraulic detention time of 32 hours. The BAC column removed 85% from activated sludge reactor effluents with a final effluent COD of 34 mg/L.

Bioaugmentation processes with addition of bacterial culture product have been used to improve the removal efficiency of organic pollutants and to reduce the amount of sludge in municipal wastewater treatment systems, particularly in activated sludge treatment processes. Three different systems, namely, extended aeration, aerated lagoon, and oxidation ditch have been used. In all three cases, bioaugmentation improved sludge settleability and BOD and COD removal efficiency [42].

Bioaugmentation with addition of bacterial culture product LLMO (live liquid micro-organisms) to the activated sludge reactor was investigated for treatment of potato wastewater [38]. Influent with 2381 mg/L COD was decreased to 200 mg/L in the bioaugmented activated sludge reactor and to 236 mg/L in the nonbioaugmented activated sludge reactor. The bioaugmented reactor can operate at a higher F/M ratio and a lower MLVSS level than the nonbioaugmented reactor and achieves a better COD removal efficiency. Effect of types of bacterial culture product addition to the activated sludge reactors on reactor performance have been studied [39]. Types of LLMO used included S1, G1, E1, N1, and New 1 LLMO. S1 LLMO was found to be the most effective, and removed 98% TOC (total organic carbon) and reduced 67% VSS (volatile suspended solids). The effect of bioaugmentation on the treatment performance of a two-stage treatment system using an anaerobic filter followed by an activated sludge process for treating combined potato and sugar wastewater was investigated [40]. The combined wastewater had 435 mg/L TOC. The bioaugmented two-stage treatment system had a better TOC removal efficiency and at a shorter hydraulic detention time of the aeration tank than the nonbioaugmented treatment system. The final effluent TOC was 75 mg/L and 89 mg/L at a hydraulic detention time of aeration tank of 12 hours and 24 hours for the bioaugmented and nonbioaugmented treatment systems, respectively.

Research on the treatment of potato processing wastewater showed that the major disadvantages of full-scale aerobic treatment are high power consumption, the large amount of sludge needing handling, and maintenance, in addition to the costs of sludge dewatering and sludge disposal (dumping and incineration), increasing substantially over the years. As a result, most potato processing companies have turned to the use of anaerobic treatment with various type of reactors followed by aerobic treatment.

Design Example 3. Continuing design example 2, a municipal extended aerobic activated sludge plant receives potato processing wastewater and has a combined BOD_5 of 450 mg/L. The return sludge has a concentration of 7000 mg/L from the secondary clarifier. Determine the required recycle ratio to the activated sludge reactor with an organic loading of 0.10 g BOD/g VSS, in order to produce an effluent meeting national discharge limits.

Solution: The organic loading (OL) can be expressed by:

$$OL = \frac{QS_o}{Q_R X_{vr}} = 0.10 \text{ g BOD/g VSS} \qquad (6.1)$$

where Q is the flow, S_o the influent BOD, Q_R the recycle flow, and X_{vr} the volatile suspended solids concentration in the recirculation line expressed in g VSS/L.

Assuming 85% VSS for the recirculation, $X_{vr} = 0.85$, $X_r = 0.85 \times 7000 = 5950$ mg VSS/L $= 5.95$ g VSS/L. The required recycle ratio can be calculated from Eq. (6.1).

$$Q_R = \frac{QS_o}{OL \cdot X_{vr}} = \frac{450 \text{ mg BOD/L} \times Q}{100 \text{ mg BOD/g VSS} \times 5.95 \text{ g VSS/L}}$$

$$= 0.756Q$$

Design Example 4. A municipal conventional activated sludge treatment plant is planning to receive the potato processing wastewater given in design example 2, without pretreatment (in an aerobic lagoon). Determine what changes need to be made in the processing conditions of the plant to avoid filamentous bulking. Assume: $T = 20°C$, $a' = 0.55$, $b' = 0.15/$day, $X = 0.6$, $N_b = 1.5$ lb $O_2/(hp.hour)$.

For the potato processing wastewater (example 2): BOD concentration = 2400 mg/L, Flow = 1150 gal/ton $\times$ 150 ton/day = 172,500 gal/day or = 4.35 m^3/ton $\times$ 150 ton/day = 652.5 m^3/day.

Solution: The municipal activated sludge treatment plant before potato processing discharge has the following characteristics: $Q_{bef.} = 2.5$ MG/day (9450 m^3/day), $S_{inf.} = 300$ mg/L, $S_e = 10$ mg/L, $S_{r,b} = 300 - 10 = 290$ mg/L, $t_b = 6$ hours = 0.25 day, $X_{v,b} = 3000$ mg/L, (F/M) = 0.3/day.

The dissolved oxygen required can be taken from reference (International water pollution control, Figs. 6.6–6.15): $DO_b = 1.7$ mg/L. The oxygen needed can be calculated by equation:

$$O_{R,b} = (a'S_{r,b} + b'XX_{v,b}t_b)Q_b$$
$$= (0.55 \times 290 + 0.15 \times 0.60 \times 3000 \times 0.25)\text{mg/L}$$
$$\times 2.5\,\text{MGD} \times 8.34(\text{lb/MG})/(\text{mg/L})$$
$$= 2733\,\text{lb/day}\ (1241\,\text{kg/day})$$
$$= 113.9\,\text{lb/hour}\ (51.71\,\text{kg/hour})$$

The power requirement is:

$$HP_b = O_{R,b}/N_b = \frac{113.9\,\text{lb/hour}}{1.5\,\text{lb/(hp.hour)}} = 76\,\text{HP}\ (57\,\text{kW})$$

After the potato industry discharge in the municipal activated sludge plant, the following will apply. Assume for the MLVSS, the value $X_{v,a} = 4000$ mg/L.

$$Q_{after} = Q_{before} + Q_{ind} = 2.5 + 0.1725 = 2.6725\,\text{MG/day (m}^3\text{/day)}$$
$$S_{inf.a} = \frac{Q_b S_{inf.b} + Q_{ind} S_{ind}}{Q_a}$$
$$= \frac{(2.5 \times 300) + (0.1725 \times 2400)}{2.6725} = 43,505\,\text{mg/L}$$

The BOD removed will be:

$$S_{r,a} = 435.5 - 10 = 425.5\,\text{mg/L}$$

The new retention time will be:

$$t_a = t_b \frac{Q_b}{Q_a} = 0.25\,\text{day}\,\frac{2.5}{2.6725} = 0.234\,\text{day}$$

The new F/M ratio can be computed using the equation:

$$(\text{F/M})_a = \frac{S_{inf.a}}{X_{v,a} \cdot t_a} = \frac{435.5}{4000 \times 0.234} = 0.465\,\text{day}$$

From the reference mentioned above, the dissolved oxygen required is: $DO_a = 3.6$ mg/L. Assuming the same values for a', b' and X, the oxygen required can be computed:

$$O_{R,a} = (0.55 \times 425.5 + 0.15 \times 0.60 \times 4000 \times 0.234)\,\text{mg/L}$$
$$\times 2.6725\,\text{MGD} \times 8.34\,(\text{lb/MG})/(\text{mg/L})$$
$$= 7093.7\,\text{lb/day}\ (3220.5\,\text{kg/day})$$
$$= 295.6\,\text{lb/hour}\ (134.2\,\text{kg/hour})$$

The oxygen saturation at $20°C$ is: $C_s = 9.2$ mg/L. The new N_a:

$$N_a = N_b \frac{(C_s - DO_a)}{(C_s - DO_b)} = \frac{1.5\,\text{lbO}_2}{(\text{hp.hour})} \times \frac{9.2 - 3.6}{9.2 - 1.7}$$
$$= 1.12\,\text{lb/(hp.hour)}\ (0.68\,\text{kg/kW.hour})$$

The power required is:

$$HP_a = O_{R,a}/N_a = \frac{295.6\,\text{lb/hour}}{1.12\,\text{lb/(hp.hour)}} = 264\,\text{HP}\ (197\,\text{kW})$$

The additional power required is:

$$HP_{add} = HP_a - HP_b = 264 - 76 = 188\,\text{HP}\ (140\,\text{kW})$$

Remark: To avoid the filamentous bulking in the conventional activated sludge plant, the following modifications are needed:

- increasing the MLVSS from 3000 to 4000 mg/L;
- increasing the power required from 76 HP (57 kW) to 264 HP (197 kW), in addition to the necessity to control the bulking.

Rotating Biological Contactors. The rotating biological contactor (RBC) is an aerobic fixed-film biological treatment process. Media in the form of large, flat discs mounted on a horizontal shaft are rotated through specially contoured tanks in which wastewater flows on a continuous basis. The media consist of plastic sheets ranging from 2 to 4 m in diameter and up to 10 mm thick. Spacing between the flat discs is approximately 30–40 mm. Each shaft, full of medium, along with its tanks and rotating device, forms a reactor module. Several modules may be arranged in parallel and/or in series to meet the flow and treatment requirements (Fig. 6.14). The contactor or disc is slowly rotated by power supplied to the shaft, with about 40% of the surface area submerged in wastewater in the reactor.

A layer of 1–4 mm of slime biomass is developed on the media (equivalent to 2500–10,000 mg/L in a mixed system) [24], according to the wastewater strength and the rotational speed of the disc. The discs, which develop a slime layer over the entire wetted surface, rotate through the wastewater and contact the biomass with the organic matter in the waste stream and then with the atmosphere for absorption of oxygen. Excess biomass on the media is stripped off by rotational shear forces, and the stripped solids are held in suspension with the wastewater by the mixing action of the discs. The sloughed solids (excess biomass) are carried with the effluent to a clarifier, where they are settled and separated from the treated wastewater.

The RBC system is a relatively new process for wastewater treatment; thus full-scale applications are not widespread. This process appears to be well suited to both the treatment of industrial and municipal wastewater. In the treatment of industrial wastewaters with high BOD levels or low reactivity, more than four stages may be desirable. For high-strength wastewaters, the first stage can be enlarged to maintain aerobic conditions. An intermediate clarifier may be

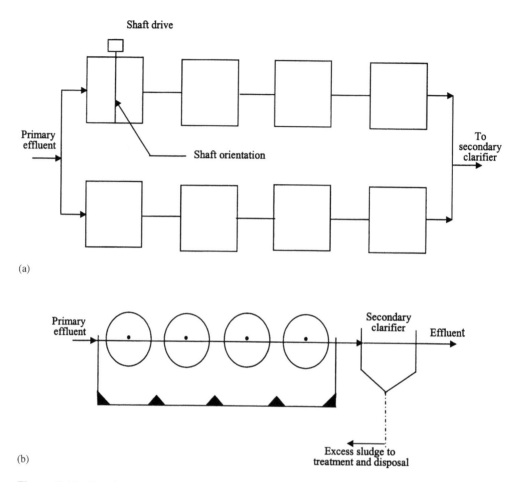

Figure 6.14 Rotating biological contactor system. (a) Flow-sheet of typical staged rotating biological contactors (RBCs). (b) Schematic diagram of the RBCs.

employed where high solids are generated to avoid anaerobic conditions in the contactor basins. Currently used media consist of high-density polyethylene with a specific surface of 37 ft^2/ft^3 (121 m^2/m^3). One module or unit, 17 ft (3.7 m) in diameter by 25 ft (7.6 m) long, contains approximately 10,000 m^2 of surface area for biofilm growth. This large amount of biomass permits a short contact time, maintains a stable system under variable loading, and should produce an effluent meeting secondary-treatment limits or standards.

Recirculating effluent through the reactor is not necessary. The sloughed solids (biomass) are relatively dense and settle well in the secondary clarifier. Low power requirement and simple operating procedure are additional advantages. A 40-kW motor is sufficient to turn the 3.7 × 7.6 m unit previously described [43]. Therefore, it can be clearly realized that the RBC can be applied successfully for treatment of potato processing effluents, in particular for values of BOD$_5$ and COD concentrations not exceeding, in the main, 5000 to 6000 mg/L in the wastewater stream. Depending on these properties, the data taken from case studies for treating contaminated wastewater with BOD$_5$ and COD concentrations close to those found in wastewater from potato processing, can be of much benefit. These data are based on the experience published by USEPA [44]. Table 6.8 summarizes the experience represented in design criteria and performance of the applied RBC for treating landfill leachate, which can be successfully applied to the potato processing industry within the range of pollutant concentrations mentioned above. However, an optimum design can be achieved by a pilot-plant study of the RBC.

Design Example 5. Design a rotating biological contactor (RBC). Determine the surface area required for an RBC system to treat preclarified potato processing wastewater with a flow of 150,000 gal/day (567 m^3/day) and BOD concentration of 4000 mg/L, with a maximum system effluent of 20 mg BOD/L. Minimum temperature is expected to be 32°C (90°F). The selected plastic medium is manufactured in 8 m shaft lengths, with each shaft containing 1.2×10^4 m^3 of surface area.

Solution: RBC performance:

$$\frac{4000 - 20}{4000} \times 100 = 99.5\%$$

No temperature correction in loading is needed, because the wastewater temperature is >55°F (13°C). Based on the hydraulic surface loading, the selected design value of Table 6.8 is: Hydraulic loading rate = 1.2 gal/ft^2/day (49 L/m^2/day).

Table 6.8 Design Criteria and Performance of Rotating Biological Contactors [44]

Parameter	Range	
(a) Design criteria		
MLSS (mg/L)	3000–4000	
MLVSS (mg/L)	1500–3000	
F/M (lb BOD/lb MLVSS/day)	0.05–0.3	
Maximum BOD volumetric loading (lb BOD/1000 ft^3/day)	15–60	
Maximum BOD surface loading (lb BOD/1000 ft^2/day)	0.05–0.7 (4–8 g BOD$_5$/m^2/day according to German experience)	
Number of stages per train	1–4	
Hydraulic surface loading (gal/day/ft^2)	0.3–1.5	
HRT (days)	1.5–10	
Compound	Influent (mg/L)	Removal (%)
(b) Performance		
SCOD	800–5200	55–99
SBOD$_5$	100–2700	95–99
TBOD$_5$	3000	99+
TOC	2100	99
DOC	300–2000	63–99
NH$_4$-N	100	80–99

Remark: These design and performance data are based on results of different references including EPA publications that handle landfill leachate treatment.

Disc area is calculated directly in a simple form:

$$A_d = \frac{150,000\,\text{gal/day}}{1.2\,\text{gal/ft}^2/\text{day}} = 125,000\,\text{ft}^2$$

$$= \frac{567\,\text{m}^3/\text{day}}{0.049\,\text{m}^3/\text{m}^2/\text{day}} = 11,600\,\text{m}^2 = 1.16 \times 10^4\,\text{m}^2$$

Based on the organic surface loading, normally adopted in Germany, the selected design value of Table 6.8 is: Organic loading rate $= 4$ g BOD/m^2/day.

$$\text{Influent BOD loading} = \frac{567\,\text{m}^3/\text{day} \times 4000\,\text{mg/L}}{1000} = 2268\,\text{kg/day}$$

Disc area is:

$$A_d' = \frac{2268\,\text{kg BOD/day}}{4\,\text{g/m}^2/\text{day}} \times \frac{1000\,\text{g}}{1\,\text{kg}} = 567,000\,\text{m}^2 = 5.67 \times 10^5\,\text{m}^2$$

In comparing A_d and A_d', it is clear that the required disc area will be:

$$A_d' = 5.67 \times 10^5$$

$$\text{Modules number} = \frac{5.67 \times 10^5\,\text{m}^2}{1.2 \times 10^4\,\text{m}^2/\text{Module}} = 47\,\text{Modules}$$

On average, 50 modules are required for the first stage of wastewater treatment.

For potato industrial wastewater, a minimum of four stages (200 modules) in series will be required. These can be placed in two lines, each line to contain four stages.

Anaerobic Treatment Systems. With more than 1800 plants worldwide using different applications (food processing, chemical industry, pulp and paper industry), anaerobic treatment has gained widespread use as a reliable and efficient means for reduction of COD [45]. Of all anaerobic processes, those technologies based on high-rate, compact, granular biomass technology, such as upflow anaerobic sludge blanket (UASB) and expended granular sludge bed (EGSB), have a leading position (more than 750 plants) [14].

A large number of analyses have been carried out since 1958, when the first full-scale anaerobic wastewater treatment plants were introduced. In Germany alone there are currently 125 methane reactors treating industrial wastewater. Forty-three plants are working with a contact process, 38 plants run sludge blanket reactors, and 33 plants work with fixed-film methane reactors. The other 11 plants have completely stirred tank reactors (CSTR), self-made contribution, hybrid reactors, or other unnamed reactor types [13].

Table 6.9 gives an overview of the typical problems and solutions in various food and beverage industries, including potato processing and potato starch industries, for all kinds of anaerobic reactor systems. This experience gathered by German researchers reveals that each industry has its own specific problems. Therefore, specific investigations should be undertaken to find the relevant solutions. Furthermore, these data show that it is possible to treat several different industrial wastewaters together in one plant, which is particularly beneficial for small factories, especially in the food industry [13].

Batch mesophilic anaerobic digestion processes for potato wastewater treatment have been conducted [16]. After 33 days of anaerobic digestion at a reactor pH of 6.5–7.3 and at a temperature of 22°C the batch treatment process removed 84, 82, and 90% COD from potato

Table 6.9 Several Food and Beverage Industries with Their Special Problems and Solutions (*Source*: Ref. 13)

Industry	Special problem	Solution
Potato processing industry	Solids	Sieve, acidification tank, EGSB methane reactor
Potato and wheat starch industry	Precipitation of MAP (magnesium ammonium phosphate)	pH regulation
Beet sugar factories	Lime precipitation	Cyclone
	pH lower than 5 in the pond system	Lowering the pH in the circuit system
Pectin factories	High nitrate concentrations over 1000 mg NO_3-N/L	Denitrification stage before methane reactor
Breweries	Considerable pH variations	Equalizing tanks, pH regulation
	Kieselguhr contents	Treatment together with municipal sludge
	Aluminum precipitation in the acidification stage	Settling tank
Distilleries (alcohol production from molasses slops)	Discontinuous production	Equalizing tanks and pH regulation
Anaerobic pretreatment of wastewater from different industries in one plant	Different small factories with high loaded wastewater and campaign processing	Anaerobic pretreatment of the wastewater mixture of a brewery, two vegetable, and one fish processing factory at the municipal sewage treatment plant
Anaerobic/aerobic treatment	Carbon : nitrogen relation bulking sludge	Bypassing the anaerobic stage, pretreatment

Source: Ref. 13.

juice, mashed potato, and potato starch wastewater, respectively. Hydrolysis played an important role in the anaerobic digestion process by converting the particulate substrate in the mashed potato and potato starch wastewaters to soluble substrate, which was subsequently utilized by anaerobes for production of organic acids and methane production.

Based on the wastewater composition (average data of settled samples: COD 4000 mg/L; total N 120 mg/L; total P 60 mg/L), wastewater from the potato processing industry is very well suited for anaerobic treatment. Accordingly, there are over 50 anaerobic plants in this sector of the industry worldwide, the majority of which consist of UASB reactors. More recently, the EGSB process (high-performance UASB), developed from the UASB, has been implemented. In the potato processing industry, several UASB plants have been built by Biothane Systems Inc. and its worldwide partners for customers such as McCain Foods (French fries) and Pepsico (potato crisps). Recently, other Biothane UASB plants have joined the Pepsico network, such as Greece (Tasty Foods, Athens), Turkey (Ozay Gida, Istanbul) and Poland (E. Wedel, Warsaw) [14].

An important prerequisite is that the influent to the UASB reactor must be virtually free of suspended solids, since the solids would displace the active pellet sludge in the system. The newly developed EGSB reactors are operated with a higher upflow velocity, which causes a partial washout of the suspended solids [14]. EGSB technology is capable of handling wastewater of fairly low temperatures and considerable fluctuations in COD composition and load throughout the year.

A description of the first large-scale EGSB (Biobed reactor) in Germany will be presented in case studies to follow.

Comparison Between Biothane UASB Reactors and Biobed EGSB Reactors [14]. The UASB technology (Fig. 6.15) and the EGSB technology (Fig. 6.16) both make use of granular anaerobic biomass. The processes have the same operation principles, but differ in terms of geometry, process parameters, and construction materials.

In both processes, wastewater is fed into the bottom of the reactor through a specially designed influent distribution system. The water flows through a sludge bed consisting of anaerobic bacteria, which develop into a granular form. The excellent settleability (60–80 m/hour) of these anaerobic granules enables high concentrations of biomass in a small reactor volume. The granules do not contain an organic carrier material, such as sand or basalt.

In the sludge bed, the conversion from COD to biogas takes place. In both reactor types, the mixture of sludge, biogas, and water is separated into three phases by means of a specially designed three-phase, separator (or settler) at the top of the reactor. The purified effluent leaves the reactor via effluent laundries, biogas is collected at the top, and sludge settles back into the active volume of the reactor.

One of the most important design parameters for both types of reactors is the maximum allowable superficial upflow liquid velocity in the settler. Upflow velocities in excess of this maximum design value result in granular sludge being washed out of the reactor. The Biobed EGSB settler allows a substantially higher upstream velocity (10 m/hour) than the Biothane UASB settler (1.0 m/hour).

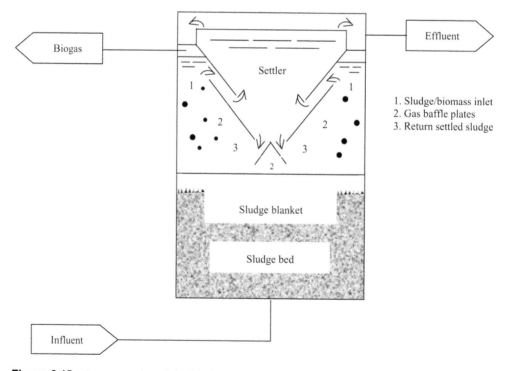

Figure 6.15   A cross-section of the Biothane UASB reactor (from Ref. 14).

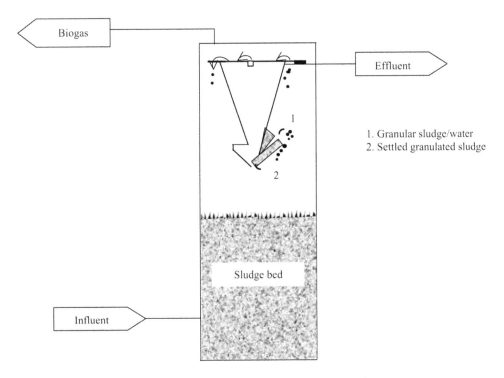

1. Granular sludge/water
2. Settled granulated sludge

Figure 6.16   A cross-section of the Biogas EGSB reactor (from Ref. 14).

Another important design parameter is the maximum COD load allowed. The Biobed EGSB process operates under substantial higher COD loads ($30 \, kg/m^3/day$) than the Biothane UASB process ($10 \, kg/m^3/day$). The result of this is that for a given COD load, the Biobed EGSB reactor volume is smaller than for a Biothane UASB reactor. Biothane UASB reactors are typically rectangular or square, with an average height of 6.0 m and are usually constructed of concrete. Biobed EGSB reactors have a substantially smaller footprint. These high and narrow tanks are built in FRP (fiber glass reinforced plastic) or stainless steel and have a typical height of 12–18 m. The height of the granular sludge bed in the Biothane UASB reactor varies between 1 and 2 m and in the Biobed EGSB between 7 and 14 m. A Biobed EGSB reactor is normally built as a completely closed reactor resulting in a system with zero odor emission. Additionally, a Biobed EGSB reactor can be operated under overpressure, thereby making any use of gas-holders and biogas compressors redundant. The general differences between the processes are shown in Table 6.10.

Wastewater in the potato processing industry contains substantial amounts of suspended solids. The Biothane UASB process is characterized by longer hydraulic retention times than the Biobed EGSB process. As a consequence, use of the Biothane UASB process results in a greater removal of suspended solids and, therefore, higher overall COD removal efficiencies. The Biobed EGSB process has been designed mainly for removal of soluble COD. Therefore, the use of Biobed EGSB in the potato processing industry is emphasized for those applications where the anaerobic effluent will be discharged to a sewer or to a final aerobic post-treatment.

Thermophilic UASB Reactors.   In general, hot wastewater streams discharge from food industries including vegetable processing. These streams are generated from high temperature unit operations and are highly concentrated due to enhanced dissolution of organic material at

Table 6.10 Comparison of the Main Characteristic Parameters of Biothane UASB and Biobed EGSB (*Source*: Ref. 14)

Parameter	Unit	Biothane UASB	Biobed EGSB
Load	kg COD/m^3/day	10	30
Height	m	5.5–6.5	12–18
Toxic Components		$+/-$	$++$
V_{liquid} settler	m/hour	1.0	10
V_{liquid} reactor	m/hour	<1.0	<6.0
V_{gas} reactor	m/hour	<1.0	<7.0

Source: Ref. 14.

elevated temperatures. Anaerobic treatment, especially the thermophilic process, offers an attractive alternative for the treatment of high-strength, hot wastewater streams [46].

In the thermophilic process, the most obvious benefits compared with the mesophilic anaerobic process involve increased loading rate and the elimination of cooling before treatment. Furthermore, the heat of the wastewater could be exploited for post-treatment, which, for example, if realized and mixed with sewage water could assist in obtaining nitrification with a normally low sewage temperature (less than 10°C) [46].

Loading rates of up to 80 kg COD/m^3/day and more have been reached in laboratory-scale thermophilic reactors treating volatile fatty acids (VFA) and glucose [47,48], acetate and sucrose [49,50] and thermomechanical pulping white water [51].

As mentioned before, during the past half century, anaerobic treatment of food processing wastewaters has been widely studied and applied using mesophilic processes. In many cases, compared with single aerobic treatment, anaerobic treatment of food industry wastewaters is economical due to decreased excess sludge generation, decreased aeration requirement, compact installation, and methane energy generation. Thermophilic anaerobic treatment of food industry wastewaters, such as vinasse [52] and beer brewing [53] wastewaters, has been studied on laboratory and pilot scales.

The removal efficiencies of pollutants in these thermophilic reactors have been found to be very satisfactory. For example, in UASB reactors treating brewery wastewater and volatile fatty acids (VFA) at 55°C with loading rates of 20–40 kg COD/m^3/day, the COD removals reached over 80% in 50–60 days.

Thermophilic anaerobic processes have been used for the treatment of high solids content in vegetable waste (slop) from distillery [24–29 kg total solids (TS)/m^3] [54] and potato sludge [42 kg suspended solids (SS)/m^3] [55]. This technology has also been applied on a laboratory scale for the treatment of vegetable processing wastewaters in UASB reactors at 55°C, where the wastewater streams result from steam peeling and blanching of different processed vegetables (carrot, potato, and swede) [46]. For further information about this application, refer to the case studies.

Case Studies

Case Study I. This study examines the first EGSB operating in a German potato processing factory [13]. A wastewater flow of 1700 m^3/day passed through a screen and a fat separator into a 3518 m^3 balancing tank (weekly balance 30% constant retention) that also served as an acidification tank. Owing to the high retention time, it may be assumed that a nearly complete acidification took place, between 40 and 50% related to filtered COD. The methane reactor had a height of 14 m with a water volume of 750 m^3. The feeding of the reactor occurred at a constant rate from a conditioning tank (pump storage reservoir), where the recirculation flow mixed with the influent and the pH was adjusted to 6.6, using sodium hydroxide. The effluent from the methane reactor passed through a lamella separator for the removal of solids, which could also be placed between the acidification and methane reactor. The anaerobically treated wastewater was fed into the municipal wastewater treatment plant.

With an average filtered COD of 3500 mg/L in the influent, the efficiency of the anaerobic treatment was 70–85%, resulting in a biogas production with about 80% methane content. The concentration of filterable solids in the influent fluctuated between 500 and 2500 mg/L. According to operational experience in this anaerobic system, these values have not caused any considerable deterioration of the pellet sludge structure during operation.

Case Study II. This study addresses the anaerobic treatment of wastewater from the potato processing industry. A Biothane UASB reactor and Biobed EGSB reactor were installed at two different potato processing facilities in the Netherlands [14]. The first example is Smiths Food, which produces potato chips. They chose the Biothane UASB anaerobic treatment process for bulk COD removal from their wastewater and aerobic final treatment to meet the discharge limits. Figure 6.17 shows the flow scheme of this process. Coarse solids are removed in a parabolic screen (mesh size 1 mm). After this screen, the water enters a preclarifier designed at a surface load of 1 m/hour for removal of suspended solids and residual fat, oil, and grease. The settled solids are dewatered in a decanter and the water flows by gravity into a buffer tank of 400 m^3. From the buffer tank, the water is pumped to a conditioning tank for pH and temperature correction. Conversion of COD takes place in the Biothane UASB reactor. The total anaerobic plant has a COD removal efficiency of approximately 80%. The remaining COD and kjeldahl nitrogen is removed in the aerobic post-treatment.

The final COD concentration is less than 100 mg/L and the K$_j$-N concentration is less than 10 m/L. The final effluent is discharged to the municipal sewer. The performance of the combined UASB anaerobic-carousel aerobic wastewater treatment plant of Smiths Food is specified in Table 6.11.

The second example is Peka Kroef, which produces potato and vegetable-based half products for the salad industry in Europe. Owing to the specific characteristics of the resulting wastewater (low temperature, COD load fluctuations, COD composition fluctuations, high suspended solids concentration) an alternative for the conventional UASB, the EGSB technology, was tested. Extensive laboratory research showed good results with this type of anaerobic treatment at temperatures of 20–25°C.

Figure 6.18 shows the flow scheme of the EGSB process at Peka Kroef. The wastewaters from the potato and the vegetable processing plants follow similar but separate treatment lines. Coarse solids are removed in parabolic screens and most of the suspended solids in a preclarifier. The settled solids are dewatered in a decanter and the overflow is fed into a buffer tank of 1000 m^3. The anaerobic plant consists of two identical streets, giving Peka Kroef a high degree of operational flexibility. From the buffer tank the water is pumped to the conditioning tanks where the pH of the wastewater is controlled. Wastewater is then pumped to the Biobed EGSB reactors where the COD conversion takes place. The conditioning tanks and the anaerobic reactors operate under 100 mbar pressure and are made from FRP. It is possible to operate without a gasholder or a compressor. In addition, the EGSB reactor guarantees operating under a "zero odor emission" and supports the aerobic post-treatment in order to increase nitrogen and phosphorus removal for final discharge to the sewer. Initial results of this Biobed reactor in the potato processing industry are very promising.

Case Study III. In this study, vegetable processing wastewaters were subjected to thermophilic treatment in UASB reactors at 55°C [46]. The high-strength wastewater streams, coming from steam peeling and balancing of carrot, potato, and swede were used. The

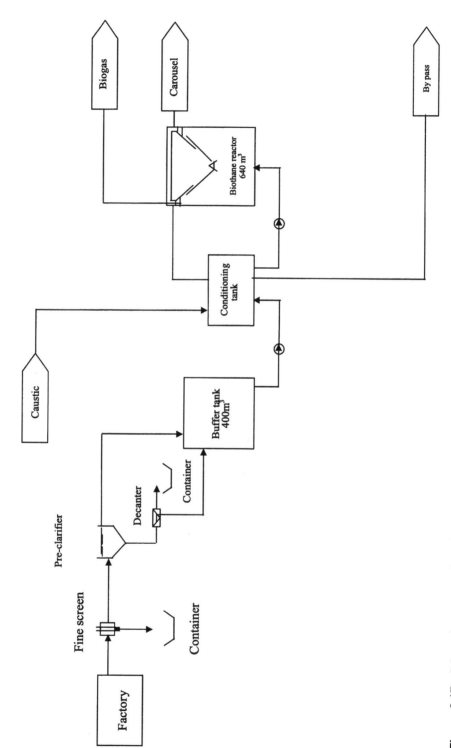

Figure 6.17 Schematic representation of the pretreatment stage and anaerobic treatment stage at Smiths Food (from Ref. 14).

Table 6.11 Performance Data of Wastewater Treatment Plant at Smiths Food (*Source*: Ref. 14)

Parameter	Unit	Value	Efficiency
Influent (data after primary clarifier)			
Flow	m³/day	517	
t-COD	mg/L	4566	
s-COD	mg/L	2770	
SS	mg/L	890	
Anaerobic effluent			
t-COD	mg/L	926	80%
s-COD	mg/L	266	90%
SS	mg/L	600	
TKN	mg/L	196	
Aerobic (final) effluent			
t-COD	mg/L	165	96%
s-COD	mg/L	60	98%
BOD	mg/L	17	
SS	mg/L	82	
TKN	mg/L	4	

Source: Ref. 14.

wastewater characteristics are summarized in Table 6.12. Carbohydrates contributed 50–60% of the COD in different wastewaters.

The reactors were inoculated with mesophilic granular sludge. Stable thermophilic methanogenesis with about 60% COD removal was reached within 28 days. During the 134 day study period, the loading rate was increased up to 24 kg COD/m³/day. High treatment efficiency of more than 90% COD removal and concomitant methane production of 7.3 m³ CH₄/m³/day were achieved.

The highest specific methanogenic activity (SMA) reported in this study was 1.5 g CH₄-COD/g VSS/day, while SMA$_s$ of 2.0 and 2.1 g COD/g VSS/day have been reported with sludge from 55°C UASB reactors treating other food industry wastewaters [52,53].

Key points of interest that can be drawn from this case study are as follows:

- The results support the previous finding that 55°C UASB reactors can be started with mesophilic granular sludge as inoculum.
- The anaerobic process performance was not affected by the changes in the wastewater due to the different processing vegetables.
- The achieved loading rates and COD removals demonstrated that the thermophilic high-rate anaerobic process is a feasible method to treat hot and concentrated wastewaters from vegetable processing.

Design Example 6. Design an anaerobic process reactor to achieve 85% removal of COD from a preclarified wastewater flow 360 m³/day (95,100 gal/day) resulting from a potato factory, depending on the steam peeling method, where total influent COD = 5000 mg/L, COD to be removed = 85%, pH = 6.2, and temperature = 30°C. The anaerobic process parameters are: sludge age (SRT) = 20 days (minimum), temperature = 35°C, a = 0.14 mg VSS/mg COD, b = 0.021 mg VSS/(mg VSS/day), K = 0.0006 L/(mg VSS/day), X_v = 5500 mg/L.

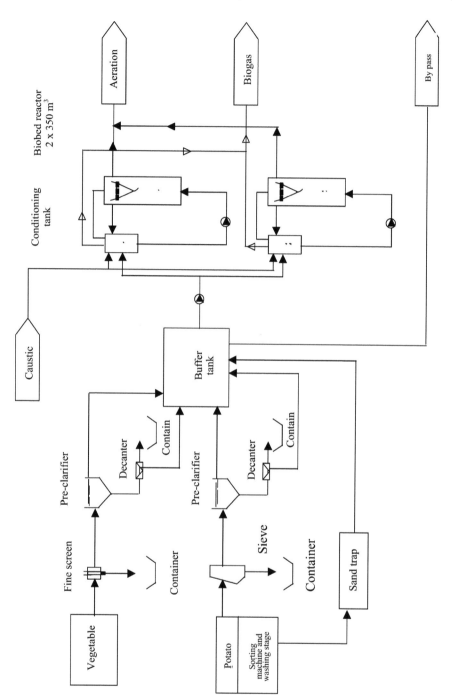

Figure 6.18 Schematic representation of the pretreatment stage and anaerobic treatment stage at Peka Kroef (from Ref. 14).

Table 6.12 Characteristics of Vegetable Processing Wastewaters after Removing Solids Through Settling and Drum

Unit	Raw material	Total COD (g/L)		Soluble COD (g/L)	
		Average	Range	Average	Range
Steam peeling	Carrot	19.4	17.4–23.6	17.8	15.1–22.6
	Potato	27.4	13.7–32.6	14.2	11.7–17.5
Blanching	Carrot	45.0	26.3–71.4	37.6	22.1–45.8
	Potato	39.6	17.0–79.1	31.3	10.9–60.6
	Swede	49.8	40.5–59.1	49.4	40.5–58.3

Source: Ref. 46.

Solution: Prior to anaerobic treatment of potato processing wastewater, it is important to provide favorable conditions for the anaerobic process through equalization and neutralization of the influent. Because the preclarified wastewater is almost neutral, there is no need for neutralization, and accordingly no need for correction of pH and temperature. Buffering of the wastewater is necessary here, to guarantee constant or near-constant flow. Total daily flow (average) = 360 m³/day. Flow (average after buffering) = 15 m³/hour, assuming that retention time is approximately 1 day in the buffer tank (balancing tank), with volume = 350 m³. Influent COD (average) = 5000 mg/L. (Exact calculation of the buffer tank requires data plotted as the summation of inflow vs. time of day.)

Digester volume from the kinetic relationship:

$$\text{Detention time: } t = \frac{S_r}{X_v \cdot K \cdot S} = \frac{5000 \times 0.85}{5500 \times 0.0006 \times 750} = 1.72 \, \text{day}$$

The digester volume is therefore:

$$\forall = (1.72 \, \text{day})(360 \, \text{m}^3/\text{day}) = 620 \, \text{m}^3 \, (0.1638 \, \text{MG})$$

Check SRT from the equation:

$$\text{SRT} = \frac{X_v t}{\Delta X_v} = \frac{X_v t}{a S_r - b X_v t}$$

$$= \frac{5500 \times 1.72}{0.14 \times 4250 - 0.021 \times 5500 \times 1.72} = 24 \, \text{day}$$

This is in excess of the recommended SRT of 20 days to ensure the growth of methane formers.

$$\text{Daily COD load} = 5000 \, \text{mg/L} \times 360 \, \text{m}^3/\text{day} \times \frac{1}{1000} = 1800 \, \text{kg COD/day}$$

$$\text{Design volumetric loading} = \frac{1800 \, \text{kg/day}}{620 \, \text{m}^3} = 3.0 \, \text{kg/m}^3 \cdot \text{day}$$

This value is acceptable for a conventional anaerobic contact process. In the case of a UASB reactor, the organic loading can be easily increased to 10 kg/m³/day, that is, it is sufficient to have only one-third or less of the calculated volume (about 200 m³), to achieve the same performance.

In the case of the expanded granular sludge bed (EGSB) reactor, the organic loading can be increased up to $30 \, \text{kg/m}^3/\text{day}$, where the required volume becomes only:

$$\frac{1800 \, \text{kg/day}}{30 \, \text{kg/m}^3 \cdot \text{day}} = 60 \, \text{m}^3$$

The sludge yield from the process is:

$$\Delta X_v = aS_r - bX_v t$$
$$= (0.14)(4250) - (0.021)(5500)(1.72) = 396.34 \, \text{mg/L}$$

$$\Delta X_v = 396.34 \, \text{mg/L} \times 360 \, \text{m}^3/\text{day} \times \frac{1}{1000}$$

$$= 142.7 \, \text{kg/day} \, (314 \, \text{lb/day})$$

Gas production

$$G = 0.351(S_r - 1.42\Delta X_v)$$

where $G = \text{m}^3$ of CH_4 produced/day

$$G = 0.351[(4250)(360) - (1.42)(142.7)]$$
$$= 0.351 \, (1530 - 202.63) = 465 \, \text{m}^3 CH_4/\text{day}$$

or

$$G = 5.62(S_r - 1.42\Delta X_v)$$

where $G = \text{ft}^3$ of CH_4 produced/day

$$G = 5.62[(4250)(0.0951 \, \text{MG/day})(8.34) - (1.42)(314)]$$
$$= 16{,}433.5 \, \text{ft}^3/\text{day} \, (465 \, \text{m}^3/\text{day})$$

Heat required can be estimated by calculating the energy required to raise the influent wastewater temperature to $35°C$ ($95°F$) and allowing $1°F$ ($0.56°C$) heat loss per day of detention time. Average wastewater temperature $= 30°C$ ($86°F$) and heat transfer efficiency $= 50\%$.

$$\text{BTU}_{\text{req.}} = \frac{W(T_i - T_e)}{E} \times (\text{specific heat})$$
$$= \frac{(95{,}100 \, \text{gal/day})(8.34 \, \text{lb/gal})(95° + 1.72°F - 86°)}{0.5} \times \left(\frac{1 \, \text{B}_{\text{tu}}}{1 \, \text{lb.°F}}\right)$$
$$= 17{,}004{,}792 \, \text{BTU} \, (17{,}940{,}055 \, \text{KJ})$$

The heat available from gas production is $\text{BTU}_{\text{avail.}} = (16{,}433.5 \, \text{ft}^3 \, CH_4/\text{day}) \, (960 \, \text{BTU} \, \text{ft}^3 \, CH_4) = 15{,}776{,}160 \, \text{BTU/day}$ ($16{,}643{,}850 \, \text{kJ/day}$). External heat of $17{,}004{,}792 - 15{,}776{,}160 = 1{,}228{,}832 \, \text{BTU/day}$ ($1{,}296{,}207 \, \text{kJ/day}$) should be supplied to maintain the reactor at $35°C$ ($95°F$).

Nutrients required: the nitrogen required is:

$$N = 0.12 \, \Delta X_v = 0.12 \times 142.7 \, \text{kg/day} = 17.124 \, \text{kg/day} \, (37.673 \, \text{lb/day})$$

The phosphorus required is:

$$P = 0.025\Delta X_v = 0.025 \times 142.7 \, \text{kg/day} = 3.568 \, \text{kg/day} \, (7.85 \, \text{lb/day})$$

Remarks:

1. The effluent from the anaerobic plant alone does not meet the national minimum discharge limits because of the high values of residual COD (15% = 750 mg/L). Therefore, it is recommended here to handle the anaerobic process effluent in an aerobic post-treatment (such as activated sludge). The final effluent of this combination of anaerobic and aerobic treatment processes can certainly be discharged to the central sewerage system or reused within the factory.

2. The equalization (buffering) was indicated in this example to dampen the fluctuations in potato processing wastewater flow that occur on a daily or longer term basis. It must be noted that optimum equalization of both flow and concentration are not achievable in a single process. To equalize flows, the buffer tank at certain times should be empty. To equalize concentration, the tank should always be full. Nevertheless, a tank that equalizes flows will also produce some reduction in peak concentration. Optimally, the organic loading to the anaerobic process reactor is constant over a 24-hour period. Equalization of flow was intended to be considered and simplified in this design example.

Advanced Treatment

Advanced wastewater treatment comprises a large number of individual treatment processes that can be utilized to remove organic and inorganic pollutants from secondary treated wastewater. The following treatment processes presented can be used to meet the effluent discharge requirements for potato processing plants. These may include suspended solids, BOD, nutrients, and COD.

Microstraining. Microstrainers consist of motor-driven drums that rotate about a horizontal axis in a basin, which collects the filtrate. The drum surface is covered by a fine screen with openings ranging from 23–60 μm. It has been reported that effluent suspended solids and BOD from microstrainers following an activated sludge plant have a ranges of 6–8 mg/L and 3.5–5 mg/L, respectively [56].

The head loss of the drum is less than 12–18 in (30–46 cm) of water. Peripheral drum speeds vary up to 100 ft/min (30.5 m/min) with typical hydraulic loadings of 0.06–0.44 m/min (1.5–10 gal/ft^2-min) on the submerged area; the backwash flow is normally constant and ranges up to 5% of the product water [57]. Periodic cleaning of the drum is required for slime control.

Granular Media Filtration. Granular filtration employing mixed media or moving bed filters plays an important role in improving the secondary effluent quality, where most of the BOD is found in bacterial solids. Therefore, removal of the suspended solids greatly improves the effluent quality. Granular filtration is generally preferred to microstraining, which is associated with greater operational problems and lower solids removal efficiencies.

Effective filter media sizes are generally greater than 1 mm. Filtration rates range from 0.06 to 0.5 m/min (1.5 to 12 gal/ft^2-min) with effluent suspended solids from 1–10 mg/L. This represents a reduction of 20 to 95% from the concentration in the filter influent [57,58]. Secondary effluent should contain less than 250 mg/L of suspended solids in order to make filtration more suitable [11]. In the case of higher concentrations of suspended solids, the secondary effluent should be first led to polishing ponds (maturation ponds) or subjected to chemical coagulation and sedimentation.

Chemical Coagulation Followed by Sedimentation. Phosphorus is a nutrient of microscopic and macroscopic plants, and thus can contribute to the eutrophication of surface waters. Phosphorus may be removed biologically or chemically. In some cases, chemicals may be added to biological reactors instead of being used in separate processes while in others, biologically concentrated phosphorus may be chemically precipitated. Chemical phosphorus removal involves precipitation with lime, iron salts, or alum. Lime should be considered for this purpose if ammonia removal is also required for pH adjustment. For low effluent phosphorus concentrations, effluent filtration may be required due to the high phosphorus content of the effluent suspended solids.

Whatever coagulant is employed, a large quantity of sludge is produced. Sludge lagoons can be considered as an economical solution to sludge disposal, although this treatment requires considerable land area.

Improved removal of phosphorus without any chemical addition can be obtained by a biological process that employs an anoxic or anaerobic zone prior to the aeration zone. When this process is used to maximize phosphate removal (sometimes called a sequencing batch reactor), it is possible to reduce the phosphorus content to a level of about 1 mg/L, with no chemical addition.

The principle of bio-P removal is the exposure of the organisms to alternating anaerobic and aerobic conditions. This can be applied with or without nitrogen removal. The alternating exposure to anaerobic and to aerobic conditions can be arranged by recirculation of the biomass through anaerobic and aerobic stages, and an anoxic stage if nitrogen removal is also required. General flowsheets of these processes are shown in Figure 6.19.

As for potato processing wastewater, which often contains high concentrations of nutrients (N and P compounds), it is recommended here to apply biological phosphorus removal including an anoxic stage for the advanced treatment.

The abovementioned role of chemical coagulation may be followed by sedimentation in the reduction of nutrients. This method can also be applied to treat potato processing wastes in general [59]. Coagulating and flocculating agents were added to wastewater from abrasive-peeled, lye-peeled, and steam-peeled potato processing. Total suspended solid and COD concentrations were significantly reduced with chemical and polymer combination treatments, at adjusted pH levels.

Nitrification–Denitrification. Based on water quality standards and point of discharge, municipal treatment plants may be: (a) free from any limits on nitrogen discharges, (b) subject to limits on ammonia and/or TKN, (c) subject to limits on total nitrogen. Nitrogen can be removed and/or altered in form by both biological and chemical techniques. A number of methods that have been successfully applied can be found in many publications. Biological removal techniques include assimilation and nitrification–denitrification. Occasionally, nitrification is adequate to meet some water quality limitations where the nitrogenous oxygen demand (NOD) is satisfied and the ammonia (which might be toxic) is converted to nitrate. According to USEPA publications, the optimum pH range for nitrification has been identified as between 7.2 and 8.0. Regarding the effect of temperature, it has been noted that nitrification is more affected by low temperature than in the case of BOD removal [60].

Nitrification can be achieved in separate processes after secondary treatment or in combined processes in which both BOD and NOD are removed. In combined processes the ratio of BOD to TKN is greater than 5, while in separate processes the ratio in the second stage is less than 3 [57].

Denitrification is a biological process that can be applied to nitrified wastewater in order to convert nitrate to nitrogen. The process is anoxic, with the nitrate serving as the electron acceptor for the oxidation of organic material.

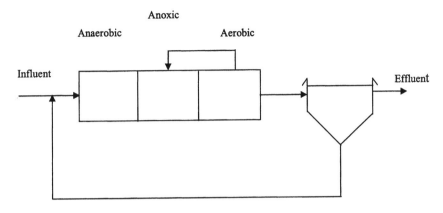

With nitrification-denitrification

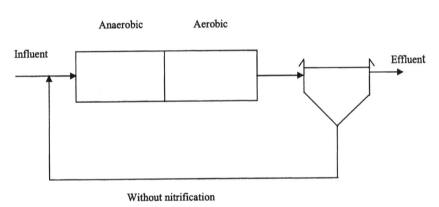

Without nitrification

Figure 6.19 General flow sheets of biological phosphorus removal with and without nitrification–denitrification (from Ref. 24).

There is a variety of alternatives for the denitrification process such as suspended growth and attached growth systems with and without using methanol as a carbon source. Chemical nitrogen-removal processes generally involve converting the nitrogen to a gaseous form (N_2) and ammonia (NH_3). The processes of major interest include break-point chlorination, ion exchange, and air stripping. Natural zeolitic tuffs play an important role as ion exchange media for ammonium and phosphate removal through columns or batch reactors [61], where the total volume treated between generation cycles depends on the ammonium concentration in the wastewater and the allowed concentration in the effluent. The wastewater itself can be stripped of ammonia if it is at the requisite pH (10.5–11.5) and adequate air is provided. The feasibility of stripping the wastewater itself depends on whether the necessary pH can be achieved at moderate cost. The air stream carries with it the stripped ammonia to be released to the atmosphere. When the ammonia is dissolved in the solution, it forms the ammonium salt of the acid, which has an economic value as a fertilizer to the soil.

Regarding land-application systems for treatment of potato processing wastewaters, they may be satisfactory regarding nitrogen removal with no need for additional biological or chemical treatment.

Membrane Technology. Membrane technology encompasses a wide range of separation processes from filtration and ultrafiltration to reverse osmosis. Generally, these processes produce a very high quality effluent defined as membrane filtration and refer to systems in which discrete holes or pores exit the filter media, generally in the order of 10^2–10^4 nm or larger. The difference in size between the pore and the particle to be removed determines the extent of filtration efficiency. The various filtration processes in relation to molecular size can be found in Ref. 24.

The criteria for membrane technology performance are related to the degree of impermeability (the extent of membrane's detention of the solute flow) or the degree of permeability (the extent of membrane's allowance of the solute flow). The design and operating parameters for a reverse osmosis system are presented in detail in Ref. 62.

Regarding potato processing wastewaters, reverse osmosis and ultrafiltration have been used for treating wastewater for the recovery of sweet potato starch [63]. They may also be successful for application within in-plant treatment and recycling systems. Other advanced treatment methods used for various industrial wastewaters such as activated carbon adsorption, deep well injection, and chlorination, are not suitable for potato processing wastewater treatment due to their high costs of application.

It is worth mentioning that important research has been carried out regarding the treatment of potato processing wastewaters by the activated carbon adsorption process used as an advanced treatment method. It was reported that activated carbon adsorption treatment following complete mix activated sludge treatment removed 97% COD from primary settled potato processing wastewaters with an effluent COD of 24 mg/L [17]. In addition, it was concluded that powdered activated carbon was more effective than granular activated carbon in removing COD from activated sludge treated effluents.

6.4.2 Bases of Potato Processing Effluent Treatment

For an existing plant, it is necessary to measure the flow of all waste streams and determine the quantity and character of the pollutants found in these flows. The reduction of wastewater discharge into the final plant effluent and the reduction of water flow throughout the plant is of major importance. For a proposed new plant for which the waste treatment units must be designed, information may be found in the literature for a similar installation. In most cases, however, a reasonable estimate of the waste flow may be determined from the estimated capacity of the plant, the recovery of product expected, and the type of screening and clarification equipment to be installed. It is necessary to have accurate estimates of water usage and methods of reuse in application. For preliminary estimates, it can be assumed that a lb (or 1 kg) of dry potato solids exerts a BOD of 0.65 lb (or 0.65 kg) and a COD of 1.1 lb (or 1.1 kg) [11].

6.5 BYPRODUCT USAGE

6.5.1 In-Plant Usage of Potato Scraps

Plants processing French fries have developed additional product lines to utilize small potatoes (chopped or sliced), cutter scraps, slivers, and nubbins. These are processed similarly to French fries and include potato patties, mashed or whipped potatoes, diced potatoes, potato puffs, and hash browns [64].

6.5.2 Potato Peels

Approximately two million tons per year of potato peels are produced from potato processing as byproducts [65]. Potato peels provide a good source of dietary fiber, particularly when processed by a lye-peeling technique [66]. Potato peels contain 40 g dietary fiber/100 g dry matter, depending on the variety of potato processed and the method of peeling [67]. Application of extruded and unextruded potato peels as a source of dietary fiber in baked goods has been evaluated [1]. Acceptable muffins were made with a 25% replacement potato peel for wheat flour. Potato peels were also found to prolong muffin shelf-life by controlling lipid oxidation [65]. Extrusion cooking of potato peels affects the color of baked goods, and some physical and chemical properties of the peels [67]. Potato peels have also been used in limited quantities in a commercial snack food potato skin type product.

6.5.3 Potato Processing Wastes as Soil Conditioner

Potato processing solid wastes are often applied to agricultural land as a disposal medium. Research supports this method [68]. Solid potato processing wastes containing nitrogen are obtained by filtering or centrifuging the settled solids from the primary clarifiers. Wastes are applied to land and used for crops, which utilize the applied nitrogen. The soil does not accumulate the nitrogen or other organic waste and becomes increasingly fertile with continued wastewater application. Additionally, potato processing wastewater was found to be effective in promoting corn growth as effectively as commercial ammonium nitrate fertilizers, when applied at optimum nitrogen levels [69]. Applying wastewater and solid wastes from potato processing provides an effective method of applying reusable nutrients that would be otherwise wasted, and thus reduces pollution levels in municipal waterways.

6.5.4 Potato Wastes as Substrate for Organic Material Production

Potato wastes have also been evaluated as a potential source from which to produce acetone, butanol, and ethanol by fermentation techniques [70]. This application of biotechnology in membrane extraction resulted in a procedure to extract a biofuel that utilizes potato wastes as a renewable resource.

6.5.5 Cattle Feed

Filter cakes and dry potato peels are used as an excellent carbohydrate source in cattle feed. Using potato wastes instead of corn in cattle feed does not affect the metabolic state or milk status of the cattle [71]. Typically, potato wastes are fed in a dry, dewatered form. The use of wet potato wastes in cattle feed has been investigated to reduce drying expenditures. Wet potato processing wastes can be introduced into cattle feed up at to 20% without negative results.

The issue of dry vs. wet application of potato processing wastes was also explored. Again, dry potato wastes are expensive due to the drying processes used to stabilize the wastes. Wet wastes must be used quickly and within a close proximity to the potato processing wastes site due to microbial and enzymatic spoilage of the waste. Barley straw has been investigated as silage material to be mixed with wet potato wastes to absorb excess moisture [72]. Problems encountered with this procedure are due to elevated pH levels being attained following five weeks of storage. Elevated pH levels can permit growth of toxigenic bacteria.

Carbohydrate-rich potato wastes can also be converted to protein for additional nutrients for animal feed [1]. Research indicates that starchy substances such as potato wastes can be converted to "microbial biomass protein" by digestion with a amylolytic, acidophilic, thermophilic fungus. The fungus hydrolyzes starch, under specific high-temperature/low-pH conditions. Utilizing nitrogen in the potato wastes, the fungus produces protein which is filtered, and has been shown to be nutritionally effective in animal feeding trials if supplemented with methionine. Limitations of this process include the short time that wastes are viable for this treatment. Wastes can become toxic to fungus during storage. Potato and corn single-cell protein was also used in place of soybean meal as a source of supplemental protein in cattle feed. Results indicate the substitution can be made, if in conjunction with soybean meal protein for growing steers [73].

6.5.6 Potato Pulp Use

Processing potato starch results in potato pulp as a major byproduct, particularly in Europe. Research indicates that potato pulp can be fractionated to produce several commercially viable resources. Pectin and starch can be isolated, as well as cellulase enzyme preparation [74]. It was hypothesized that ethanol production would be feasible, but low sugar concentration prevented this. Potato pulp may also have applications for reuse in the following industries: replacement of wood fiber in paper making, and as a substrate for yeast production and B_{12} production [74]. Potato pulp isolated from potato starch production can be isolated and sold as pomace [75]. Protein can also be isolated from the starch processing wastewater and sold as fractionated constituents [74].

In summary, new technologies have served to minimize potato processing wastes and appropriate means of utilizing the rich byproducts are still under research. The vast quantities of wastes will continue to be minimized and byproducts have found new applications as renewable resources and potential energy sources. All of these goals will continue to be realized as research leads to the development of unique technologies to treat wastes, minimize the impact on the environment, reduce use of valuable natural resources, and reduce the impact of waste effluent.

REFERENCES

1. Stevens, C.A.; Gregory, K.F. Production of microbial biomass protein from potato processing wastes by cephalosporium eichhorniae. Appl. Environ. Microbiol. **1987**, *53*, 284–291.
2. Vegt, A.; Vereijken. M. Eight year full-scale experience with anaerobic treatment of potato processing effluent, In *Proceedings of the 46th Industrial Waste Conference*, Purdue University, West Lafayette, IN, 1992; 395–404.
3. Guttormsen, K.G.; Carlson, D.A. *Current Practice in Potato Processing Waste Treatment*, Water Pollution Research Series, Report No. DAST-14; Federal Water Pollution Control Federation, U.S. Department of the Interior: Washington, DC, 1969.
4. Talburt, W.F.; Smith, O. *Potato Processing*; Van Nostrand Reinhold Company: New York, 1967.
5. Gray, H.F.; Ludwig, H.F. Characteristics and treatment of potato dehydration wastes. Sewage Works, **1943**, *15*, 1.
6. Cooley, A.M.; Wahl, E.D.; Fossum, G.O. Characteristics and amounts of potato wastes from various process stream. In *Proceedings of the 19th Industrial Waste Conference*, Purdue University, West Lafayette, IN, 1964; 379–390.
7. Abeling, U.; Seyfried, C.F. Anaerobic-aerobic treatment of potato-starch wastewater. Water Sci. Technol. **1993**, *28* (2), 165–176.
8. Hadjivassilis, I.; Gajdos, S.; Vanco, D.; Nicolaou, M. Treatment of wastewater from the potato chips and snacks manufacturing industry. Water Sci. Technol. **1997**, *36* (2–3), 329–335.

9. Kadlec, R.H. Deterministic and stochastic aspecting constructed wetland performance and design. Water Sci. Technol. **1997**, *35* (*5*), 149–156.

10. Hung, Y.T. Tertiary treatment of potato processing waste by biological activated carbon process. Am. Potato J. **1983**, *60* (*7*), 543–555.

11. Pailthorp, R.E.; Filbert, J.W.; Richter, G.A. Treatment and disposal of potato wastes. In *Potato Processing*; Talburt, W.F., Smith, O., Eds.; Van Nostrand Reinhold Co.: New York, 1987; 747–788.

12. USEPA. *Development Document for Proposed Effluent Limitation Guidelines and New Source Performance Standards for the Citrus, Apple and Potato Segment of the Canned and Preserved Fruits and Vegetables Processing Plant Source Category*, EPA-440/1–73/027; U.S. Environmental Protection Agency: Washington, DC, 1973.

13. Austerman-Haun, U.; Mayer, H.; Seyfried, C.F.; Rosenwinkel, K.H. Full scale experiences with anaerobic/aerobic treatment plants in the food and beverage industry. Water Sci. Technol. **1999**, *40*(*1*), 305–325.

14. Zoutberg, G.R.; Eker, Z. Anaerobic treatment of potato processing wastewater. Water Sci. Technol. **1999**, *40* (*1*), 297–304.

15. Kadlec, R.H.; Burgoon, P.S.; Henderson, M.E. Integrated natural systems for treating potato processing wastewater. Water Sci. Technol. **1997**, *35* (*5*), 263–270.

16. Hung, Y.T. Batch mesophilic anaerobic digestions of potato wastewaters. *Am. Potato J.* **1989**, *66* (*7*), 437–447.

17. Hung, Y.T. Treatment of potato processing wastewaters by activated carbon adsorption process. Am. Potato J. **1984**, *61* (*1*), 9–22.

18. Olson, O.O.; Van Heuvelen, W.; Vennes, J.W. Experimental treatment of potato wastes in North Dakota. In *Proceeding of the International Symposium, Utilization and Disposal of Potato Wastes*; New Brunswick Research and Productivity Council: New Brunswick, Canada, 1965; 316–344.

19. Dickey, H.C.; Brugman, H.H.; Highlands, M.E.; Plummer, B.E. The use of by-products from potato starch and potato processing. In *Proceeding of the International Symposium, Utilization and Disposal of Potato Wastes*; New Brunswick Research and Productivity Council: New Brunswick, Canada, 1965; 106–121.

20. Hung, Y.T.; Priebe, B.D. *Biological Activated Carbon Process for Treatment of Potato Processing Wastewater for In-Plant Reuse*, Report No. 81-10-EES-01; Engineering Experimental Station, University of North Dakota: Grand Forks, North Dakota, 1981.

21. Loehr, R.C. Biological processes. In *Agricultural Wastes Management Problems, Processes and Approaches*; Academic Press: New York, 1974; 129–182.

22. Bertola, N.; Palladino, L.; Bevilacqua, A.; Zaritzky, N. Optimisation of the design parameters in an activated sludge system for the wastewater treatment of a potato processing plant. Food Eng. **1999**, *40*, 27–33.

23. Davis, M.L.; Cornwell, D.A. *Introduction to Environmental Engineering*, 2nd Ed.; McGraw-Hill, International: New York, 1991.

24. Eckenfelder, W.W. *Industrial Water Pollution Control*, 2nd Edition; McGraw-Hill, International: New York, 1989.

25. Fossum, G.O.; Cooley, A.M.; Wahl, G.D. Stabilization ponds receiving potato wastes with domestic sewage. In *Proceedings of the 19th Industrial Waste Conference*, Purdue University, West Lafayette, IN, 1964; 96–111.

26. Bastian, R.K.; Hammer, D.A. The use of constructed wetlands for wastewater treatment and recycling. In *Constructed Wetlands for Water Quality Improvement*; Moshiri, G.A; Ed.; Lewis Publishers: Boca Raton, FL 1993; 59–68.

27. de Zeeuv, W.; Heijnen, G.; de Vries, J. Reed bed treatment as a wastewater (post) treatment alternative in the potato starch industry. In *Constructed Wetlands in Water Pollution Control (Adv. Water Pollut, Control No. 11)*; Cooper, P.F., Findlate, B.C., Eds.; Pergamon Press: Oxford, UK 1990; 551–554.

28. Van Oostrom, A.J. Nitrogen removal in constructed wetlands treating nitrified meat processing effluent. Water Sci. Technol. **1995**, *33* (*3*), 137–148.

29. Van Oostrom, A.J.; Russel, J.M. Denitrification in constructed wastewater wetlands receiving high concentrations of nitrate. Water Sci. Technol. **1992**, *29* (*4*), 7–14.

30. Van Oostrom, A.J.; Cooper, R.N. Meat processing effluent treatment in surface-flow and gravel-bed constructed wastewater wetlands. In *Constructed Wetlands in Water Pollution Control (Adv. Water Pollut. Control No. 11)*; Cooper, P.F., and Findlater, B.C., Eds.; Pergamon Press: Oxford, UK, 1990; 321–332.

31. Haberl, R.; Partler, R.; Mayer, H. Constructed wetlands in Europe. Water Sci. Technol. **1995**, *33* (*3*), 305–315.

32. Burka, U.; Lawrence, P.C. A new community approach to wastewater treatment with higher plants. In *Constructed Wetlands for Water Pollution Control (Adv. Water Pollut, Control No. 11)*; Cooper, P.F., Findlater, B.C., Eds.; Pergamon Press: Oxford, UK, 1990; 359–371.

33. Bahlo, K.E.; Wach, F.C. Purification of domestic sewage with and without faeces by vertical intermittent filtration in reed and rush beds. In *Constructed Wetlands for Water Pollution Control (Adv. Water Pollut, Control No. 11)*; Cooper, P.F., Findlater, B.C., Eds.; Pergamon Press: Oxford, UK, 1990; 215–221.

34. Burgoon, P.S.; Kadlec, R.H.; Henderson, M. Treatment of potato processing wastewater with engineered natural systems. Water Sci. Technol. **1999**, *40* (*3*), 211–215.

35. Cocci, A.A.; Page, I.C.; Grant, S.R.; Landine, R.C. Low-rate anaerobic treatment of high-strength industrial wastewater: ADI-BVF case histories. In *Seminar of Anaerobic Treatment for Industrial Wastes*; East Syracuse, New York, 1997.

36. Malina, J.F.; Pohland, F.C. *Design of Anaerobic Processes for the Treatment of Industrial and Municipal Wastes*; Technomic Publishing Company, Inc.: Lancaster, PA, 1992.

37. Metcalf and Eddy, Inc. *Wastewater Engineering*; McGraw-Hill: New York, 1991.

38. Liyah, R.Y.; Hung, Y.T. Bio-augmented activated sludge treatment of potato wastewaters. Acta Hydrochim Hydrobiol. **1988**, *16* (*2*), 223–230.

39. Hung, Y.T.; Howard, H.L.; Javaid, A.M. Effect of bio-augmentation on activated sludge treatment of potato wastewater. Environ. Stud. **1994**, *45*, 98–100.

40. Hung, Y.T.; Jen, P.C. Anaerobic filter followed by activated sludge process with bio-augmentation for combined potato and sugar wastewater treatment. In *Proceedings of 1987 Food Processing Waste Conference*, Atlanta, Georgia, September 1–2, 1987.

41. Shih, J.K.C.; Hung, Y.T. Biological treatment of potato processing wastewaters. Am. Potato J. **1987**, *64* (*9*), 493–506.

42. Chambers, D.A. Improving removal performance reliability of a wastewater treatment system through bio-augmentation. In *Proceedings of the 36th Industrial Waste Conference*, Purdue University, West Lafayette, IN, 1981; 631.

43. Peavy, H.S.; Rowe, D.R.; Tchobanoglous, C. *Environmental Engineering* 1st Ed.; McGraw-Hill, International: New York, 1985.

44. USEPA. *Ground-Water and Leachate Treatment Systems*, EPA/625/R-94/005; Environmental Protection Agency: Washington, DC, 1995.

45. Hulshoffpol, L.; Hartlieb, E.; Eitner, A.; Grohganz, D. GTZ sectorial project promotion of anaerobic technology for the treatment of municipal and industrial sewage and wastes. In *Proceedings of the 8th International Conference on Anaerobic Digestions*; Sendai, Japan, **1997**, *2*, 285–292.

46. Lepisto, S.S.; Rintala, J.A. Start-up and operation of laboratory-scale thermophilic upflow anaerobic sludge blanket reactors treating vegetable processing wastewaters. Chem. Technol. Biotechnol. **1997**, *68*, 331–339.

47. Wiegant, W.M.; de Man, A.W.A. Granulation of biomass in the thermophilic upflow anaerobic sludge blanket reactor treating acidified wastewaters. Biotechnol. Bioeng. **1986**, *28*, 718–727.

48. Wiegant, W.M.; Lettinga, G. Thermophilic anaerobic digestion of sugars in upflow anaerobic sludge blanket reactors. Biotechnol. Bioeng. **1985**, *27*, 1603–1607.

49. Van Lier, J.B.; and Lettinga, G. Limitations of thermophilic anaerobic wastewater treatment and the consequences for process design. In *Proceedings of International Meeting on Anaerobic Processes for Bioenergy and Environment*, Copenhagen, Denmark, January 25–27, 1995, 1995; Section 16.

50. Uemura, S.; Harada, H. Microbial characteristic of methanogenic sludge consortia developed in thermophilic UASB Reactors. Appl. Microbiol. Biotechnol. **1995**, *39*, 654–660.

51. Rintala, J.; Lepistö, S. Anaerobic treatment of thermomechanical pulping whitewater at 35–70°C. Water Res. 1992, *26*, 1297–1305.

52. Souza, M.E.; Fuzaro, G.; Polegato, A.R. Thermophilic anaerobic digestion of vinasse in pilot plant UASB reactor. Water Sci. Technol. **1992**, *25* (*7*), 213–222.

53. Ohtsuki, T.; Tominaga, S.; Morita, T.; Yoda, M. Thermophilic UASB system start-up and management-change in sludge characteristics in the start-up procedure using mesophilic granular sludge. In *Proceedings of Seventh International Symposium on Anaerobic Digestion*, Cape Town, South Africa, January 23–27, 1994; 348–357.

54. Garavini, B.; Mercuriali, L.; Tilche, A.; Xiushan, Y. Performance characteristics of a thermophilic full scale hybrid reactor treating distillery slops. In *Poster-Papers of the Fifth International Symposium on Anaerobic Digestion,* Bologna, Italy, 22–26 May, 1988; Tilche, A; Rozzi, A; Eds.; 509–515.

55. Trösch, W.; Chmiel, H. Two-stage thermophilic anaerobic digestion of potato wastewater-experience with laboratory, pilot scale and full-scale plants. In *Poster-Papers of the Fifth International Symposium on Anaerobic Digestion,* Bologna, Italy, 22–26 May, 1988; Tilche, A; Rozzi, A; Eds.; 599–602.

56. Lynam, B.; Ettelt, G.; McAloon, T. Tertiary treatment at Metro Chicago by means of rapid sand filtration and microstrainers. Water Pollut. Control Feder. **1969**, *41*, 247.

57. McGhee, T.J. *Water Supply and Sewerage*, 6th Ed.; McGraw-Hill, International: New York, 1991.

58. Ripley, P.G.; Lamb, G. Filtration of effluent from a biological-chemical system. Water Sewage Works. **1973**, *12* (*2*), 67.

59. Karim, M.I.A.; Sistrunk, W.A. Treatment of potato processing wastewater with coagulating and polymeric flocculating agents. Food Sci. **1985**, *50*, 1657–1661.

60. Sutton, P.M. et al. Efficacy of biological nitrification. Water Pollut. Control Feder. **1975**, *47*, 2665.

61. Awad, A.; Garaibeh, S. Nutrients removal of biological treated effluent through natural zeolite. In *Proceedings of the Second Syrian-Egyptian Conference in Chemical Engineering*; Al-Baath University: Homs, Syria, 20–22 May 1997; 616–640.

62. Agardi, F.J. Membrane processes. In *Process Design in Water Quality Engineering*; Tackston, E.L.; and Eckenfelder, W.W., Eds.; Jenkins Publishing Co.: Austin, Texas, 1972.

63. Chiang, B.H.; Pan, W.D. Ultrafiltration and reverse osmosis of the wastewater from sweet potato starch process. Food Sci. **1986**, *51* (*4*), 971–974.

64. Talburt, W.F.; Weaver, M.L.; Renee, R.M.; Kueneman, R.W. Frozen french fries and other frozen potato products. In *Potato Processing*; Talburt, W.F, Smith, O; Eds.; Van Nostrand Reinhold Co.: New York, 1987; 491–534.

65. Arora, A.; Camire, M.E. Performance of potato peels in muffins and cookies. Food Res. Inter. **1994**, *27*, 15–22.

66. Smith, O. Potato Chips. In *Potato Processing*; Talburt, W.F, and Smith, O; Eds.; Van Nostrand Reinhold Co.: New York, 1987; 371–474.

67. Arora, A.; Jianxin, Z.; Camire, M.E. Extruded potato peel functional properties affected by extrusion conditions. Food Sci. **1993**, *58* (*2*), 335–337.

68. Smith, J.H. Decomposition of potato processing wastes in soil. *Environ. Qual.* **1986**, *15*(*1*), 13–15.

69. Smith, J.H.; Hayden, C.W. Nitrogen availability from potato processing wastewater for growing corn. Environ. Qual. **1984**, *13* (*1*), 151–158.

70. Grobben, N.G.; Egglink, G.; Cuperus, F.P.; Huizing H.J. Production of acetone, butanol and ethanol (ABE) from potato wastes: fermentation with integrated membrane extraction. Appl. Microbiol. Biotechnol. **1993**, *39*, 494–498.

71. Onwubuwmell, C.; Huber, J.T.; King, K.J.; Johnson, C.O.L.E. Nutritive value of potato processing wastes. Dairy Sci. **1985**, *68* (*5*), 1207–1214.

72. Sauter, E.A.; Hinman, D.D.; Parkinson, J.F. The lactic acid and volatile fatty acid content and in vitro organic matter digestibility of silages made from potato processing residues and barley. Anim. Sci. **1985**, *60* (*5*), 1087–1094.

73. Hsu, J.C.; Perry, T.W.; Mohler, M.T. Utilization of potato-corn biosolids single-cell protein and potato-corn primary waste by beef cattle. Anim Sci. **1984**, *58* (*5*), 1292–1299.
74. Kingspohn, U.; Bader, J.; Kruse, B.; Kishore, P.V.; Schugerl, K.; Kracke-Helm, H.A.; Likidis, Z. Utilization of potato pulp from potato starch processing. Proc. Biochem. **1993**, *28*, 91–98.
75. Treadway, R.H. Potato Starch. In *Potato Processing*; Talburt, W.F, Smith, O. Eds.; Van Nostrand Reinhold Co.: New York, New York, 1987; 647–666.

7
Soft Drink Waste Treatment

J. Paul Chen and Swee-Song Seng
National University of Singapore, Singapore

Yung-Tse Hung
Cleveland State University, Cleveland, Ohio, U.S.A.

7.1 INTRODUCTION

The history of carbonated soft drinks dates back to the late 1700s, when seltzer, soda, and other waters were first commercially produced. The early carbonated drinks were believed to be effective against certain illnesses such as putrid fevers, dysentery, and bilious vomiting. In particular, quinine tonic water was used in the 1850s to protect British forces abroad from malaria.

The biggest breakthrough was with Coca-Cola, which was shipped to American forces wherever they were posted during World War II. The habit of drinking Coca-Cola stayed with them even after they returned home. Ingredients for the beverage included coca extracted from the leaves of the Bolivian Coca shrub and cola from the nuts and leaves of the African cola tree. The first Coca-Cola drink was concocted in 1886. Since then, the soft drink industry has seen its significant growth.

Table 7.1 lists the top 10 countries by market size for carbonated drinks, with the United States leading the pack with the largest market share. In 1988 the average American's consumption of soft drinks was 174 L/year; this figure has increased to approximately 200 L/year in recent years. In 2001, the retail sales of soft drinks in the United States totaled over $61 billion. The US soft drink industry features nearly 450 different products, employs

Table 7.1 Top Ten World Market Size in Carbonated Soft Drinks, 1988

Rank	Country	1000 million liters
1	United States	42.7
2	Mexico	8.4
3	China	7.0
4	Brazil	5.1
5	West Germany	4.6
6	United Kingdom	3.5
7	Italy	2.6
8	Japan	2.5
9	Canada	2.4
10	Spain	2.3

Source: Ref. 1.

more than 183,000 nationwide and pays more than \$18 billion annually in state and local taxes.

The soft drink industry uses more than 12 billion gallons of water during production every year. Therefore, the treatment technologies for the wastewater resulting from the manufacturing process cannot be discounted. This chapter reviews the technologies that are typically used to treat soft drink wastewater.

7.1.1 Composition of Soft Drinks

The ingredients of soft drinks can vary widely, due to different consumer tastes and preferences. Major components include primarily water, followed by carbon dioxide, caffeine, sweeteners, acids, aromatic substances, and many other substances present in much smaller amounts. Table 7.2 lists calories and components of major types of soft drinks.

Water

The main component of soft drinks is water. Regular soft drinks contain 90% water, while diet soft drinks contain up to 99% water. The requirement for water in soft drink manufacturing is that it must be pure and tasteless. For this reason, some form of pretreatment is required if the tap water used has any kind of taste. The pretreatment can include coagulation–flocculation, filtration, ion exchange, and adsorption.

Carbon Dioxide

The gas present in soft drinks is carbon dioxide. It is a colorless gas with a slightly pungent odor. When carbon dioxide dissolves in water, it imparts an acidic and biting taste, which gives the drink a refreshing quality by stimulating the mouth's mucous membranes. Carbon dioxide is delivered to soft drink factories in liquid form and stored in high-pressure metal cylinders.

Carbonation can be defined as the impregnation of a liquid with carbon dioxide gas. When applied to soft drinks, carbonation makes the drinks sparkle and foam as they are dispensed and consumed. The escape of the carbon dioxide gas during consumption also enhances the aroma since the carbon dioxide bubbles drag the aromatic components as they move up to the surface of the soft drinks. The amount of the carbon dioxide gas producing the carbonation effects is specified in volumes, which is defined as the total volume of gas in the liquid divided by the volume of the liquid. Carbonation levels usually vary from one to a few volumes of carbon dioxide. Figure 7.1 shows the typical carbonation levels for a range of well-known drinks [1].

In addition, the presence of carbon dioxide in water inhibits microbiological growth. It has been reported that many bacteria die in a shorter time period in carbonated water than in noncarbonated water.

Caffeine

Caffeine is a natural aromatic substance that can be extracted from more than 60 different plants including cacao beans, tea leaves, coffee beans, and kola nuts. Caffeine has a classic bitter taste that enhances other flavors and is used in small quantities.

Table 7.2 List of Energy and Chemical Content per Fluid Ounce

Flavor types	Calories	Carbohydrates (g)	Total sugars (g)	Sodium (mg)	Potassium (mg)	Phosphorus (mg)	Caffeine (mg)	Aspartame (mg)
Regular								
Cola or Pepper	12–14	3.1–3.6	3.1–3.6	0–2.3	0–1.5	3.3–6.2	2.5–4.0	0
Caffeine-free cola or Pepper	12–15	3.1–3.7	3.1–3.7	0–2.3	0–1.5	3.3–6.2	0	0
Cherry cola	12–15	3.0–3.7	3.0–3.7	0–1.2	0–1.0	3.9–4.5	1.0–3.8	0
Lemon-lime (clear)	12–14	3.0–3.5	3.0–3.5	0–4.6	0–0.3	0–0.1	0	0
Orange	14–17	3.4–4.3	3.4–4.3	1.1–3.5	0–1.4	0–5.0	0	0
Other citrus	10–16	2.5–4.1	2.5–4.1	0.8–4.1	0–10.0	0–0.1	0–5.3	0
Root beer	12–16	3.1–4.1	3.1–4.1	0.3–5.1	0–1.6	0–1.6	0	0
Ginger ale	10–13	2.6–3.2	2.6–3.2	0–2.3	0–0.3	0–trace	0	0
Tonic water	10–12	2.6–2.9	2.6–2.9	0–0.8	0–0.3	0–trace	0	0
Other regular	12–18	3.0–4.5	3.0–4.5	0–3.5	0–2.0	0–7.8	0–3.6	0
Juice added	12–17	3.0–4.2	3.0–4.2	0–1.8	2.5–10.0	0–6.2	0	0
Diet								
Diet cola or pepper	<1	0–0.1	0	0–5.2	0–5.0	2.1–4.7	0–4.9	0–16.0
Caffeine-free diet cola, pepper	<1	0–0.1	0	0–6.0	0–10.0	2.1–4.7	0	0–16.0
Diet cherry cola	<1	0–<0.04	0–trace	0–0.6	1.5–5.0	2.3–3.4	0–3.8	15.0–15.6
Diet lemon-lime	<1	0–0.1	0	0–7.9	0–6.9	0–trace	0	0–16.0
Diet root beer	<2	0–0.4	0	3.3–8.5	0–3.0	0–1.6	0	0–17.5
Other diets	<6	0–1.5	0–1.5	0–8.0	0.3–10.1	0–trace	0–5.8	0–17.0
Club soda, Seltzer, sparkling water	0	0	0	0–8.1	0–0.5	0–0.1	0	0
Diet juice added	<3	0.1–0.5	0.1–0.5	0–1.8	0–9.0	0–5.0	0	11.4–16.0

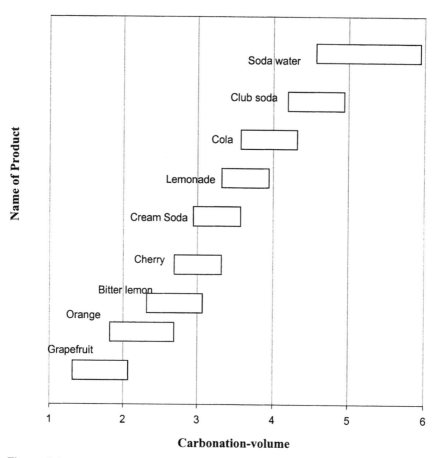

Figure 7.1 Carbonation levels of various popular soft drinks.

Sweeteners

Nondiet and diet soft drinks use different types of sweeteners. In nondiet soft drinks, sweeteners such as glucose and fructose are used. Regular (nondiet) soft drinks contain about 7–14% sweeteners, the same as fruit juices such as pineapple and orange. Most nondiet soft drinks are sweetened with high fructose corn syrup, sugar, or a combination of both. Fructose is 50% sweeter than glucose and is used to reduce the number of calories present in soft drinks.

In diet soft drinks, "diet" or "low calorie" sweeteners such as aspartame, saccharin, suralose, and acesulfame K are approved for use in soft drinks. Many diet soft drinks are sweetened with aspartame, an intense sweetener that provides less than one calorie in a 12 ounce can. Sweeteners remain an active area in food research because of the increasing demand in consumer's tastes and preferences.

Acids

Citric acid, phosphoric acid, and malic acid are the common acids found in soft drinks. The function of introducing acidity into soft drinks is to balance the sweetness and also to act as a preservative. Its importance lies in making the soft drink fresh and thirst-quenching. Citric acid is naturally found in citrus fruits, blackcurrants, strawberries, and raspberries. Malic acid is found in apples, cherries, plums, and peaches.

Other Additives

Other ingredients are used to enhance the taste, color, and shelf-life of soft drinks. These include aromatic substances, colorants, preservatives, antioxidants, emulsifying agents, and stabilizing agents.

7.1.2   Manufacturing and Bottling Process of Soft Drinks

The manufacturing and bottling process for soft drinks varies by region and by endproducts. Generally, the process consists of four main steps: syrup preparation; mixing of carbonic acid, syrup and water; bottling of the soft drink; and inspection.

Syrup Preparation

The purpose of this step is to prepare a concentrated sugar solution. The types of sugar used in the soft drinks industry include beet sugar and glucose. For the production of "light" drinks, sweeteners or a combination of sugar and sweeteners is used instead. After the preliminary quality control, other minor ingredients such as fruit juice, flavorings, extracts, and additives may be added to enhance the desired taste.

Mixing of Carbonic Acid, Syrup, and Water

In this second step, the finished syrup, carbonic acid, and water of a fixed composition are mixed together in a computer-controlled blender. This is carried out on a continuous basis. After the completion of the mixing step, the mixed solution is conveyed to the bottling machine via stainless steel piping. A typical schematic diagram of a computer-controlled blender is shown in Figure 7.2.

Bottling of Soft Drinks

Empty bottles or cans enter the soft drinks factory in palletized crates. A fully automated unpacking machine removes the bottles from the crates and transfers them to a conveyer belt. The unpacking machines remove the caps from the bottles, then cleaning machines wash the bottles repeatedly until they are thoroughly clean. The cleaned bottles are examined by an inspection machine for any physical damage and residual contamination.

Inspection

This step is required for refillable plastic bottles. A machine that can effectively extract a portion of the air from each plastic bottle is employed to detect the presence of any residual foreign substances. Bottles failing this test are removed from the manufacturing process and destroyed.

A typical bottling machine resembles a carousel-like turret. The speed at which the bottles or cans are filled varies, but generally the filling speed is in excess of tens of thousands per hour. A sealing machine then screws the caps onto the bottles and is checked by a pressure tester machine to see if the bottle or can is properly filled. Finally, the bottles or cans are labeled, positioned into crates, and put on palettes, ready to be shipped out of the factory.

Before, during, and after the bottling process, extensive testing is performed on the soft drinks or their components in the laboratories of the bottling plants. After the soft drinks leave the manufacturing factory, they may be subjected to further testing by external authorities.

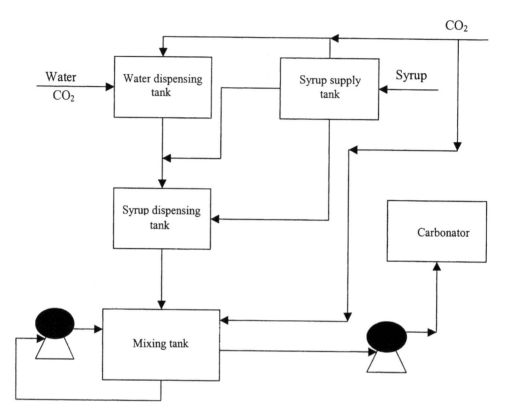

Figure 7.2 Schematic diagram of a computer-controlled blender.

7.2 CHARACTERISTICS OF SOFT DRINK WASTEWATER

Soft drink wastewater consists of wasted soft drinks and syrup, water from the washing of bottles and cans, which contains detergents and caustics, and finally lubricants used in the machinery. Therefore, the significant associated wastewater pollutants will include total suspended solids (TSS), 5-day biochemical oxygen demand (BOD_5), chemical oxygen demand (COD), nitrates, phosphates, sodium, and potassium (Table 7.2). Table 7.3 gives a list of typical wastewater parameters. As shown, higher organic contents indicate that anaerobic treatment is a feasible process.

7.3 BIOLOGICAL TREATMENT FOR SOFT DRINK WASTEWATER

Biological treatment is the most common method used for treatment of soft drink wastewater because of the latter's organic content (Table 7.3). Since BOD_5 and COD levels in soft drink wastewaters are moderate, it is generally accepted that anaerobic treatment offers several advantages compared to aerobic alternatives. Anaerobic treatment can reduce BOD_5 and COD from a few thousands to a few hundreds mg/L; it is advisable to apply aerobic treatment for further treatment of the wastewater so that the effluent can meet regulations. High-strength wastewater normally has low flow and can be treated using the anaerobic process; low-strength wastewater together with the effluent from the anaerobic treatment can be treated by an aerobic process.

Table 7.3 Soft Drink Wastewater
Characteristics

Item	Value (mg/L)
COD	1200–8000
BOD_5	600–4500
Alkalinity	1000–3500
TSS	0–60
VSS	0–50
NH_3-N	150–300
PO_4-P	20–40
SO_4	7–20
K	20–70
Fe	10–20
Na	1500–2500
Ni	1.2–2.5
Mo	3–8
Zn	1–5
Co	3–8

A complete biological treatment includes optional screening, neutralization/equalization, anaerobic and aerobic treatment or aerobic treatment, sludge separation (e.g., sedimentation or dissolved air flotation), and sludge disposal. Chemical and physical treatment processes (e.g., coagulation and sedimentation/flotation) are occasionally used to reduce the organic content before the wastewater enters the biological treatment process. Since the wastewater has high sugar content, it can promote the growth of filamentous bacteria with lower density. Thus, dissolved air flotation may be used instead of the more commonly used sedimentation.

7.4 AEROBIC WASTEWATER TREATMENT

Owing to the high organic content, soft drink wastewater is normally treated biologically; aerobic treatment is seldom applied. If the waste stream does not have high organic content, aerobic treatment can still be used because of its ease in operation. The removal of BOD and COD can be accomplished in a number of aerobic suspended or attached (fixed film) growth treatment processes. Sufficient contact time between the wastewater and microorganisms as well as certain levels of dissolved oxygen and nutrients are important for achieving good treatment results. An aerobic membrane bioreactor (MBR) for organic removal as well as separation of biosolids can be used in the wastewater treatment.

7.4.1 Aerobic Suspended Growth Treatment Process

Aerobic suspended growth treatment processes include activated sludge processes, sequencing batch reactors (SBR), and aerated lagoons. Owing to the characteristics of the wastewater, the contact time between the organic wastes and the microorganisms must be higher than that for domestic wastewater. Processes with higher hydraulic retention time (HRT) and solids retention time (SRT), such as extended aeration and aerated lagoon, are recommended to be used.

O'Shaughnessy et al. [2] reported that two aerobic lagoons with volume of 267,800 gallons each were used to treat a wastewater from a Coca Cola bottling company. Detention time was 30 days; the design flow was 20,000 gpd. A series of operational problems occurred in the early phase, including a caustic spill incident, continuous clogging of air diffusers, and bad effluent quality due to shock loading (e.g., liquid sugar spill). Failure to meet effluent standards was a serious problem in the treatment plant. It was observed that the effluent BOD_5 and COD were above 100 and 500 mg/L, respectively. This problem, however, was solved by addition of potassium; the effluent BOD_5 decreased to 60 mg/L.

Tebai and Hadjivassilis [3] used an aerobic process to treat soft drink wastewater with a daily flow of 560 m^3/day, BOD_5 of 564 mg/L, and TSS of 580 mg/L. Before beginning biological treatment, the wastewater was first treated by physical and chemical treatment processes. The physical treatment included screening and influent equalization; in the chemical treatment, pH adjustment was performed followed by the traditional coagulation/flocculation process. A BOD_5 and COD removal of 43.2 and 52.4%, respectively, was achieved in the physical and chemical treatment processes. In the biological treatment, the BOD_5 loading rate and the sludge loading rate were 1.64 kg BOD_5/day m^3 and 0.42 kg BOD_5/kg MLSS day; the BOD_5 and COD removal efficiencies were 64 and 70%, respectively. The biological treatment was operated at a high-rate mode, which was the main cause for the lower removal efficiencies of BOD_5 and COD.

7.4.2   Attached (Fixed Film) Growth Treatment Processes

Aerobic attached growth treatment processes include a trickling filter and rotating biological contactor (RBC). In the processes, the microorganisms are attached to an inert material and form a biofilm. When air is applied, oxidation of organic wastes occurs, which results in removal of BOD_5 and COD.

In a trickling filter, packing materials include rock, gravel, slag, sand, redwood, and a wide range of plastic and other synthetic materials [4]. Biodegradation of organic waste occurs as it flows over the attached biofilm. Air through air diffusers is provided to the process for proper growth of aerobic microorganisms.

An RBC consists of a series of closely placed circular discs of polystyrene or polyvinyl chloride submerged in wastewater; the discs are rotated through the wastewater. Biodegradation thus can take place during the rotation.

A trickling filter packed with ceramic tiles was used to treat sugar wastewater. The influent BOD_5 and COD were 142–203 mg/L and 270–340 mg/L; the organic loading was from 5 to 120 g BOD_5/m^2 day. Removal efficiencies of BOD_5 of 88.5–98% and COD of 67.8–73.6% were achieved. The process was able to cope effectively with organic shock loading up to 200 g COD/L [5].

An RBC was recommended for treatment of soft drink bottling wastewater in the Cott Corporation. The average wastewater flow rate was 60,000 gpd; its BOD_5 was 3500 mg/L; and TSS was of the order of 100 mg/L. Through a laboratory study and pilot-plant study, it was found that RBC demonstrated the capability of 94% BOD_5 removal at average loading rate of 5.3 lb BOD_5 applied per 1000 square feet of media surface [6].

7.5   ANAEROBIC WASTEWATER TREATMENT

The anaerobic process is applicable to both wastewater treatment and sludge digestion. It is an effective biological method that is capable of treating a variety of organic wastes. Because the anaerobic process is not limited by the efficiency of the oxygen transfer in an aerobic process, it is more suitable for treating high organic strength wastewaters (≥ 5 g COD/L). Disadvantages of the process include slow startup, longer retention time, undesirable odors from production of hydrogen sulfite and mercaptans, and a high degree of difficulty in operating as compared to aerobic processes. The microbiology of the anaerobic process involves facultative and anaerobic microorganisms, which in the absence of oxygen convert organic materials into mainly gaseous carbon dioxide and methane.

Two distinct stages of acid fermentation and methane formation are involved in anaerobic treatment. The acid fermentation stage is responsible for conversion of complex organic waste (proteins, lipids, carbohydrates) to small soluble product (triglycerides, fatty acids, amino acids, sugars, etc.) by extracellular enzymes of a group of heterogeneous and anaerobic bacteria. These small soluble products are further subjected to fermentation, β-oxidations, and other metabolic processes that lead to the formation of simple organic compounds such as short-chain (volatile) acids and alcohols. There is no BOD_5 or COD reduction since this stage merely converts complex organic molecules to simpler molecules, which still exert an oxygen demand. In the second stage (methane formation), short-chain fatty acids are converted to acetate, hydrogen gas, and carbon dioxide in a process known as acetogenesis. This is followed by methanogenesis, in which hydrogen produces methane from acetate and carbon dioxide reduction by several species of strictly anaerobic bacteria.

The facultative and anaerobic bacteria in the acid fermentation stage are tolerant to pH and temperature changes and have a higher growth rate than the methanogenic bacteria from the second stage. The control of pH is critical for the anaerobic process as the rate of methane fermentation remains constant over pH 6.0–8.5. Outside this range, the rate drops drastically. Therefore, maintaining optimal operating conditions is the key to success in the anaerobic process [7]. Sodium bicarbonate and calcium bicarbonate can be added to provide sufficient buffer capacity to maintain pH in the above range; ammonium chloride, ammonium nitrate, potassium phosphate, sodium phosphate, and sodium tripolyphosphate can be added to meet nitrogen and phosphorus requirements.

A number of different bioreactors are used in anaerobic treatment. The microorganisms can be in suspended, attached or immobilized forms. All have their advantages and disadvantages. For example, immobilization is reported to provide a higher growth rate of methanogens since their loss in the effluent can be diminished; however, it could incur additional material costs. Typically, there are three types of anaerobic treatment processes. The first one is anaerobic suspended growth processes, including complete mixed processes, anaerobic contactors, anaerobic sequencing bath reactors; the second is anaerobic sludge blanket processes, including upflow anaerobic sludge blanket (UASB) reactor processes, anaerobic baffled reactor (ABR) processes, anaerobic migrating blanket reactor (AMBR) processes; and the last one is attached growth anaerobic processes with the typical processes of upflow packed-bed attached growth reactors, upflow attached growth anaerobic expanded-bed reactors, attached growth anaerobic fluidized-bed reactors, downflow attached growth processes. A few processes are also used, such as covered anaerobic lagoon processes and membrane separation anaerobic treatment processes [4].

It is impossible to describe every system here; therefore, only a select few that are often used in treating soft drink wastewater are discussed in this chapter. Figure 7.3 shows the schematic diagram of various anaerobic reactors, and the operating conditions of the corresponding reactors are given in Table 7.4.

7.5.1 Upflow Anaerobic Sludge Blanket Reactor

The upflow anaerobic sludge blanket reactor, which was developed by Lettinga, van Velsen, and Hobma in 1979, is most commonly used among anaerobic bioreactors with over 500 installations

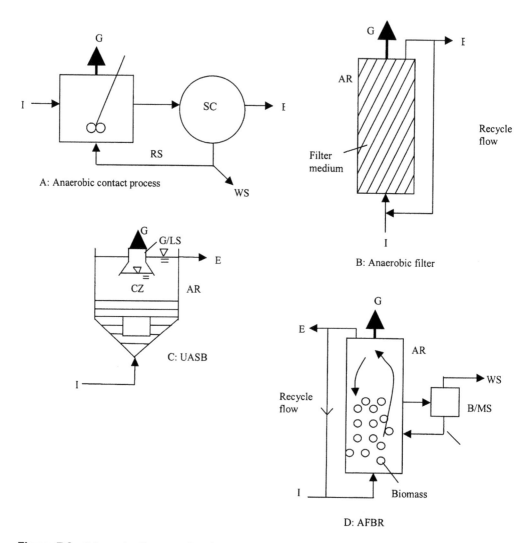

Figure 7.3 Schematic diagram of various anaerobic wastewater treatment reactors. AR: anaerobic reactor; B/MS: biofilm/media separator; CZ: clarification zone; E: effluent; G: biogas; G/LS: gas-liquid separator; I: influent; RS: return sludge; SC: secondary clarifier; SZ: sludge zone; WS: waste sludge.

treating a wide range of industrial wastewaters [4]. The UASB is essentially a suspended-growth reactor with the fixed biomass process incorporated. Wastewater is directed to the bottom of the reactor where it is in contact with the active anaerobic sludge solids distributed over the sludge blanket. Conversion of organics into methane and carbon dioxide gas takes place in the sludge blanket. The sludge solids concentration in the sludge bed can be as high as 100,000 mg/L. A gas–liquid separator is usually incorporated to separate biogas, sludge, and liquid. The success of UASB is dependent on the ability of the gas–liquid separator to retain sludge solids in the system. Bad effluent quality occurs when the sludge flocs do not form granules or form granules that float.

The UASB can be used solely or as part of the soft drink wastewater treatment process. Soft drink wastewater containing COD of 1.1–30.7 g/L, TSS of 0.8–23.1 g/L, alkalinity of

Table 7.4 Operating Conditions of Common Anaerobic Reactors

Reactor type	AC	UASB	AF	AFBR
Organic loading (kg COD/ m³-day)	0.48–2.40	4.00–12.01	0.96–4.81	4.81–9.61
COD removal (%)	75–90	75–85	75–85	80–85
HRT (hour)	2–10	4–12	24–48	5–10
Optimal temperature (°C)		30–35 (mesophilic) 49–55 (thermophilic)		
Optimal pH		6.8–7.4		
Optimal total alkalinity (mgCaCO₃/L)		2000–3000		
Optimal volatile acids (mg/L as acetic acid)		50–500		

AC, anaerobic contactor; UASB, upflow anaerobic sludge bed; AF, anaerobic filter; AFBR, anaerobic fluidized bed reactor.
Source: Ref. 7.

1.25–1.93 g CaCO₃/L, nitrogen of 0–0.05 g N/L and phosphate of 0.01–0.07 gP/L was treated by a 1.8 L UASB reactor [8]. The pH of wastewater was 4.3–13.0 and temperature was between 20 and 32°C. The highest organic loading reported was 16.5 kg COD m^{-3} day^{-1}. A treatment efficiency of 82% was achieved.

The "Biothane" reactor is a patented UASB system developed by the Bioethane Corporation in the United States. Its industrial application in wastewater treatment systems was described by Zoutberg and Housley [9]. The wastewater mainly consists of waste sugar solution, product spillage, and wastewater from the production lines. The flow rate averages about 900 m³/day with an average BOD and COD load of 2340 and 3510 kg/day, respectively. The soft drink factory was then producing 650×10^6 L of product annually, with three canning lines each capable of producing 2000 cans/min and three bottling lines each capable of filling 300 bottles/min. A flow diagram of the "Biothane" wastewater treatment plant is shown in Figure 7.4. Monitoring of the plant could be performed on or off site. A supervisory control and data acquisition system (SCADA) was responsible for providing continuous monitoring of the process and onsite equipment. In normal operation, COD removal of 75–85% was reported with 0.35 m³ of biogas produced per kg COD.

7.5.2 Anaerobic Filters

The anaerobic filter was developed by Yong and McCarty in the late 1960s. It is typically operated like a fixed-bed reactor [10], where growth-supporting media in the anaerobic filter contacts wastewater. Anaerobic microorganisms grow on the supporting media surfaces and void spaces in the media particles. There are two variations of the anaerobic filters: upflow and downflow modes. The media entraps SS present in wastewater coming from either the top (downflow filter) or the bottom (upflow filter). Part of the effluent is recycled and the magnitude of the recycle stream determines whether the reactor is plug-flow or completely mixed. To prevent bed clogging and high head loss problems, backwashing of the filter must be periodically performed to remove biological and inert solids trapped in the media [7]. Turbulent fluid motion that accompanies the rapid rise of the gas bubbles through the reactor can be helpful to remove solids in the media [10].

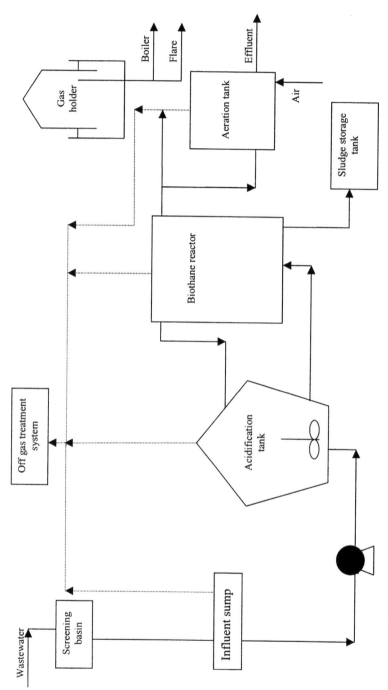

Figure 7.4 Flow diagram of the "Biothane" wastewater treatment plant.

Siino *et al.* [11] used an anaerobic filter to treat soluble carbohydrate waste (soft drink wastewater). At an HRT of 1.7 days, organic loading of 44–210 lb. $COD/1000 \, ft^3/day$, and SRT of 137 days, removal of 85–90% of COD ranging from 1200 to 6000 mg/L can be achieved. The percentage of methane ranged from 60 to 80%; its product was $0.13–0.68 \, ft^3/day$. COD removal efficiency (E %) can be estimated by the following equation:

$$E = 93(1 - 1.99/HRT) \tag{7.1}$$

7.5.3 Anaerobic Fluidized Bed Reactor

Soft drink wastewater can also be treated by an anaerobic fluidized bed reactor (AFBR), which is similar in design to the upflow expanded-bed reactor. Influent wastewater enters the reactor from the bottom. Biomass grows as a biolayer around heavy small media particles. At a certain upflow velocity, the weight of the media particles equals the drag force exerted by the wastewater. The particles then become fluidized and the height of the fluidized bed is stabilized.

Packing size of 0.3–0.8 mm and upflow liquid velocities of 10–30 m/hour can be used in order to provide 100% bed expansion. The high flow velocity around the media particles provides good mass transfer of the dissolved organic matter from the bulk liquid to the particle surface. The bed depth normally ranges from 4 to 6 m. Sand, diatomaceous earth, anion and cation exchange resins, and activated carbon can be used as packing materials [4]. The overall density of media particles decreases as the biomass growth accumulates on the surface areas. This can cause the biomass attached media particles to rise in the reactor and eventually wash out together with the effluent. To prevent this from occurring, a portion of the biomass attached particles is wasted and sent to a mechanical device where the biomass is separated from the media particles. The cleaned particles are then returned to the reactor, while the separated biomass is wasted as sludge [7,12]. Owing to the high turbulence and thin biofilms developed in an AFBR, biomass capture is relatively weak; therefore, an AFBR is better suited for wastewater with mainly soluble COD.

Borja and Banks [13] reported that bentonite, saponite, and polyurethane were respectively used as the suspended support materials for three AFBRs. The composition and parameters of the soft drink wastewater were: total solids (TS) of 3.7 g/L; TSS of 2.9 g/L; volatile suspended solids (VSS) of 2.0 g/L; COD of 4.95 g/L; volatile acidity (acetic acid) of 0.12 g/L; alkalinity of 0.14 g $CaCO_3$/L; ammonium of 5 mg/L; phosphorus of 12 mg/L; pH of 4.8. The average COD removal efficiencies for the three reactors were 89.9% for bentonite, 93.3% for saponite, 91.9% for polyurethane. The amount of biogas produced decreases with increasing HRT. The percentages of methane were 66.0% (bentonite), 72.0% (saponite), and 69.0% (polyurethane).

Borja and Banks [14] used zeolite and sepiolite as packing materials in AFBRs to treat soft drink wastewater. On average, the COD removal of 77.8% and yield coefficient of methane was 0.325 L CH_4/g COD destroyed. The effluent pH was around 7.0–7.3 in all reactors. The content of methane in the biogas ranges from 63 to 70%.

Hickey and Owens [15] conducted a pilot-plant study on the treatment of soft drink bottling wastewater using an AFBR. Diluted soda syrup was used as the substrate, and nitrogen and phosphorus were added with a COD : N : P ratio of 100 : 3 : 0.5. An organic loading rate of 4.0–18.5 kg COD/m^3 day results in BOD_5 and COD removal of 61–95% and 66–89%, respectively. Within this organic loading range, the solids production varies from 0.029 to 0.083 kg TSS/kg COD removed. Methane gas was produced at a rate of 0.41 L/g COD destroyed. The composition of the biogas consists of 60% methane and 40% CO_2.

7.5.4 Combined Anaerobic Treatment Process

A combination of different anaerobic reactors has been used to treat soft drink wastewater. It has been reported that treatment efficiency and liability for combined reactors are better than those of a single type of reactor. Several examples are given below.

Stronach *et al.* [16] reported that a combination of upflow anaerobic sludge blanket reactor, anaerobic fluidized-bed reactor, and anaerobic filter was used to treat fruit processing and soft drink wastewater with TSS, COD, and pH of 160–360 mg/L, 9–15 g/L, and 3.7–6.7, respectively. The organic loadings were 0.75–3.00 kg COD m^{-3} day^{-1} for all three different reactors. COD removal efficiency >79% was achieved. The AFBR performed better than the UASB and the AF in terms of COD removal efficiency and pH stability; however, the methane production was the greatest in the UASB.

Vicenta *et al.* [17] reported that a 68 L semipilot scale AF installed in series with a UASB was used to treat bottling wastes (bottling washing water and spent syrup wastewater). At an organic loading of 0.59 and 0.88 kg COD m^{-3} day^{-1} for the AF and UASB respectively, an overall COD removal of 75% was achieved. The hydraulic retention time (HRT) for the AF and UASB was maintained at 3.4 and 2.2 days, respectively. An average gas yield of 0.83 L per L of influent was produced.

Silverio *et al.* [18] used a series of UASB and upflow AF and trickling filter to treat bottling wastewater with pH of 7.6, COD of 7500 mg/L, TSS of 760 mg/L, and alkalinity of 370 mg $CaCO_3$/L, respectively. The total capacity of the reactors in series is 239 L. An organic loading of 2.78 kg COD m^{-3} day^{-1} and HRT of 2.5 days achieved COD removal of 73% and gas yield of 1 L per L of wastewater in the UASB. The COD level of the effluent from the AF after the UASB further dropped to 550 mg/L and corresponded to a removal efficiency of 87%. The HRT and organic loading in the AF were 2.2 days and 0.88 kg COD m^{-3} day^{-1}, respectively. Incorporation of the trickling filter further reduced the COD level of the effluent to 100 mg/L [18]. All biological treatment processes are discussed in detail in Wang et al. [19] and Wang et al. [20].

REFERENCES

1. Mitchell, A.J. *Formulation and Production of Carbonated Soft Drinks*; Blackie: Glasgow and London, 1990.
2. O'Shaughnessy, J.C.; Blanc, F.C.; Corr, S.H.; Toro, A. Enhanced treatment of high strength soft drink bottling wastewaters, *42nd Annual Purdue Industrial Waste Conference*, 1987; 607–618.
3. Tebai, L.; Hadjivassilis, I. Soft drinks industry wastewater treatment. *Water Sci. Technol.*, **1992**, *25*, 45–51.
4. Metcalf and Eddy. *Wastewater Engineering: Treatment Disposal Reuse*, 4th ed.; McGraw-Hill, 2003.
5. Hamoda, M.F.; Al-Sharekh, H.A. Sugar wastewater treatment with aerated fixed-film biological systems. *Water Sci. Technol.*, **1999**, *40*, 313–321.
6. Blanc, F.C.; O'Shaughnessy, J.C.; Miller, C.H. Treatment of bottling plant wastewater with rotating biological contactors, *33rd Annual Purdue Industrial Waste Conference*, 1978; 614–623.
7. Liu, H.F. Wastewater treatment. In *Environmental Engineers's Handbook*, 2nd ed.; Lewis Publishers: Boca Raton, New York, 1997; 714–720.
8. Kalyuzhnyi, S.V.; Saucedo, J.V.; Martinez, J.R. The anaerobic treatment of soft drink wastewater in UASB and hybrid reactors. *Appl. Biochem. Biotech.*, **1997**; *66*, 291–301.
9. Housley, J.N.; Zoutberg, G.R. Application of the "Biothane" wastewater treatment system in the soft drinks industry. *J. Inst. Water. Env. Man.* **1994**; *8*, 239–245.
10. Rittmann, B.E.; and McCarty, P.L. Anaerobic treatment by methanogenesis. In *Environmental Biotechnology: Principles and Applications*; McGraw Hill: New York, 2001; 573–579.

11. Siino, F.J.; Blanc, F.C.; O'Shaughnessy, J.C. Performance of an anaerobic filter treating soluble carbohydrate waste, *40th Annual Purdue Industrial Waste Conference*, 1985; 785–793.

12. Heijnen, J.J.; Mulder, A.; Enger, W.; Hoeks, F. Review on the application of anaerobic fluidized bed reactors in wastewater treatment. *Chem. Eng. J. & Biochem. Eng. J.*, **1989**; *41*, B37–50.

13. Borja, R.; Banks, C.J. Semicontinuous anaerobic digestion of soft drink wastewater in immobilized cell bioreactors. *Biotechnol. Lett.*, **1993**; *15*, 767–772.

14. Borja, R.; Banks, C.J. Kinetics of anaerobic digestion of soft drink wastewater in immobilized cell bioreactors. *J. Chem. Technol. Biotechnol.*, **1994**; *60*, 327–334.

15. Hickey, R.F.; Owens, R.W. Methane generation from high-strength industrial wastes with the anaerobic biological fluidized bed. *Biotechnol. Bioeng. Symp.*, **1981**; *11*, 399–413.

16. Stronach, S.M.; Rudd, T.; Lester, J.N. Start-up of anaerobic bioreactors on high strength industrial wastes. *Biomass*, **1987**; *13*, 173–197.

17. Vicenta, M.; Pacheco, G.; Anglo, P.G. Anaerobic treatment of distillery slops, coconut water, and bottling waste using an upflow anaerobic filter reactor. In *Alternative Energy Sources 8: Solar Energy Fundamentals & Applications*, Vol 1, Hemisphere Publication: 1989; 865–875.

18. Silverio, C.M.; Anglo, P.G.; Luis, Jr, V.S.; Avacena, V.P. Anaerobic treatment of bottling wastewater using the upflow anaerobic reactor system. In *Alternative Energy Sources 8: Solar Energy Fundamentals & Applications* Vol 1; Hemisphere Publication: 1989; 843–853.

19. Wang, L.K.; Pereira, N.C.; Hung, Y.T. (Eds.) *Biological Treatment Processes*. Humana Press: Totowa NJ, 2004.

20. Wang, L.K.; Hung, Y.T.; Shammas, N.K. (Eds.) *Advanced Biological Treatment Processes*. Humana Press: Totowa NJ, 2004..

8

Bakery Waste Treatment

J. Paul Chen, Lei Yang, and Renbi Bai
National University of Singapore, Singapore

Yung-Tse Hung
Cleveland State University, Cleveland, Ohio, U.S.A.

8.1 INTRODUCTION

The bakery industry is one of the world's major food industries and varies widely in terms of production scale and process. Traditionally, bakery products may be categorized as bread and bread roll products, pastry products (e.g., pies and pasties), and specialty products (e.g., cake, biscuits, donuts, and specialty breads). In March 2003, there were more than 7000 bakery operations in the United States (Table 8.1) with more than 220,000 employees. More than 50% of bakery businesses are small, having fewer than 100 employees [1].

The bakery industry has had a relatively low growth rate. Annual industry sales were $14.7 billion, $16.6 billion, and $17.7 billion in 1998, 2000, and 2002, respectively; the average weekly unit sales were $9,890, $10,040, and $10,859 during the same periods. Industry sales increased 6.5%, only 1.6% ahead of the compounded rate of inflation, according to *www.bakery-net.com*. Production by large plant bakers contributes more than 80% of the market's supply, while master bakers sell less than 5% [1].

The principles of baking bread have been established for several thousand years. A typical bakery process is illustrated in Figure 8.1. The major equipment includes miller, mixer/kneading machine, bun and bread former, fermentor, bake ovens, cold stage, and boilers [2–4]. The main processes are milling, mixing, fermentation, baking, and storage. Fermentation and baking are normally operated at $40°C$ and $160–260°C$, respectively. Depending on logistics and the market, the products can be stored at $4–20°C$.

Flour, yeast, salt, water, and oil/fat are the basic ingredients, while bread improver (flour treatment agents), usually vitamin C (ascorbic acid), and preservatives are included in the commercial bakery production process.

Flour made from wheat (e.g., hard wheats in the United States and Canada) contains a higher protein and gluten content. Yeast is used to introduce anaerobic fermentation, which produces carbon dioxide. Adding a small amount of salt gives the bread flavor, and can help the fermentation process produce bread with better volume as well as texture. A very small quantity of vegetable oil keeps the products soft and makes the dough easier to pass through the

Table 8.1 Bakery Industry Market in the United States

Number of employees	Number of businesses	Percentage of businesses	Total employees	Total sales	Average employees/ businesses
Unknown	1,638	23.65	N/A	N/A	N/A
1	644	9.30	644	487	1
2–4	1,281	18.50	3,583	505.5	3
5–9	942	13.60	6,138	753	7
10–24	1,117	16.13	16,186	1,208.1	14
25–49	501	7.23	17,103	1,578.7	34
50–99	287	4.14	18,872	23,51.7	66
100–249	305	4.40	45,432	10,820.5	149
250–499	130	1.88	43,251	6,909.1	333
500–999	70	1.01	45,184	3,255	645
1,000–2,499	7	0.10	8,820	N/A	1,260
2,500–4,999	2	0.03	7,295	760.2	3,648
10,000–14,999	1	0.01	11,077	N/A	11,077
Total/Average	6,925	100.00	223,585	28,628.8	32

Note: data include bread, cake, and related products (US industry code 2051); cookies and crackers (US industry code 2052); frozen bakery products, except bread (US industry code 2053); sales are in $US.
Source: Ref. 1.

manufacturing processes. Another important component in production is water, which is used to produce the dough. Good bread should have a certain good percentage of water. Vitamin C, a bread improver, strengthens the dough and helps it rise. Preservatives such as acetic acid are used to ensure the freshness of products and prevent staling. The ratio of flour to water is normally 10 : 6; while others are of very small amounts [3–6].

During the manufacturing process, 40–50°C hot water mixed with detergents is used to wash the baking plates, molds, and trays. Baking is normally operated on a single eight-hour shift and the production is in the early morning hours.

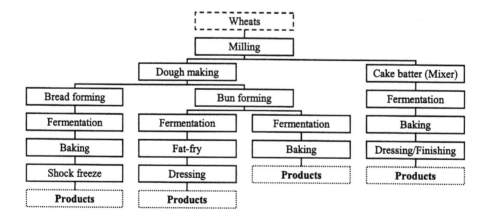

Figure 8.1 General production process diagram of bakery industry.

8.2 BAKERY INDUSTRY WASTE SOURCES

The bakery industry is one of the largest water users in Europe and the United States. The daily water consumption in the bakery industry ranges from 10,000 to 300,000 gal/day. More than half of the water is discharged as wastewater. Facing increasing stringent wastewater discharge regulations and cost of pretreatment, more bakery manufacturers have turned to water conservation, clean technology, and pollution prevention in their production processes.

As shown in Figure 8.1, almost every operation unit can produce wastes and wastewaters. In addition, other types of pollution resulting from production are noise pollution and air pollution.

8.2.1 Noise

Noise usually comes from the compressed air and the running machines. It not only disturbs nearby residents, but can harm bakery workers' hearing. It is reported that sound more than 5 dB(A) above background can be offensive to people. A survey of bakery workers' exposure showed that the average range is 78–85 dB(A), with an average value of 82 dB(A). Ear plugs can help to effectively reduce the suffering. Other noise control measures include the reduction of source noise, use of noise enclosures, reduction of reverberation, and reduction of exposure time [2,7].

8.2.2 Air Pollution

The air pollution is due to emission of volatile organic compounds (VOC), odor, milling dust, and refrigerant agent. The VOC can be released in many operational processes including yeast fermentation, drying processes, combustion processes, waste treatment systems, and packaging manufacture. The milling dust comes from the leakage of flour powder. The refrigerant comes from the emissions leakage of the cooling or refrigeration systems. All of these can cause serious environmental problems. The controlling methods may include treatment of VOC and odor, avoidance of using the refrigerants forbidden by laws, and cyclic use of the refrigerants.

8.2.3 Wastewater

Wastewater in bakeries is primarily generated from cleaning operations including equipment cleaning and floor washing. It can be characterized as high loading, fluctuating flow and contains rich oil and grease. Flour, sugar, oil, grease, and yeast are the major components in the waste.

The ratio of water consumed to products is about 10 in common food industry, much higher than that of 5 in the chemical industry and 2 in the paper and textiles industry [3,6]. Normally, half of the water is used in the process, while the remainder is used for washing purposes (e.g., of equipment, floor, and containers).

Typical values for wastewater production are summarized in Tables 8.2–8.4 [3,8,9]. Different products can lead to different amounts of wastewater produced. As shown in Table 8.2, pastry production can result in much more wastewater than the others. The values of each item can vary significantly as demonstrated in Table 8.3. The wastewater from cake plants has higher strength than that from bread plants. The pH is in acidic to neutral ranges, while the 5-day biochemical oxygen demand (BOD_5) is from a few hundred to a few thousand mg/L, which is much higher than that from the domestic wastewater. The suspended solids (SS) from cake plants is very high. Grease from the bakery industry is generally high, which results from the production operations. The waste strength and flow rate are very much dependent on the operations, the size of the plants, and the number of workers. Generally speaking, in the plants with products of bread, bun, and roll, which are termed as dry baking, production equipment (e.g., mixing vats and baking pans) are cleaned dry and floors are swept before washing down. The wastewater from cleanup

Table 8.2 Summary of Waste Production from the Bakery Industry

Manufacturer	Products	Wastewater production (L/tonne-production)	COD (kg/tonne-production)	Contribution to total COD loading (%)
Bread and bread roll	Bread and bread roll	230	1.5	63
Pastry	Pies and sausage rolls	6000	18	29
Specialty	Cake, biscuits, donuts, and Persian breads	74	–	–

Source: Ref. 3.

has low strength and mainly contains flour and grease (Table 8.3). On the other hand, cake production generates higher strength waste, which contains grease, sugar, flour, filling ingredients, and detergents.

Due to the nature of the operation, the wastewater strength changes at different operational times. As demonstrated in Table 8.3, higher BOD_5, SS, total solids (TS), and grease are observed from 1 to 3 AM, which results from lower wastewater flow rate after midnight.

Bakery wastewater lacks nutrients; the low nutrient value gives $BOD_5 : N : P$ of $284 : 1 : 2$ [8,9]. This indicates that to obtain better biological treatment results, extra nutrients must be added to the system. The existence of oil and grease also retards the mass transfer of oxygen. The toxicity of excess detergent used in cleaning operations can decrease the biological treatment efficiency. Therefore, the pretreatment of wastewater is always needed.

8.2.4 Solid Waste

Solid wastes generated from bakery industries are principally waste dough and out-of-specified products and package waste. Solid waste is the loss of raw materials, which may be recovered by cooking waste dough to produce breadcrumbs and by passing cooked product onto pig farmers for fodder.

8.3 BAKERY WASTE TREATMENT

Generally, bakery industry waste is nontoxic. It can be divided into liquid waste, solid waste, and gaseous waste. In the liquid phase, there are high contents of organic pollutants including chemical oxygen demand (COD), BOD_5, as well as fats, oils, and greases (FOG), and SS. Wastewater is normally treated by physical and chemical, biological processes.

Table 8.3 Wastewater Characteristics in the Bakery Industry

Type of bakery	pH	BOD_5 (mg/L)	SS (mg/L)	TS (mg/L)	Grease (mg/L)
Bread plant	6.9–7.8	155–620	130–150	708	60–68
Cake plant	4.7–8.4	2,240–8,500	963–5,700	4,238–5,700	400–1,200
Variety plant	5.6	1,600	1,700	–	630
Unspecified	4.7–5.1	1,160–8,200	650–13,430	–	1,070–4,490

Source: Refs. 8 and 9.

Table 8.4 Average Waste Characteristics at Specified Time Interval in a Cake Plant

Time interval	pH	BOD$_5$ (mg/L)	SS (mg/L)	TS (mg/L)	Grease (mg/L)
3 am–8 am	7.9	1480	834	3610	428
9 am–12 am	8.6	2710	1080	5310	457
1 pm–6 pm	8.1	2520	795	4970	486
7 pm–12 pm	8.6	2020	953	3920	739
1 am–3 am	8.9	2520	1170	4520	991

Source: Ref. 9.

8.4 PRETREATMENT SYSTEMS

Pretreatment or primary treatment is a series of physical and chemical operations, which precondition the wastewater as well as remove some of the wastes. The treatment is normally arranged in the following order: screening, flow equalization and neutralization, optional FOG separation, optional acidification, coagulation–sedimentation, and dissolved air flotation. The pretreatment of bakery wastewater is presented in Figure 8.2.

In the bakery industry, pretreatment is always required because the waste contains high SS and floatable FOG. Pretreatment can reduce the pollutant loading in the subsequent biological and/or chemical treatment processes; it can also protect process equipment. In addition, pretreatment is economically preferable in the total process view as compared to biological and chemical treatment.

8.4.1 Flow Equalization and Neutralization

In bakery plants, the wastewater flow rate and loading vary significantly with the time as illustrated in Table 8.4 [8,9]. It is usually economical to use a flow equalization tank to meet the peak discharge demand. However, too long a retention time may result in an anaerobic environment. A decrease in pH and bad odors are common problems during the operations.

8.4.2 Screening

Screening is used to remove coarse particles in the influent. There are different screen openings ranging from a few μm (termed as microscreen) to more than 100 mm (termed as coarse screen). Coarse screen openings range from 6–150 mm; fine screen openings are less than 6 mm. Smaller opening can have a better removal efficiency; however, operational problems such as clogging and higher head lost are always observed.

Fine screens made of stainless material are often used. The main design parameters include velocity, selection of screen openings, and head loss through the screens. Clean operations and waste disposal must be considered. Design capacity of fine screens can be as high as 0.13 m^3/sec; the head loss ranges from 0.8–1.4 m. Depending on the design and operation, BOD$_5$ and SS removal efficiencies are 5–50% and 5–45%, respectively [8,9].

8.4.3 FOG Separation

As wastewater may contain high amount of FOG, a FOG separator is thus recommended for installation. Figure 8.3 gives an example of FOG separation and recovery systems [4]. The FOG

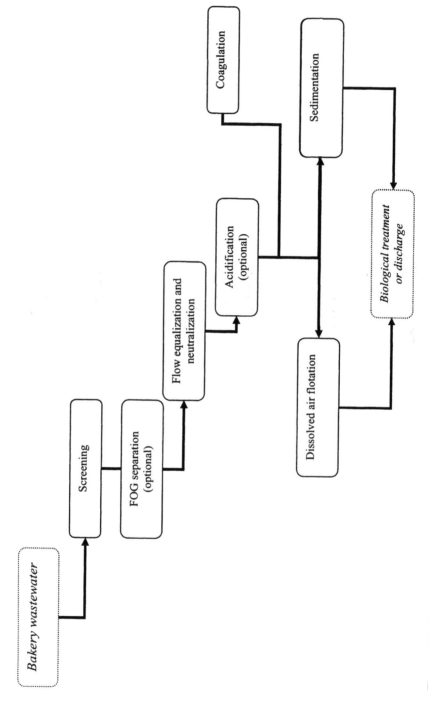

Figure 8.2 Bakery wastewater pretreatment system process flow diagram.

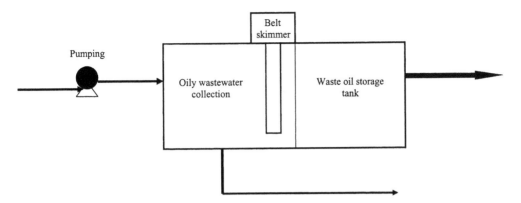

Figure 8.3 Fats, oils, and grease (FOG) separation unit.

can be separated and recovered for possible reuse, as well as reduce difficulties in the subsequent biological treatment.

8.4.4 Acidification

Acidification is optional, depending on the characteristics of the waste. Owing to the presence of FOG, acid (e.g., concentrated H_2SO_4) is added into the acidification tank; hydrolysis of organics can occur, which enhances the biotreatability. Grove et al. [10] designed a treatment system using nitric acid to break the grease emulsions followed by an activated sludge process. A BOD_5 reduction of 99% and an effluent BOD_5 of less than 12 mg/L were obtained at a loading of 40 lb $BOD_5/1000$ ft^3 and detention time of 87 hour. The nitric acid also furnished nitrogen for proper nutrient balance for the biodegradation.

8.4.5 Coagulation–Flocculation

Coagulation is used to destabilize the stable fine SS, while flocculation is used to grow the destabilized SS, so that the SS become heavier and larger enough to settle down. The Coagulation–flocculation process can be used to remove fine SS from bakery wastewater. It normally acts as a preconditioning process for sedimentation and/or dissolved air flotation.

The wastewater is preconditioned by coagulants such as alum. The pH and coagulant dosage are important in the treatment results. Liu and Lien [11] reported that 90–100 mg/L of alum and ferric chloride were used to treat wastewater from a bakery that produced bread, cake, and other desserts. The wastewater had pH of 4.5, SS of 240 mg/L, and COD of 1307 mg/L. Values of 55% and 95–100% for removal of COD and SS, respectively, were achieved. The optimum pH for removal of SS was 6.0, while that for removal of COD was 6.0–8.0. It was also found that $FeCl_3$ was relatively more effective than alum. Yim et al. [8] used coagulation–flocculation to treat a wastewater with much higher waste strength. Table 8.5 gives the treatment results. Owing to the higher organic content, SS, and FOG, coagulants with high dosage of 1300 mg/L were applied [8,9]. The optimal pH was 8.0. As shown, removal for the above three items was fairly high, suggesting that the process can also be used for high-strength bakery waste. However, the balance between the cost of chemical dosage and treatment efficiency should be justified.

8.4.6 Sedimentation

Sedimentation, also called clarification, has a working mechanism based on the density difference between SS and the water, allowing SS with larger particle sizes to more easily settle

Table 8.5 Comparison of Different Bakery Waste Pretreatment Methods

	BOD$_5$		SS		FOG	
Coagulant	Influent (mg/L)	Removal (%)	Influent (mg/L)	Removal (%)	Influent (mg/L)	Removal (%)
Ferric sulfate	2780	71	2310	94	1450	93
Alum	2780	69	2310	97	1450	96

Source: Ref. 9.

down. Rectangular tanks, circular tanks, combination flocculator–clarifiers, and stacked multilevel clarifiers can be used[6].

8.4.7 Dissolved Air Flotation (DAF)

Dissolved air flotation (DAF) is usually implemented by pumping compressed air bubbles to remove fine SS and FOG in the bakery wastewater. The wastewater is first stored in an air pressured, closed tank. Through the pressure-reduction valves, it enters the flotation tank. Due to the sudden reduction in pressure, air bubbles form and rise to the surface in the tank. The SS and FOG adhere to the fine air bubbles and are carried upwards. Dosages of coagulant and control of pH are important in the removal of BOD$_5$, COD, FOG, and SS. Other influential factors include the solids content and air/solids ratio. Optimal operation conditions should be determined through the pilot-scale experiments. Liu and Lien [11] used a DAF to treat a wastewater from a large-scale bakery. The wastewater was preconditioned by alum and ferric chloride. With the DAF treatment, 48.6% of COD and 69.8% of SS were removed in 10 min at a pressure of 4 kg/cm^2, and pH 6.0. Mulligan [12] used DAF as a pretreatment approach for bakery waste. At operating pressures of 40–60 psi, grease reductions of 90–97% were achieved. The BOD$_5$ and SS removal efficiencies were 33–62% and 59–90%, respectively.

8.5 BIOLOGICAL TREATMENT

The objective of biological treatment is to remove the dissolved and particulate biodegradable components in the wastewater. It is a core part of the secondary biological treatment system. Microorganisms are used to decompose the organic wastes [6,8–15].

With regard to different growth types, biological systems can be classified as suspended growth or attached growth systems. Biological treatment can also be classified by oxygen utilization: aerobic, anaerobic, and facultative. In an aerobic system, the organic matter is decomposed to carbon dioxide, water, and a series of simple compounds. If the system is anaerobic, the final products are carbon dioxide and methane.

Compared to anaerobic treatment, the aerobic biological process has better quality effluent, easies operation, shorter solid retention time, but higher cost for aeration and more excess sludge. When treating high-load influent (COD > 4000 mg/L), the aerobic biological treatment becomes less economic than the anaerobic system. To maintain good system performance, the anaerobic biological system requires more complex operations. In most cases, the anaerobic system is used as a pretreatment process.

Suspended growth systems (e.g., activated sludge process) and attached growth systems (e.g., trickling filter) are two of the main biological wastewater treatment processes. The activated sludge process is most commonly used in treatment of wastewater. The trickling filter is easy to control, and has less excess sludge. It has higher resistance loading and low energy cost. However, high operational cost is its major disadvantage. In addition, it is more sensitive to temperature and has odor problems. Comprehensive considerations must be taken into account when selecting a suitable system.

8.6 AEROBIC TREATMENT

8.6.1 Activated Sludge Process

In the activated sludge process, suspended growth microorganisms are employed. A typical activated sludge process consists of a pretreatment process (mainly screening and clarification), aeration tank (bioreactor), final sedimentation, and excess sludge treatment (anaerobic treatment and dewatering process). The final sedimentation separates microorganisms from the water solution. In order to enhance the performance result, most of the sludge from the sedimentation is recycled back to the aeration tank(s), while the remaining is sent to anaerobic sludge treatment. A recommended complete activated sludge process is given in Figure 8.4.

The activated sludge process can be a plug-flow reactor (PFR), completely stirred tank reactor (CSTR), or sequencing batch reactor (SBR). For a typical PFR, length–width ration should be above 10 to ensure the plug flow. The CSTR has higher buffer capacity due to its nature of complete mixing, which is a critical benefit when treating toxic influent from industries. Compared to the CSTR, the PFR needs a smaller volume to gain the same quality of effluent. Most large activated sludge sewage treatment plants use a few CSTRs operated in series. Such configurations can have the advantages of both CSTR and PFR.

The SBR is suitable for treating noncontinuous and small-flow wastewater. It can save space, because all five primary steps of fill, react, settle, draw, and idle are completed in one tank. Its operation is more complex than the CSTR and PFR; in most cases, auto operation is adopted.

The performance of activated sludge processes is affected by influent characteristics, bioreactor configuration, and operational parameters. The influent characteristics are wastewater flow rate, organic concentration (BOD_5 and COD), nutrient compositions (nitrogen and phosphorus), FOG, alkalinity, heavy metals, toxins, pH, and temperature. Configurations of the bioreactor include PFR, CSTR, SBR, membrane bioreactor (MBR), and so on. Operational parameters in the treatment are biomass concentration [mixed liquor volatile suspended solids concentration (MLVSS) and volatile suspended solids (VSS)], organic load, food to microorganisms (F/M), dissolved oxygen (DO), sludge retention time (SRT), hydraulic retention time (HRT), sludge return ratio, and surface hydraulic flow load. Among them, SRT and DO are the most important control parameters and can significantly affect the treatment results. A suitable SRT can be achieved by judicious sludge wasting from the final clarifier. The DO in the aeration tank should be maintained at a level slightly above 2 mg/L. The typical design parameters and operational results are listed in Table 8.6.

Owing to the high organic content, it is not recommended that bakery wastewater be directly treated by aerobic treatment processes. However, there are a few cases of this reported in the literature, including a study from Keebler Company [4]. The company produces crackers and cookies in Macon, Georgia. The FOG and pH of the wastewater from the manufacturing facility were observed as higher than the regulated values. Wastewater was treated by an aerobic activated sludge process, which included a bar screen, nutrient feed system, aeration tank, clarifier, and sludge storage tank. Because of the large quantities of oil in the water (Table 8.7), two FOG separators as shown in Figure 8.3 (discussed previously) were installed in the oleo/lard

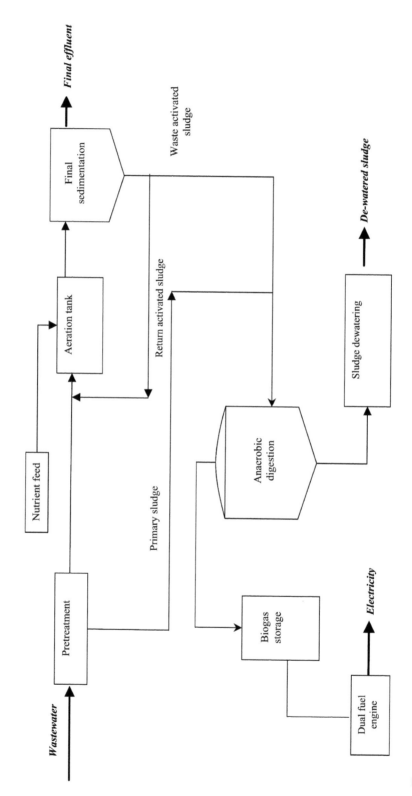

Figure 8.4 Process flow diagram of activated sludge treatment of bakery wastewater.

Table 8.6 Design and Performance of Activated Sludge Processes

Activated sludge processes	Extended	Conventional	High rate
F/M (kg BOD_5/kg MLSS · day)	0.06–0.2	0.3–0.6	0.5–1.9
MLSS (g/L)	4–7.5	1.9–4	5–12
HRT (hour)	18–36	4–10	2–4
SRT (day)	20–30	5–15	3–8
BOD_5 removal (%)	> 95	95	70–75
VLR (kg BOD_5/m³ · day)	0.2–0.4	0.4–1.0	2–16

Note: F/M, food to microorganisms ratio; MLSS, mixed liquid suspended solids; SRT, sludge retention time; HRT, hydraulic retention time; BOD_5, five-day biochemical oxygen demand; VLR, volumetric loading rate.

storage area and the coconut oil spray machines. Characteristics of influent and effluent as well as design parameters are given in Table 8.7. As shown, the company had favorable treatment results; the effluent was good enough for direct discharge to a nearby watercourse. Owing to the poor nutrient content in the influent, nutrient was fed directly into the aeration tank. Not all the added nitrogen was consumed in the treatment, thus the total Kjedahl nitrogen (TKN) concentration in the effluent was higher than that in the influent. The high HRT in Table 8.7 shows that the process was not in fact economical. The bakery wastewater treatment can be more cost-effective if the waste is first treated by an anaerobic process and then an aerobic process.

8.6.2 Trickling Filter Process

Aerobic attached-growth processes include tricking filters (biotower) and rotating biological contactors (RBC). In these processes, microorganisms are attached onto solid media and form a layer of biofilm. The organic pollutants are first adsorbed to the biofilm surface, oxidation reactions then occur, which break the complex organics into a group of simple compounds, such as water, carbon dioxide, and nitrate. In addition, the energy released from the oxidation together with the organics in the waste is used for maintenance of microorganisms as well as synthesis of new microorganisms.

Table 8.7 Summary of Wastewater Treatment in the Keebler Company

Parameter	Influent: Design basis[a]	Influent: Operation[b]	Effluent[b]
Flow rate (gpd)	51,200	37,000	–
PH	5.6	6.0	6.8
TCOD (mg/L)	1620	830	65
SCOD (mg/L)	–	290	40
$TBOD_5$ (mg/L)	891	500	39
$SBOD_5$ (mg/L)	–	175	24
TS (mg/L)	756	–	11[b]
FOG (mg/L)	285	–	3[b]
TKN (mg/L)	–	2	5
PO_4-P (mg/L)	–	3	3

[a]Based on historical pretreatment program monitoring data. [b]Based on operation in August 1988. Operational parameters: HRT = 2.8 day; MLSS = 3300 mg/L; VSS = 2600 mg/L; DO = 2.2 mg/L; F/M = 0.07 1b BOD/1b VSS/day. Yield = 0.32; clarifier overflow rate = 118 gpd/ft²; clarifier solids loading rate = 5 1b/ft²/day.
Source: Ref. 4.

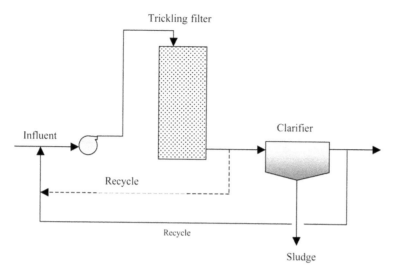

Figure 8.5 Flow diagram of trickling filter for bakery wastewater treatment.

The tricking filter can be used to treat bakery wastewater. Solid media such as crushed rock and stone, wood, and chemical-resistant plastic media are randomly packed in the reactor. Figure 8.5 shows a typical trickling filter, which can be used for the bakery wastewater treatment. Surface area and porosity are two important parameters of filter media. A large surface area can cause accumulation of a large amount of biomass and result in high treatment efficiency; large porosity would lead to higher oxygen transfer rate and less blockage. A common problem in trickling filter systems is the excess growth of microorganisms, which can cause serious blockage in the medium and reduce the porosity. Typical design parameters and performance data for aerobic trickling filters are listed in Table 8.8. Keenan and Sabelnikov [14] demonstrated that a biological system containing a mixing-aeration tank and biological filter (trickling filter) was able to eliminate grease and oil in bakery waste. A dramatic reduction of FOG content from 1500 mg/L to less than 30 mg/L was achieved. This system was fairly stable during 20 months of continuous operation.

8.7 ANAEROBIC BIOLOGICAL TREATMENT

Bakery waste contains high levels of organics, FOG, and SS, which are treated using the preferred method of anaerobic treatment processes. There are different types of anaerobic

Table 8.8 Design and Performance of Trickling Filter

Type of filter	BOD$_5$ loading (kg/m^3/day)	Hydraulic loading (m^3/m^2/day)	Depth (m)	BOD$_5$ removal (%)	Medium
Low rate	0.07–0.4	1–3	1.8–2.4	95	Rock, slag
Mid-range rate	0.2–0.45	3–7	1.8–2.4	–	Rock, slag
High rate	0.5–1	6–20	1–1.8	50–70	Rock

processes available on the market, such as CSTR, AF, UASB, AFBR, AC, and ABR. The most obvious operational parameters are high SRT, HRT, and biomass concentration. Anaerobic processes have been widely used in treatment of a variety of food processing and other wastes since they were first developed in the early 1950s. Figure 8.6 illustrates a typical anaerobic treatment process for bakery wastewater.

In addition to accommodating organic waste treatment, anaerobic treatment can produce methane, which can be used for production of electricity (Fig. 8.6). The disadvantages, however, include complexity in operation, sensitivity to temperature and toxicity, time-consuming in startup, and susceptibility to process upset. Table 8.9 gives a summary of design and performance of typical anaerobic treatment processes.

Anaerobic processes are suitable for a variety of bakery wastewater. For example, an anaerobic contactor was successfully used to treat wastewater from a production facility of snack cake items [13]. The waste strength was extremely high as demonstrated in Table 8.10. The BOD_5 to COD ratio of the raw wastewater was 0.44. An anaerobic contact reactor was used, similar to that in Figure 8.6, except that two bioreactors were operated in series. As shown in Table 8.10, the system provides good treatment results. The removal efficiencies for BOD_5, COD, TSS, and FOD were above 96%. The treated stream can be directly discharged to the domestic sewage systems. Alternatively, a subsequent aerobic treatment can be used to further reduce the waste strength and the effluent can then be discharged to a watercourse.

8.8 AIR POLLUTION CONTROL

While air pollution in the bakery industry may be not serious, it can become a concern if not properly handled. Dust, VOC, and refrigerant are three main types of air pollutants.

8.8.1 Dust

Flour production workers are usually harmed by dust pollution. Lengthy exposure time at a high exposure level can cause serious skin and respiration diseases. The control approaches include prevention of the leakage of flour power, provision of labor protection instruments, and post treatment. Filters and scrubbers are commonly used.

8.8.2 Refrigerant

In the chilling, freezing storage or transport of bakery products, a large amount of refrigerant is used. Chlorofluorocarbons (CFCs) and hydrochlorofluorocarbons (HCFCs) are the common refrigerants and can damage the ozone layer. They can be retained in the air for approximately 100 years. Owing to the significantly negative environmental effects, replacement chemicals such as hydrofluorocarbons (HFC) have been developed and used. Another measure is the prevention of the refrigerant leakage.

8.8.3 VOC

Several measures can be used to control VOC pollution, including biological filters and scrubbers.

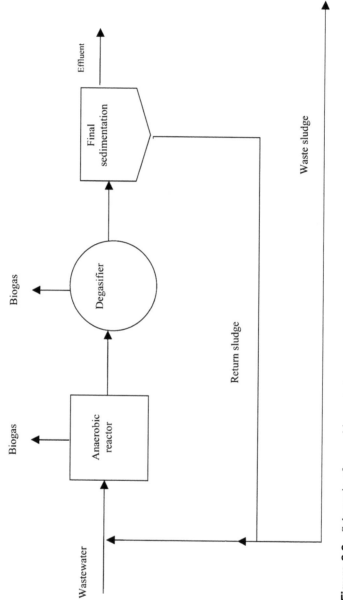

Figure 8.6 Schematic of anaerobic contact process.

Table 8.9 Design and Performance of Anaerobic Treatment Processes

Reactor	Influent COD (g/L)	HRT (day)	VLR (kg COD/m³/day)	Removal (%)
AF	3–40	0.5–13	4–15	60–90
AC	3–10	1–5	1–3	40–90
AFBR	1–20	0.5–2	8–20	80–99
UASB	5–15	2–3	4–14	85–92

8.9 SOLID WASTE MANAGEMENT

Bakery solid waste includes stale bakery products, dropped raw materials (e.g., dough), and packages. The most simple and common way is to directly transport these to landfill or incineration. Landfill can cause the waste to decompose, which eventually leads to production of methane (a greenhouse gas) and groundwater pollution (organic compounds and heavy metals). Incineration of bakery waste can also release nitrogen oxide gases.

Reclamation of the bakery waste can play an important role in its management. The waste consists primarily of stale bread, bread rolls, and cookies – all of which contain high energy and can be fed directly to animals, such as swine and cattle. Another application is to use the waste for production of valuable products. For example, Oda *et al.* [15] successfully used bakery waste to produce lactic acid with a good conversion efficiency of 47.2%.

8.10 CLEANER PRODUCTION IN THE BAKERY INDUSTRY

8.10.1 Concepts

The production of bakery products involves many operation units that may cause a variety of wastes. Most bakery industries are of small or medium size, and are often located in densely populated areas, which makes environmental problems more critical. Nevertheless, the conventional "end-of-pipe" treatment philosophy has its restrictions in dealing with these problems. It only addresses the result of inefficient and wasteful production processes, and should be considered only as a final option.

Table 8.10 Performance of Anaerobic Contact Process

Parameter	Raw water (mg/L) Range	Average	Clarifier effluent (mg/L) Range	Average	Average removal (%)
BOD₅	906–24,000	9,873	65–267	145	98.5
COD	2,910–50,400	23,730	315–1,340	642	97.3
TS	848–36,700	15,127	267–1,260	502	96.7
FOG	429–10,000	5,778	9–113	41	99.3

Operational parameters: Bioreactor: HRT = 7.8 day; SRT = 50 day; volumetric BOD₅ loading = 1.3 kg BOD₅/m³/day, volumetric COD loading = 3.0 kg COD/m³/day. Clarifier: Overflow rate = 3.7 m³/m²/day; HRT = 16 hour, solids loading = 20.5 m³/m²/day, clarification efficiency = 91%.
Source: Ref. 13.

Manufacturing will always cause direct or indirect pollution of the environment. It is hard to realize "zero discharge," and waste treatment is always expensive. Cleaner production (CP) has two key components: maximization of waste reduction and minimization of raw material usage and energy consumption. The United Nations Environment Program (UNEP) defines CP as [7]:

> The continuous application of an integrated preventive environmental strategy to processes, products, and services to increase overall efficiency, and reduce risks to humans and the environment. Cleaner Production can be applied to the processes used in any industry, to products themselves and to various services provided in society.

Cleaner production results from one or a combination of conserving raw materials, water, and energy; eliminating toxic and dangerous raw materials; and reducing the quantity and toxicity of all emissions and wastes at source during the production process. It aims to reduce the environmental, health, and safety impacts of products over their entire life-cycles, from raw materials extraction, through manufacturing and use, to the "ultimate" disposal of the product. It implies the incorporation of environmental concerns into designing and delivering services [3,7].

In the CP process, raw materials, water, and energy should be conserved, their emission or wastage should be reduced, and application of toxic raw materials must be avoided. It is also important to reduce the negative impacts during the whole production life-cycle, from the design of the production to the final waste disposal. The main steps of a CP assessment are outlined in Figure 8.7. The CP can be illustrated by the following example.

8.10.2   A Case Study in Country Bake Pty. Ltd.

Country Bake Pty. Ltd. [3] is a well-known bakery in Queensland, Australia, that produces mainly bread and bread rolls, as well as pastry products and cakes. Production is highly automated, and CP was carried out at the bakery to improve its operational efficiency.

Staff Awareness and Management Expectation

An initial brief survey showed that general awareness of CP at the manufacturing facility was fairly low before its implementation. The staff felt that changes were most likely to be in the areas of general housekeeping and minor process improvements. However, workers were keen on voluntary improvements to their operations as CP could lead to reduction of environmental and health risk liability, less operating costs through better waste and energy management, and reduction of environmental impact. In addition, both management and labor believed that higher business profitability as well as improvement of the company's public image could be achieved through exercising CP.

Assessment of Waste

Areas of waste generation were identified and characterized. It was found that water usage was 719,000 L/week, with about 59% used in production, while the remainder was ultimately discharged as wastewater from cleaning and other ancillary uses. The pastry area and bread and bread rolls area contributed 35 and 36% of wastewater volume, respectively. Other wastewater arose from the boiler, the crate wash, and the staff amenities. In terms of COD loading, the pastry area, bread and bread rolls area, and night cleaning contributed 29, 25, and 38%, respectively. The characterization of wastes can be found in Table 8.2.

Approximately 1.7 tons of dough per week was lost in the waste stream, leading to a loss of 0.5% of the total mass of ingredients (or a loss of $4000/month). Pancoat oil and white oil

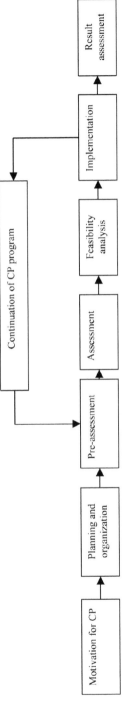

Figure 8.7 Outline of CP assessment process.

were used in production, most of which were lost and became the main contributors to the FOG in the waste stream. Monthly cost for their purchase was $13,140. Prevention of oil loss therefore could lead to significant savings for the bakery.

CP Strategies

Three CP strategies were proposed. The first was to reduce the COD load of wastewater discharged from the bread/bread roll area. Some dough material still fell on the floor and ultimately found its way to the drains. The following approaches were used for reclaiming and recycling the material: relocation of drains for easier collection of dough and installation of screens at drain points to capture fallen dough. A second strategy was to reduce the volumes of wastewater discharged from the pastry area by modification of cleaning practices, elimination or reuse of water discharges from the vacuum pump, and reuse of water discharges from the blast chiller. The last strategy was to reduce the loss of oil by modifications of equipment.

Staff Involvement

Cleaner production cannot be implemented well without great enthusiasm and commitment of the staff to CP, as they are the first to fulfill the CP. The company developed 12 work teams made up of individuals from the major functional work areas. These teams met regularly to discuss issues relevant to their specific work areas. These teams assumed responsibility for driving CP in the workplaces. Team leaders who were trained by the UNEP Working Group conducted a series of training programs for the remaining staff. Finally, the staff was rewarded for their implementation of CP.

Cost-Saving Benefits

Through implementation of CP in production, it was estimated that a total monthly saving of $27,700 could be achieved.

REFERENCES

1. D&B Sales & Marketing Solutions. *Poultry Slaughtering and Processing Report.* http://www.zapdata.com/, 2003.
2. Kannan, P; Boie, W. Energy management practices in SME – case study of a bakery in Germany. Energy Conv. Mgnt 2003, *44*, 945–959.
3. Gainer, D.; Pullar, S.; Lake, M.; Pagan, R. The Country Bake Story – How a modern bakery is achieving productivity and efficiency gains through cleaner production. *Sustainable Energy and Environmental Technology – Challenges and Opportunities, Proceedings*, Gold Coast, 14–17, June, 1998. 573–578.
4. Givens, S.; Cable, J. Case study – A tale of two industries, pretreatment of confectionary and bakery wastewaters. *1988 Food Processing Waste Conference*, presented by the Georgia Tech Research Institute, Atlanta, Georgia, October 31 – November 2, 1988.
5. Dalzell, J.M. *Food Industry and the Environment in the European Union – Practical Issues and Cost Implications*, 2nd Ed., Aspen Publishers, Inc.: Gaithersburg, Maryland, 2000.
6. Metcalf and Eddy. *Wastewater Engineering: Treatment Disposal Reuse*, 4th Ed.; McGraw-Hill, 2002.
7. Nations Environment Programme (UNEP). (http://www.uneptie.org/pc/cp/home.htm). 2003.
8. Yim, B.; Young, R.H.F.; Burbank, N.C. Dugan, G.L. Bakery waste: its characteristics, Part I. Indust. Wastes **1975** *March/April*, 24–25.
9. Yim, B.; Young, R.H.F.; Burbank, N.C.; Dugan, G.L. Bakery waste: its characteristics and treatability, Part II. Indust. Wastes **1975** *September/October*, 41–44.

10. Grove, C.S. Jr.; Emerson, D.B.; Dul, E.F.; Schlesiger, H.; Brown, W. Design of a treatment plant for bakery wastes. *24th Purdue Industrial Waste Conference (PIWC)*, Lafayetle, IN; 1969; 155–178.
11. Liu, J.C.; Lien, C.S. Pretreatment of bakery wastewater by coagulation–flocculation and dissolved air flotation. Water Sci. Technol. **2001**, *43*, 131–137.
12. Mulligan, T. Bakery sewage disposal. *Proceedings of the 1967 Meeting of the American Society of Bakery Engineers*, 1967; 254–263.
13. Shin, B.S.; Eklund, C.W.; Lensmeyer, K.V. Bakery waste treatment by an anaerobic contact process. Res. J. Water Pollut. Control **1990**, *62* (*7*), 920–925.
14. Keenan, D.; Sabelnikov, A. Biological augmentation eliminates grease and oil in bakery wastewater. Water Environ. Res. **2000**, *72(2)*, 141–146.
15. Oda, Y.; Park, B.S.; Moon, K.H.; Tonomura, K. Recycling of bakery wastes using an amylolytic lactic acid bacterium. Biores. Technol. **1997**, *60*, 101–106.

9
Food Waste Treatment

Masao Ukita and Tsuyoshi Imai
Yamaguchi University, Yamaguchi, Japan

Yung-Tse Hung
Cleveland State University, Cleveland, Ohio, U.S.A.

9.1 INTRODUCTION

Food processing industries occupy an important position economically and generate large volumes of mostly biodegradable wastes. However, hazardous wastes are also occasionally generated depending on situations such as contamination by pesticides or herbicides, and pathogens. Even simply unbalanced localization may induce unsuitable accumulation or putrefaction of organic wastes. Discarded gourmet foods might also generate hazardous wastes.

Wastes derived from food industries are categorized into three groups: (a) manufacturing losses, (b) food products thrown away as municipal solid waste (MSW), and (c) discarded wrappers and containers. These groups may be further divided into liquid and solid wastes. This chapter will focus on the background and issues surrounding food wastes from a structural point of view, liquid wastes and wastewater treatment systems, and solid wastes and hazardous wastes of the US and Japanese food industries. Although both countries are in developed stages, they offer contrasting pictures of food waste treatment.

Additionally, several topics will be introduced regarding recent technologies relating to food wastes. These are (a) examples of fermentation factories, (b) cassava starch industries, (c) resource recovery by UASB (up-flow anaerobic sludge bed reactor) or EGSB (extended granular sludge bed reactor), (d) reduction and reuse of wastewater, (e) zero-emission of beer breweries, and (f) technology for garbage recycling.

9.2 STRUCTURAL POINT OF VIEW

The recycling of food wastes should be considered as part of the long-term sustainability of agriculture. As Japan is a typical island state, the undesirable influence of oversea-dependent food production has become obvious. Although free trade systems are commonly accepted in the world today, reconsideration of them may be necessary concerning food and feed from environmental aspects. Figure 9.1 shows the food and feed cycle in Japan in 1998 on the basis of nitrogen (N), in 10^3 tons/year.

From this figure, it is obvious that the nitrogen cycles originally closed, are very open, because of the consumption of chemical fertilizer and large amount of imported food and feed.

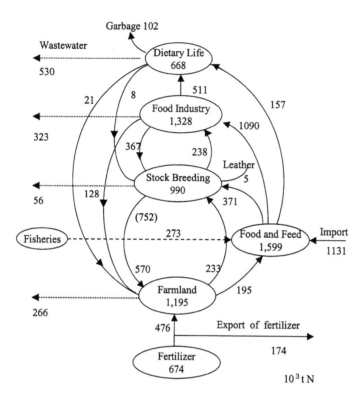

Figure 9.1 Nitrogen cycle relating to food in Japan.

The rate of Japan's self supply of domestic food was 41% in 1970, 32% in 1990, and 29% in 1998 for N, and 33% (1970), 29% (1990), and 28% (1998) for Phosphorus, excluding grass feed [1]. These facts make the recycling of food wastes difficult in various phases. We have not enough farmlands for food wastes to be recycled. The supply of composts to paddy field for rice plantation decreased from 5.07 ton/ha/year in 1965 to 1.25 ton/ha/year in 1997 [2]. Figure 9.2 shows a comparison of food balance between Japan and the United States.

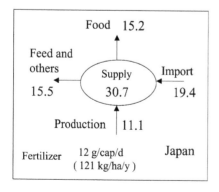
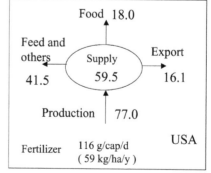

Figure 9.2 Nitrogen balance of food and feed in Japan and the United States (1992–1994).

Based on international statistics on agriculture, forestry, and fisheries [3], the United States exports food and feed of 4.2 g N/capita/day. However Japan imports food and feed at a rate as high as 19.4 g N/capita/day. The supply of food is 15.2 g/capita/day in Japan, and 18.0 g/capita/day in the United States. The ratio of the amount of food to be recycled to farmland vs. chemical fertilizer consumption is $15.5:12 = 1.3$ in Japan, $41.5:116 = 0.36$ in the United States. The consumption of chemical fertilizer on farmland is 121 kg N/ha and 59 kg N/ha for Japan and for the United States, respectively. Considering these situations, it easy to understand the difficulty of food waste recycling in Japan, which uses more than twice as much foreign farmland overseas as domestic farmland.

These realities also profoundly affect issues of eutrophication in water, not only shortage of the demand for recycled food wastes. The principle that organic wastes should return to the land needs to be enforced.

9.3 LIQUID WASTES FROM FOOD INDUSTRIES

9.3.1 Wastewater Treatment Systems for Food Processing

Different sources contribute to the generation of wastewater in food processing industries, including meat processing, dairy products, seafood and fish processing, fruits and vegetable processing, starch and gluten products, confectionery, sugar processing, alcoholic/nonalcoholic beverages, and bean products.

Wastewaters released from these industries are turbid, with high concentrations of biochemical oxygen demand (BOD), Fats, oils and grease (FOG), suspended solids (SS), and usually nitrogen and phosphorus. Hazardous chemical content is generally low. Other characteristics of food processing wastewater are (a) large seasonal variation, (b) large hourly variation and concentration in daytime, (c) factories are often of small scale, (d) sometimes unbalanced ratio of BOD:N:P that induces the bulking of sludge, and (e) colored effluent.

Usually it is desirable to group wastewater as high concentration, medium concentration, and low concentration. High-concentration wastewater may sometimes be concentrated further, treated, and recycled or disposed as solid wastes. Medium-concentration wastewater may be treated on site or discharged into public sewers. Low-concentration wastewater such as indirect cooling water may be discharged without any treatment.

Decreasing the pollutant load of wastewater requires the reduction of both water consumption and pollutant, as well as reduction of the opportunity for interaction between the two. This may be accomplished by the following measures [4]: (a) reducing the amount of washing water for raw materials and reusing the water [5]; (b) mechanical separation and obtaining concentrated wastewater by saving processing water; (c) minimization of over spills during bottling processes [6] and; (d) reducing the amount of water used to wash tanks and containers after operations.

Typically, the wastewater is subjected to pH adjustment and chemical/physical processes that cause the pollutants to form flocs for subsequent removal. Activated sludge processes are generally employed in food industries. Sometimes advanced treatment systems are used such as coagulation and filtration or other innovated technologies. Recent wastewater treatment technologies are listed in Table 9.1 [7–12].

Sequencing batch reactors (SBR) are often employed in small food processing factories and have been observed to improve the activated sludge process for the removal efficiency of nitrogen and phosphorus [13]. The conversion of aeration tanks to include anaerobic mixing capabilities increases the removal efficiency of phosphorus and is also effective in preventing bulking. One way to improve the nitrification or the removal efficiency of refractory organics is

Table 9.1 Recent Progress in Biological Treatment

Type of reactor	Characteristic point	Advantage
Aerobic treatment		
Sequencing batch reactor	Automatic sequential control	Space saving, removal of nutrients
Anerobic–aerobic method	One sludge method	Removal of P, resistant to bulking
Membrane separating method	0.04–0.1 μUF hollow fiber membrane	Space saving, removal of microorganisims
Moving-bed reactor	PP, PU, PE, Activated carbon, etc., are used	Enhance nitrification
Entrapped media method	PEG, PEG-PPG, Calcium arginate, etc., are used	Enhance nitrification
Anaerobic treatment		
UASB	Self-granulation	High efficiency
EGSB	Higher velocity of upflow	Higher efficiency for medium concentration

Source: Refs. 7–12.

to use the membrane separating method of activated sludge, MF/UF bioreactors, moving-bed or fluidized-bed bioreactors, and entrapped media bioreactors. However, this is not frequently used due to its high costs. Accompanying the development of membrane technology, MF/UF bioreactors may become popular.

Anaerobic treatment systems have been salvaged by adopting UASB [11] or EGSB [12] processes for saving energy. It should also be noted that the primary processing of agro-industries has shifted back to the site of production of raw materials, often in developing countries. These include sugar cane mills and sugar refineries, cassava starch factories, and alcohol fermentation using molasses. Developed countries then import processed raw materials or crude products for further refining or applications. The wastewater can be recycled after treatment, usually through oxidation ponds or stabilization ponds for irrigation on farms; biomass wastes may also be used as fuel for factory operations.

Labor-intensive industries such as sea-food processing also tend to shift to developing countries where cheap labor is available.

9.3.2 Effluent Guidelines and Standards for Food Processing

Table 9.2 shows the water quality of untreated wastewater from food processing industries in Japan 20 years ago [14]. The flow rate shown represents the values for standard size factories. It should be noted that the oxidation efficiency of COD_{Mn} (JIS K102) may be about one-third.

Table 9.3 shows the present state of effluent quality in food processing in Japan [15]. The regulation of BOD or COD_{Mn} has already been enacted for a long period, therefore the values for them are converged near the range of standards without large variance. On the other hand, the regulation for TN and TP began only recently, and then only for the specified enclosed water areas such as important lakes, inner sea areas like Seto Inland Sea, Tokyo Bay, and Ise Bay.

Table 9.4 shows the effluent guidelines by USEPA [16] together with Japanese guidelines [17], which were recently revised for the specified sea areas. The Japanese guidelines are set for the specified wastewater as mentioned above. The USEPA values for effluent limits are noticeably larger than the Japanese values. As a further reference, Table 9.5 shows the effluent standards for food processing industries of the Tokyo Metropolitan Government [18]. For a flow

Table 9.2 Characterization of Raw Wastewater from Food Processing in Japan

	Flow rate (m³/day)	BOD₅ (mg/L)	COD_Mn (mg/L)	SS (mg/L)	TN (mg/L)	TP (mg/L)
Meat processing	830	600	400	300	50	15
Dairy products	820	250	170	200	35	5
Seafood cans	530	2700	1700	450	210	75
Fish paste products	650	800	600	500	150	50
Vegetable pickles	440	2300	2500	1000	100	30
Animal feeds	1490	600	300	100	50	10
Starch and gluten	2160	2300	1300	800–1300	130	30–40
Bread and cookies	540	1300	800	900	30	15
Frozen cooked products	400	440	170	200	30	5
Beet sugar processing	4600	450	300	100	25	5
Beer	8500	1000	500	300	40	10
Spirit	1170	500	200	300	20	10
Seasoning chemicals	6500	1000	680	300	460	50
Soy source and amino acids	1090	1000	300	200	100	15

Note: After the investigation by Japanese EPA in 1979–1980, before treatment; flowrate: factories of standard scale. BOD₅, five-day biochemical oxygen demand; COD, chemical oxygen demand; SS, suspended solids; TN, total nitrogen; TP, total phosphorus.
Source: Ref. 14.

rate of more than 500 m³/day, the criteria of BOD, SS, TN, and TP are set to be 20, 40, 20, and 2–3 mg/L, respectively.

9.4 SOLID WASTES FROM FOOD PROCESSING

Two groups of solid wastes are generated in food industries. One group is organic residual wastes such as sludge from wastewater treatment and food wastes or garbage accompanied with consumption. Another group is solid wastes such as vessels, containers, and wrappers. Among the wastes of this group, plastic wastes should be noted in particular.

9.4.1 Organic Residual Wastes

Table 9.6 shows the estimated amount of bio-organic wastes in Japan [19]. Wastes from food processing industries amount to only 62,000 tons of N. This is smaller than the value shown in Figure 9.1 because the statistics of food wastes are not sufficiently arranged.

Table 9.3 Present State of Effluent Quality in Food Processing in Japan

Categories of specified plants	Number of samples	Flow rate (m³/day)	Relative varience	BOD₅ (mg/L)	Relative varience	COD$_{Mn}$ (mg/L)	SS (mg/L)	n-hex. extract (mg/L)	TN (mg/L)	TP (mg/L)
Meat and dairy products	528	649	1.1	12	0.2	16	13	2.0	10.9	2.77
Fishery products	443	317	2.1	19	0.3	25	28	3.8	21.7	4.01
Vegetables and fruits products	404	451	1.5	29	3.1	27	15	1.3	7.4	1.41
Soy source and amino acids	133	2,144	5.5	10	0.1	23	14	2.0	11.2	2.42
Sugar processing	36	21,850	0.8	28	0.4	19	15	0.0	4.2	0.19
Bread and cookies	53	465	2.5	7	0.1	14	11	0.9	8.3	2.43
Rice cake and kouji	46	312	0.9	17	0.2	21	21	1.8	5.2	5.26
Beverage	399	1,068	1.9	12	0.2	14	11	0.9	5.2	1.45
Animal feeds and organic fertilizer	61	952	1.6	18	0.2	20	17	1.3	36.5	0.91
Animal oil processing	56	3,852	2.3	51	2.2	20	17	7.1	11.4	2.29
Yeast	3	6,762	1.1	16	0.2	3	14	–	1.4	0.01
Starch and gluten	43	4,785	1.4	49	0.4	47	53	0.6	6.4	2.65
Glucose and maltose	12	7,239	1.3	22	0.2	16	16	0.6	3.3	1.21
Noodle products	98	305	0.9	5	0.1	9	8	1.3	4.7	1.31
Bean curd and processing	187	429	1.9	10	0.1	15	12	1.3	12.3	2.24
Instant coffee	4	2,170	0.5	7	0.0	12	3	0.8	5.8	0.64
Frozen cooked products	94	275	0.9	11	0.1	16	12	1.8	8.4	2.00

Note: Number of samples for some items is less than that for flow rate.
Source: Ref. 15.

Table 9.4 Effluent Guidelines for Food Processing in the United States and Japan

Category of industries	Japanese guidelines for effluent[a]			USEPA guidelines (BPT)[b]		USEPA guidelines (SPN)[c]	
	COD_{Mn}	TN	TP	BOD_5	TSS	BOD_5	TSS
Meat processing	30–80	10–60	1–16	370–740	450–900	370–740	450–900
Seafood products							
Tuna processing	30–120	10–55	3–12		3300–8300		
Fish meal processing				3900–7000	1500–3700	3800–6700	1500–3700
Dairy Products							
Fluid products	20–50	10–30	1–16	1350–3375	2025–5506	370–740	463–925
Dry milk				650–1625	975–2438	18–36	225–450
Canned vegetables and fruits	30–80	10–30	3–12	140–420	80–240		
Cereal processing							
Flour and other grain mill products	30–70	10–30	1.5–7.5				
Animal feeds	20–90	10–30	1–3.5				
Starch and gluten	40–80	10–30	1.5–10	2000–6000	2000–6000	100–3000	1000–3000
Bread and cookies	40–70	10–30	1.5–7.5				
Sugar processing							
Beet sugar	30–80	10–30	1.5–7.5	2200–3300			
Raw cane sugar processing				630–1140	470–1410		
Cane sugar refining	30–80	10–30	1.5–7.5	430–1190	90–270	90–180	35–110
Oil mill and processing	30–80	10–30	1–7.5				
Wines and beverage							
Beer	30–70	10–30	1.5–4				
Spirit	20–50	10–30	1.5–4				
Yeast	90–130	10–30	1.5–7.5				
Seasoning	20–100	10–145	1.5–9				

[a]Effluent limitations guidelines for specified wastewater from specified plants. Local governments select a value in the range from lower values and higher values.
[b]Effluent limitations guidelines for existing point sources attainable by the application of the best practicable control technology currently available. Lower values: maximum for any 1 day; higher values: average for concecutive 30 days.
[c]Standards of performance for new sources of effluent discharged into navigable waters.
Source: Refs. 16 and 17.

For organic food wastes, the options of feedstuff use, composting, biogas then composting, and heat recovery are adopted for treatment and recycling.

Feedstuff Use

Industrial wastes from food processing are still recycled as feed or organic fertilizer to a fair extent. As shown in Table 9.7, 5.5 million tons of food processing byproducts are used for

Table 9.5 Effluent Standards for Food Processing Industries by Tokyo Metropolitan Government

	Flow rate (m³/day)	BOD (mg/L)	COD$_{Mn}$ (mg/L)	SS (mg/L)	TN (mg/L)	TP (mg/L)	Odor intensity
Existing	50–500	25	25	50	30	6	4
	>500	20	20	40	20	3	4
New sources	50–500	20	20	40	25	3	4
	>500	20	20	40	20	2	4

Source: Ref. 18.

Table 9.6 Amount of Bio-organic Wastes in Japan

	Generation (10⁴ tons)	Dry matter		Contents		Generation of N, P	
		(parts)	(10⁴ tons)	Nitrogen	P$_2$O$_5$	N	P
Agriculture							
Rice stems	1094			0.60	0.20	6.6	0.96
Straw	78			0.40	0.20	0.3	0.07
Rice husks	232			0.60	0.20	1.4	0.20
Stockbreeding							
Manure	9430			0.79	0.13	74.9	11.96
Residues	167			5.01	7.13	8.4	5.20
Forestry							
Bark	95			0.53	0.08	0.5	0.03
Sawdust	50			0.15	0.03	0.1	0.01
Wood waste	402			0.15	0.03	0.6	0.05
Food processing							
Animal/plant residues	248	0.28	69	1.41	0.53	1.0	0.16
Sludge	1504	0.05	75	7.01	4.02	5.3	1.32
Construction waste							
Wood waste	632			0.15	0.03	0.9	0.08
Municipal solid waste							
Garbage	2028	0.29	588	1.41	0.53	8.3	1.36
Wood and bamboo	247			0.76	0.19	1.9	0.20
Others							
Sewage sludge	8550	0.02	171	5.18	5.37	8.9	4.01
Nightsoil sludge	1995			0.6	0.10	12.0	0.87
Joukasou septage	1359	0.02	27	5.18	5.37	1.4	0.63
Farm sewage sludge	32	0.02	0.6	5.18	5.37	0.0	0.01
Total	28,143					132.3	27.13

Source: Ref. 19.

Table 9.7 Feedstuff Use of Byproducts from Food Processing

Byproducts of:	Generation, a (10^3 t/year)	Use for feed, b (10^3 t/year)	b/a (%)
Fruit juice	116	98	84
Vegetable can	80	72	90
Wine	3030	2707	89
Starch processing	1162	80	7
Bean processing	795	411	52
Sugar	1858	1350	73
Fish processing	58	24	41
Bread and malt	29	16	55
Total	7128	5487	77

Source: Ref. 20.

feeding [20]. The rate of use is 77%. Other than that, rice bran, wheat bran, and plant oil residues and BMP, and others are used for general feedstuff products. Feeding use of residual food for pigs has drastically decreased from 206 kg/head/year in 1965 to 6 kg/head/year in 1997 in Japan [2]. Although this system is a good option for recycling, it is a general tendency that pig farms has gradually shifted far from residential areas and have changed to modernized farms.

Composting and Biogas Production

Composting is a traditional, reliable method of recycling food wastes. This will be discussed later together with biogas production.

Incineration with Energy Recovery

Utilization of biomass as it relates to CO_2 reduction against global warming has been focused on recently. For this purpose, woody biomass is more suitable, and organic wastes including various minerals may not be appropriate for incineration.

9.4.2 Vessels, Containers, and Wrapping Wastes

Another type of waste relating to food industries is the waste originating from containers, vessels, bottles, and wrapping materials. These wastes occupy a large portion of municipal solid waste (MSW). Among these wastes, plastic wastes in particular should be focused on from an environmental standpoint.

One company, FP Co. Ltd., has developed a good recycling system for polystylene paper (PSP) trays in Japan. They employ a whole network of transportation systems from factories through markets and back to factories again. The recycling is a tray-to-tray system. However, the rate of recycling is restricted one-third, because the efficiency of transportation for wastes of PSP tray is one-third of that for the products that can be packed compactly.

Polyethylene terephtalate (PET) bottles are the most suitable wastes for material recycling. They can be recycled as polyester fiber products through PET flake and to raw chemicals through chemical recycling. However, the amount of incinerated PET bottles has still been increasing because the rate of consumption continues to exceed the recycling effort. Chemical recycling to obtain the monomer of dimethyl terephtalate (DMT) has been conducted successfully using recycled PET bottles collected by municipalities [21]. The company Teijin Co. Ltd. plans to transport PET bottles amounting to 60,000 t/year to its factory in the Yamaguchi Prefecture.

Recently, other plastic wastes used for wrapping and vessels have been recycled by means of gasification, liquifaction to oil, or heat recovery in the blast furnaces of steel industries (substituting cokes) and cement kilns (substituting coal). However, there are complicated arguments as to whether the direct incineration for heat recovery is more environmentally friendly than options through gas and oil as mentioned above.

9.5 HAZARDOUS WASTES FROM FOOD PROCESSING

9.5.1 Management of Chemicals Based on EPCRA or PRTR

Chemicals commonly encountered in food processing are listed in Table 9.8, relating to the Emergency Planning and Community Right-to-Know Act (EPCRA) in the United States [22]. The Pollutant Release and Transfer Register (PRTR) system was also enacted in Japan from the 2001 fiscal year. Figure 9.3 describes the chemicals used in food processing [22]. Similarly, it is applicable for other materials included in the food itself.

Food that has been accidentally contaminated by pesticides, herbicides, or fumigants may also be treated as hazardous waste. Chlorine is frequently used for sanitary cleaning in food processing at the end of daily operations. Therefore chlorinated organic compounds should be noted in the wastewater treatment plants of food industries. It is very possible that wastewater contains certain levels of trihalomethane and related compounds.

9.5.2 Accidentally Contaminated Food Wastes

Food products contaminated with pathogenic microbes or food poisoning sometimes result in hazardous wastes. The incident of Kanemi rice oil contaminated by PCB in 1968 is still discussed with regards to dioxins (DXNs) as the possible cause of the Kanemi Yusho disease. Two recent examples discussed below include the treatment of contaminated milk products and the issues relating to the issues of BSE.

Table 9.8 Chemicals Commonly Encountered in Food Processing

Purposes	Chemicals used
Water treatment	Chlorine, chlorine dioxide
Refrigerant uses	Ammonia, ethylene glycol, freon gas
Food ingredients	Phosphoric acid, various food dyes, various metals, peracetic acid
Reactants	Ammonia, benzoyl peroxide, Cl_2, ClO_2, ethylene oxide, propylene oxide, phosphoric acid
Catalysts	Nickel and nickel compounds
Extraction/carrier solvents	n-Butyl alcohol, dichloromethane, n-hexane, phosphoric acid, cyclohexane, tert-butyl alcohol
Cleaning/disinfectant uses	Chlorine, chlorine dioxide, formaldehyde, nitric acid, phosphoric acid, 1,1,1-trichloroethane
Wastewater treatment	Ammonia, hydrochloric acid, sulfuric acid
Fumigants	Bromomethane, ethylene oxide, propylene oxide, bromine
Pesticides/herbicides	Various pesticides and herbicides
Byproducts	Ammonia, chloroform, methanol, hydrogen fluoride, nitrate compounds
Can making/coating	Various ink and coating solvents, various listed metals, various metal pigment compounds

Source: Ref. 22.

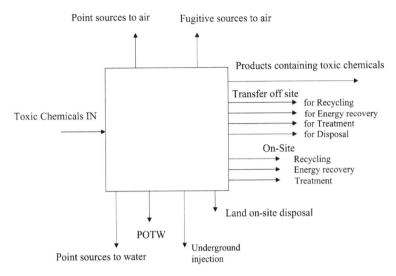

Figure 9.3 Possible release and other waste management types for EPCRA.

Contaminated Milk with Enterotoxin A

In June 2000, skim milk contaminated by *Staphylococcus aureus* led to a huge outbreak of food poisoning in Japan. The milk was contaminated in April because of an electricity outage that lasted several hours in the factory of Yukijirushi Co. Ltd. Afterwards, contaminated milk products were widely distributed and 14,780 persons exhibited food poisoning symptoms [23]. The milk products produced by the company were removed from market displays, and most of them were incinerated as hazardous wastes.

Treatment of Bone and Meat Powder Suspicious of BSE

The Japanese government has prohibited the import of bone powder from England since March 1996 and also from EU countries since January 2001 to prevent the introduction of Bovine Spongiform Encephalopathy (BSE), known as "mad cow" disease. After three infected cows were reported in Japan, bone and meat powder (BMP) rendered in domestic bone-boiling factories has been prohibited as use as feedstuff and also as fertilizer for a certain period. Although BMP is a good fertilizer, especially for fruit trees, consumers have avoided it because of the risk of BSE. Therefore BMP has been labeled a hazardous waste that is incinerated or treated in cement factories. In total, 8.5% of cow's body is bone, corresponding to 20% of meat. It is said that the amount that should be incinerated is near 1 thousand tons daily [24].

The BMP is transported to cement factories, packed in plastic bags of 15–20 kg each. The bags are thrown down into the end of the kiln at 1000°C. The contents of raw material for cement are shown in Table 9.9. For 1 ton of cement clinker 2 kg of BMP are used. The percentage of BMP among raw materials is 0.13% by dry base. The receiving toll is around 40,000 yen/t of BMP. Since prion is a kind of protein, it must be decomposed at very high temperatures in the kiln.

9.5.3 Incineration Ash of Food Wastes

Incineration is the most popular method of disposing of combustible solid wastes in Japan, especially for MSW from restaurants, hotels, and supermarkets. For a long time incineration was perceived as a progressive method and the percentage of incineration of MSW was near 80%.

Table 9.9 Percentages of Raw Materials of Various Kinds of Cement

	Portland cement original	Portland cement recent	Eco-cement
Lime stone	78	76.5	52
Clay	16	1	
Silicate	4	7.5	
Iron source	2	2	1
Pretreated ash of MSW		1	39
Others (coal ash, sludge, plastics)		12	8
Total	100	100	100

However, the evaluation of this method has changed with the discovery of dioxins (DXNs). After incineration, ash containing DXNs is inevitably generated, resulting in another type of hazardous wastes derived from food products. Recently, it has become popular in Japan to recycle the ash to cement raw material. Table 9.8 illustrates two methods for this type of recycling: (a) eco-cement which uses MSW ash at 39% of total amount of raw material, (b) ordinary cement, which uses pretreated ash at 0.5–0.7% of raw material, as shown in Figure 9.4. In Yamaguchi Prefecture 50,000 tons of MSW ash can be recycled by this second method.

This method is considered to be tentative as the final means of recycling. Essentially the goal should be to recycle back to farmland not only major nutrients like N and P but also various minor nutrients like Ca, Mg, Fe, Mn, Cu, Zn, Mo, and B. This means mineral resources derived from food are contaminated by heavy metals through incinaration of garbage together with various other wastes.

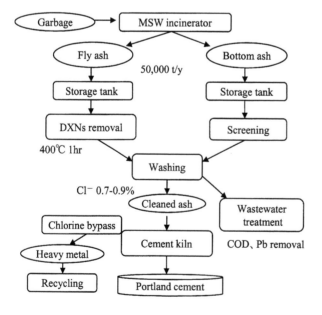

Figure 9.4   Ash recycling for Portland cement feedstock.

9.6 RECENT TECHNOLOGIES ON FOOD WASTES TREATMENT

9.6.1 Waste Management in Fermentation Industries

Fermentation industries cover a wide range of food processing from the traditional industries of breweries, soy source, miso, pickles, to yeast, alcohol, amino acids, nucleic acids, antibiotics, enzyme, and other bio-active fine chemicals. Usually, the harvest rate of these products is not high except for traditional fermentation, which typically has a large pollutant load. However, the possibility of resources recovery is also high because hazardous chemicals are rarely used.

The wastewater comes from (a) the mother liquid after harvesting the products, (b) cleaning water of cells or reactors, (c) condensates from the evaporator, (d) spent eluting solution in purifying processes, and (e) ammonium sulfate, used in salt crystallization of enzymes, and others.

The following information is cited mainly from a thesis [25] published in 1983. While details may have changed since then, the fundamental structures remain largely the same.

Alcohol

Molasses and sweet potatoes are used as the raw materials for alcohol fermentation. In Japan, to improve wastewater quality, the trend of importing crude alcohol and refining it increased in the 1970s. Here the case of alcohol production using sugar cane molasses as raw material is introduced. It is also replaced for example by acetic acid.

The raw materials necessary to produce 1 kL of 95% alcohol are 3.12 tons of molasses, 1.2 kg of urea, and 1.2 kg of ammonium sulfate. The input of N and P is 14.2 kg/kL and 2.18 kg/kL, respectively. The main part of wastewater is the evaporation residues of 10 kL/kL. The water quality of this is 30 g/L for COD_{Mn}, 1.1 g/L for TN, and 0.2 g/L for TP. Therefore the unit loading factor is 300 kg/kL for COD_{Mn}, 11 kg/kL for TN, and 2 kg/kL for TP. The loss of N likely occurs in the process. Methane fermentation was employed before the 1970s; however, it has been replaced by ocean dumping and by drying for reuse to organic fertilizer. Dried matter of distillery waste from the maize alcohol process is valuable and popular as distillery feeds in the United States. On the other hand, distillery wastes from processes using sugar cane molasses have high color and ash content, were recycled to organic fertilizer through concentration, humination with acid treatment to reduce viscosity, adding calcium and magnesium silicate, drying, and palletizing. Consequently, the cleaning water of reactors and the condensates from the evaporation process of distillery wastes are generated as the wastewater.

Bread Yeast

Bread yeast is still produced by fermentation using molasses. As shown in Figure 9.5, this process includes cultivation, separation, pressurized filtration, and the addition of baking powder. Wastewater derives mainly from centrifugal supernatant and the cell-washing process.

The estimated mass balance of this process is shown in Table 9.10 assuming that the yeast content is 2.4% of N and 0.36% of P, and the harvest vs. molasses is 75% and sugar content of molasses is 55%. The rate of loss to wastewater was estimated to be 8% for N and 3% for P. This factory in Hyogo Prefecture had sophisticated treatment systems for its wastewater. The fraction of high concentration was further concentrated and dried to make organic fertilizer. The removal rates of COD_{Mn}, TN, and TP are 99, 72, and 44%, respectively. The unit loading factor vs. product was around 3.3 kg of COD_{Mn}/ton of yeast, 0.58 kg TN/ton and 0.058 kg TP/ton, respectively.

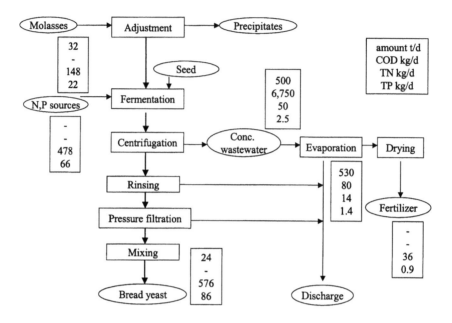

Figure 9.5 Estimated flow of N and P in a bread yeast production factory.

Mono-Sodium Glutamate

Mono-sodium glutamate (MSG) is a product that originates from Japan, and is produced mostly by fermentation. Glucose, acetic acid, or molasses are used as the raw material for the source of carbon, while ammonia and urea may be used as the source of nitrogen. Figure 9.6 shows the flow sheet for one factory in Japan. The main part of the wastewater comes from the crystal separating process.

This factory employs concentration and wet combustion processes. Through this treatment, COD was reduced from 2.4 ton/day to 0.05 ton/day with a removal rate of 98%. The unit loading factor of the discharging stage was estimated to be 15 kg COD/t of MSG, similarly 30 kg/ton for TN and 1.5 kg/ton for TP. Assuming the input of N and P of 4.6 ton/day and 0.88 ton/day, respectively, the rate of loss was 22% for N and 57% for P, as shown in Table 9.11.

Nucleotides

Through hydrolysis of RNA, nucleotides such as inosinic acid and guanyl acid are formed and used for seasonings. Ribonucleotides are produced by the combination of extraction,

Table 9.10 Wastewater Load of Bread Yeast Production

	Amount of use, a (kg/day)	Generated load, b (kg/day)*	Loss rate, b/a (%)	Discharged load, c (t/day)	Reduction rate, $1-c/b$ (%)	Unit loading factor (kg/t yeast)**
COD$_{Mn}$	–	6750	–	80	99	33
TN	626	50	8	14	72	16
TP	88	3	3	1	44	0.06

*Excluding other nontreated wastewater; **Assuming moisture of yeast is 68%.

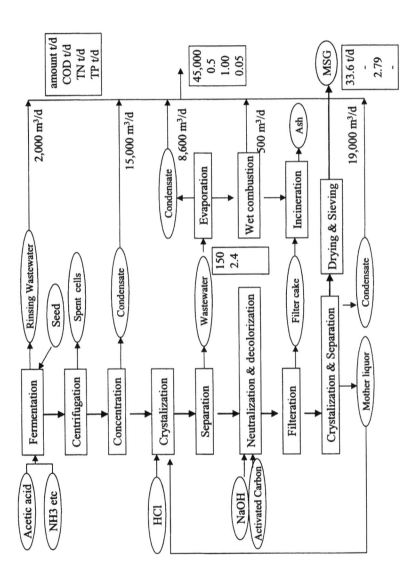

Figure 9.6 Estimated flow of wastewater in the production of MSG.

Table 9.11 Wastewater Load of MSG Production

	Amount of use, a (t/day)	Generated load, b (t/day)	Loss rate, b/a (%)	Discharged load, c (t/day)	Reduction rate, $1 - c/b$ (%)	Unit loading factor (t/t MSG)
COD$_{Mn}$	–	–	–	0.50	–	0.015
TN	4.6	1.8	39.1	1.00	44.4	0.030
TP	0.09	0.09	100.0	0.05	43.2	0.002

fermentation, and chemical synthesis. During fermentation, yeast capable of accumulating RNA is cultured and forms ribonucleotides through enzymic hydrolysis of extracted RNA from the yeast. Finally, ribonucleotides are purified through ion exchange resins, crystallized, and dried to become the products. The main part of wastewater is derived from the yeast separation and purifying processes as shown in Figure 9.7.

As these products consist of finer materials, the loss rate may be larger than in the case of MSG. Therefore, various byproducts are recovered. The spent cell of the yeast is utilized for feedstuff and the concentrated part of the wastewater from the purifying process is used as liquid fertilizer after further concentration. A fraction of the wastewater of medium concentration is treated by the activated sludge process. After separating the yeast cells, the wastewater is treated by an activated sludge process of deep shaft aeration. Molasses had been used previously in this process, but acetic acid replaced it as a countermeasure for reducing wastewater load. Table 9.12 shows the balance of N and P as the result of estimation shown in Table 9.13. The rate of loss is very large at

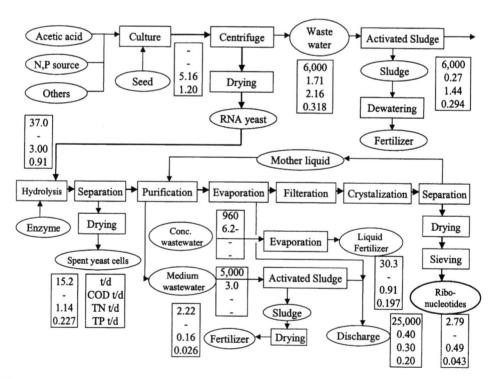

Figure 9.7 Estimated flow of N and P in the production of ribonucleotides.

Table 9.12 Wastewater Load of Ribonucleotides Production

	Amount of use, a (t/day)	Generated load, b (t/day)	Loss rate, b/a (%)	Discharged load, c (t/day)	Reduction rate, $1 - c/b$ (%)	Unit loading factor (t/t of nucleotide)
COD_{Mn}	–	11.01	–	0.67	94	0.24
TN	5.16	3.90	76	1.74	55	0.62
TP	1.24	0.74	62	0.49	33	0.17

76% for TN and 62% for TP. The unit loading factor of COD_{Mn}, TN, and TP after treatment was 240 kg/t, 620 kg/t, and 170 kg/t, respectively. The removal rate was 94% for COD_{Mn}, 56% for TN, and 33% for TP.

Example of Comprehensive Fermentation Factory

During the 1970s, a fermentation factory in Yamaguchi Prefecture produced alcohol, MSG, lysine, antibiotics, and organic combined fertilizer and used molasses to produce alcohol, MSG, and yeast. It is introduced as an example, although its present situation has changed.

Previously, the factory employed a combination of methane fermentation and biological aerobic treatment. With the advance of environmental criteria, the following countermeasures were conducted, in stages: (a) to make the substrate concentration higher, (b) reuse of wastewater to substrate solution, (c) conversion of raw material from molasses to acetic acid, and (d) improvement of extraction and purifying processes. For treatment processes, measures included, (a) classification of wastewater, and (b) possible resource recovery of valuable matters. Figure 9.8 shows the flow of wastewater treatment of this factory in 1976.

The values shown in Figure 9.8 are the flow rate, daily load of COD_{Mn}, TN, and TP. Table 9.14 summarizes the estimation process of the balance shown in Figure 9.7. In fermentation

Table 9.13 Estimation of N and P Balance of Ribonucleotides Production Starting from RNA Yeast

	Amount (t/day)[c]	Content		Amount of N and P	
		N (%)	P (%)	N (t/day)	P (t/day)
Input					
RNA yeast	37.0	8.1	2.4	3.00	0.88
Output[a]					
Nucleotides seasonings[b]	2.8	11.8	5.8	0.33	0.16
Spent yeast cells	15.2	7.5	1.5	1.14	0.23
Liquid fertilizer	30.3	3.0	0.7	0.91	0.20
Sludge fertilizer	2.2	8.0	1.3	0.16	0.03
Wastewater	28,000			0.30	0.20
Total				2.84	0.81

[a]Ignoring the input of enzyme and the output of filter cake.
[b]Taking average of N, P contents for 5'-GMP 2Na $7H_2O$ and 5'-IMP 2Na 7-$8H_2O$.
[c]Assuming 27 days/month and 330 days/year.

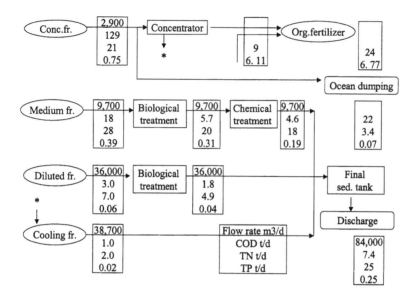

Figure 9.8 Estimated flow of wastewater in a comprehensive fermentation factory.

factories, batch processes are generally adopted, so that it is difficult to estimate the reliable balance. As shown in Table 9.15, 76 ton of N were used; 74% of it was generated as wastewater and 25 ton were discharged to the coastal water. As for P, 2.01 ton were used, excluding organic fertilizer production, and 1.2 ton, corresponding to 77% of use, were generated and

Table 9.14 Estimation of N and P Balance in a Comprehensive Fermentation Factory

	Amount (t/year)	Content		Amount of N and P[a]	
		N (%)	P (%)	N (t/day)	P (t/day)
Input					
Molasses	282,000	0.43	0.07	3.67	0.598
Soy bean husk	3,000	3.57	0.67	0.32	0.061
Ammonia	28,800	82.25	0	71.78	0
Calcium perphosphate	6,000	0	7.42	0	1.349
Phosphate and others					5.68[b]
Subtotal				75.8	7.69
Output					
Mono-sodium glutamate	15	8.28	0	5.12	0
Amino acids	4,800	15	0	2.18	0[c]
Seasonings	7,080	7	1	1.5	0.215
Organic fertilizer	85,898	9.11	2.6	23.71	6.768
Yeast	3,012	7	1	0.64	0.091
Recycled cells	195	7	1	0.51	0.072
Ocean dumping				3.4	0.07
Removal				12.3	0.22
Discharge	84,000			24.7	0.25
Subtotal				74.1	7.69

[a]Assuming 330 working days/year; [b]estimated from the balance; [c]lysine, leucine, valine.

Table 9.15　N and P Balance in a Fermentation Factory

	Use, a (t/day)	Generated load, b (t/day)	Loss rate, b/a (%)	Discharged load, c (t/day)	Reduction rate, $1 - c/b$ (%)	Recovery, d (t/day)	Recovery rate, d/b (%)
COD_{Mn}	–	150	–	7.4	95	106	71
TN	76	56	74	25	55	16	29
TP	7.7	1.2	16	0.3	79	0.7	57
	2.01[a]		60				

[a]Excluding the input to fertilizer production.

0.25 tons were discharged to the coastal water. The COD_{Mn} generated was 150 ton/day and discharged 7.4 ton/day. The production of organic fertilizer had been increasing year by year at that time, and 71, 29, and 57% of COD_{Mn}, TN, and TP, respectively, were recovered as organic fertilizer out of generation load.

9.6.2　Agro-Industries in Tropical Countries

Cassava is a key food product in many tropical countries. In 1997, 165 million tons were produced worldwide for food and feed. Cassava chips, pellet, and starch are major exports for Thailand, which produces 18 million tons annually. Eight million tons were exported annually in 1992–94 to Europe, mainly for the feedstuff, but has decreased since then [26]. The processes for starch are peeling, washing, grating, starch extraction, settling, drying, milling, and sieving.

　　Solid waste from the extraction process of cassava is known as cassava pulp [27]. The quantity of discharged pulp (60–75% moisture content) is about 15–20% of the root weight (65–75% moisture content) being crushed. This is equivalent to about 1.5–2.0 million ton of pulp discharged each year from 10 million ton of root crushed. Discharged pulp still contains a high starch content (around 50% dry basis). This is due to the inextricable starch that is trapped inside the cells. Pulp is sun dried to reduce the moisture content and used as filler in animal feed. Environmental problems from the solid waste occur only if the storage of pulp is badly managed and it becomes exposed to rain. Utilization of pulp as a substrate for industrial fermentation has been investigated, but to date there has been no success. An attempt to extract the starch from the pulp by means of enzyme hydrolysis has been reported. Treatment of pulp with a mixture of cellulase and pectinase increased starch recovery by 40%.

　　Liquid waste is discharged from the factory at about 10–30 m^3/ton of starch produced. This waste has a high BOD and COD content. In Thailand, simple treatment of the wastewater is practiced. The first stage is screening to remove insoluble debris such as peel and woody parts. Wastewater is treated in an open-type anaerobic pond followed by oxidation ponds or aerated lagoons and finally a polishing pond before final discharge. Balogoparan and Padmaja [29] reported cyanide concentration between 10.4 and 274 mg/L in the final effluent, and also a high concentration of cyanide in groundwater near the factory ranging between 1.2 and 1.6 mg/L [28].

　　Most factories prefer to build a "no discharge" system; this means more ponds are prepared for retaining the treated wastewater. A huge land area is required for wastewater because of the long retention times required. Some factories near towns have had to improve their wastewater treatment processes. These factories now employ closed-type anaerobic reactors or an activated sludge process. The composition of wastewater from five factories during the year 1997 is shown in Table 9.16. Regulation of the Ministry of Industry, Thailand, allows a BOD in discharge wastewater of 20 mg/L.

Table 9.16 Wastewater Characteristics from Representative Cassava Starch Factories

	Factories				
	1	2	3	4	5
Production, a (t/day)	70	90	155	300	141
Wastewater, b (m^3/day)	2350	2700	4100	2936	1770
b/a (m^3/t)	33	32	25	11	14
pH	4.8	5.0	5.7	8.3	5.3
COD (g/L)	13.0	16.3	15.0	19.3	19.2
BOD (g/L)	6.5	11.6	10.6	12.6	9.4
TS (g/L)	13.2	16.4	12.5	19.8	18.0
SS (g/L)	7.4	8.0	6.8	7.0	7.8

Source: Ref. 27.

Table 9.17 shows other examples of agro-industrial wastewater treatment in Thailand. Similarly to that mentioned above, the wastewaters are generally treated with lagoon systems. In the case of the palm oil mill, through 81 days of retention time, the effluent is utilized for irrigation of palm fields [1]. A type of closed system is realized in this case. However, if other systems can recover energy at low cost, they would be more preferable.

The color of the effluent has also become a serious problem of public interest. The colors of food processing wastewater are caused generally by melanoidines formed through the reaction of sugars with amino acids and polyphenols derived from lignin materials. It is not easy to decompose these materials biologically, although many researchers have tried to use effective

Table 9.17 Examples of Agro-Industry Wastewater in Thailand

	Raw material (t/day)	Treatment	Wastewater (m^3/day)	Water quality (mg/L)				
				BOD	COD	SS	TN	TP
Fishery products								
Influent	500	4 ponds (42,000 m^2)	3,000	1,020	1,950	368	46	12
Effluent				20	385	19	2	14
Criteria				20	120	30–150		
Palm oil mill								
Influent	600	22 ponds (10,120 m^2)	300	50,000				
Effluent				100				
Criteria				60				
Palm oil mill								
Influent				25,000	50,000	18,000	750	180
Effluent			3 m^3/t of crude oil					
Criteria				100		400	200	

Source: Ref. 1.

fungi. It should be noted that for the decolorization of alcohol distillery wastewater using *Mycelia sterilia*, even if melanoidines are removed, residual polyphenols cause color through oxidation [29]. Ozone, electrolysis, and coagulation with chitosan are effective [30]. However, these processes have not prevailed because of the high cost to date.

9.6.3 UASB and EGSB Treatment Systems

Anaerobic treatment, especially thermophilic treatment, offers an attractive alternative for the treatment of high-strength, hot wastewater. The thermophilic process, compared to the mesophilic anaerobic process, has the advantages of increased loading rate and the elimination of cooling before treatment. Furthermore, the heat content of the wastewater would be available for post-treatment. Loading rates up to 80 kgCOD/m^3/day and more have been reached in laboratory-scale thermophilic reactors treating glucose, acetate, and sucrose and thermomechanical pulping whitewater. Table 9.18 shows the results of food wastewater treatment by thermophilic UASB at 55°C together with the examples for pulp mill wastewater [31–40]. For alcohol distillery wastewater at the loading rate of 100 kgCOD/m^3/day, successful removal efficiencies were reported. Rintala and Lipisto [42] reported 70°C thermopilic UASB experiment using pulp mill wastewater; however, the COD removal was not high at 56% at the loading rate of 41 kgCOD/m^3/day [41].

Table 9.19 summarizes successful performances using EGSB in China. Biogas recovered at the rate of 0.4–0.6 m^3/day vs. 1 kL of 95% alcohol has been used as supplementary fuel for coal [42].

Among seven beet sugar processing factories in Hokkaido, Japan, three use UASB. Biogas is generated at the rate of 7000 N m^3/day under the condition of 15 kgCOD/m^3/day. It is used as fuel for boilers and dryers of beet pulp. Starch processing also generates wastewater of high BOD. Four factories in Japan use UASB reactors and generate biogas of 8000–9000 N m^3/day under the condition of 15 kgCOD/m^3/day.

Thermophilic anaerobic treatment of hot vegetable processing wastewaters deriving from steam peeling and blanching of carrot, potato, and swede, was studied in laboratory-scale UASB reactors at 55°C [43]. The reactors were inoculated with mesophilic granular sludge. Stable thermophilic methanogenesis with about 60% COD removal was reached within 28 days. During the 134 day study period the loading rate was increased up to 24 kgCOD/m^3/day. More than 90% COD removal and methane production of 7.3 m^3CH_4/day were achieved. The anaerobic process performance was not affected by the changes in the wastewater due to the different processed vegetables. The wastewater characteristics are summarized in Table 9.20, and the water qualities of influent and effluent in the experiments are shown in Table 9.21.

Several studies have also attempted to use membrane technology in combination with anaerobic packed bed reactors [44]. Three different methane fermentation processes were evaluated using soybean processing wastewater: (a) Process A — acidification (empty bed volume: 1 m^3) and methane reactors (empty bed volume: 2 m^3); (b) Process B — acidification and methane reactors followed by membrane (Polysulfone and PVA, MW cutoff, approx. 15,000); and (c) Process C — acidification reactor, membrane, and methane reactors. The characteristics of the wastewater are BOD 1000 mg/L, COD 1629 mg/L, VSS 693 mg/L, protein 544 mg/L, and lipid 23 mg/L.

Process B showed a COD removal of 77.7% by decreasing the free SS in the treated water. Higher acetic acid and propionic acid concentrations were found as residual in the treated water. The rate of methane conversion was 68.9%.

Process C showed a remarkable removal of COD by 92.4% and methane conversion of 83.4%. Process C gave noteworthy improvement in results compared with process A. It has been

Table 9.18 Results of Wastewater Treatment by Thermophilic UASB

Reactor volume (L)	Temperature (°C)	Inoculum sludge	Types of wastewater	Wastewater conc. (mgCOD/L)	Volumetric loading rate (kgCOD/m³/day)	COD removal rate (%)	Reference
75,000	55–57	Methophilic digested sludge and cow dung	Alcohol wastewater	31,500	26.5	71.7	31
5.5	55	Thermophilic granular sludge	Alcohol wastewater	15,400	98.3	58	32
126	55	Thermophilic digested sludge	Alcohol wastewater	10,000	28	41.5	33
0.12	55	Methophilic granular sludge	Pulp wastewater	1,900–2,200	14.1	45	34
0.12	55	Thermophilic granular sludge	Pulp wastewater	1,900–2,200	17.7	56	34
0.12	55	Thermophilic granular sludge	Pulp wastewater	1,900–2,200	40.8	56	34
71.5	50–53	—	Alcohol wastewater	19,200–25,600	9.6	91.2	35
46.2	51–54	—	Acetic acid production	6,600–8,500	5.0–6.0	93	36
1,400	55	Methophilic granular sludge	Beer wastewater	2,000	50	82	36
2	55	Methophilic granular sludge	Coffee processing	1,000–4,000	4	>70.0	37
0.25	55	Methophilic granular sludge	Pulp wastewater	1,900	2.6	26.0–42.0	38
8	55	Thermophilic granular sludge	Alcohol wastewater	3,000–10,000	100	85.0–90.0	39
11.9	55	Methophilic granular sludge	Alcohol wastewater	9,000	30	85	40

Source: Refs. 31–40.

Table 9.19 Performance of EGSB in Alcohol Wastewater in China

Place	Production (kL/year)	Alcohol (%)	Raw material	Wastewater (m³/day)	COD before EGSB (mg/L)	COD after EGSB (mg/L)	Biogas generation (m³/day)
Nin Bo	35,000	95	Cassava	1,500	25,000–30,000	1,000–1,500	20,000–23,000
Nan Hai	50,000	38	Rice	550	30,000–35,000	1,000–1,500	10,000
Nin He	60,000	95	Maize	1,500	25,000–30,000	1,000–1,500	25,000
Tan Chen	15,000	95	Potato	700	10,000–14,000	500–1,000	3,000

Source: Ref. 42.

concluded that process C was excellent and recommended when wastewater enriched with VSS is treated. The combination of membrane with UASB was also studied. However, the results are not feasible for practical use [45].

9.6.4 Zero-Emission in Beer Breweries

Waste recycling systems in beer breweries are very complete. Kirin Beer Co. Ltd. has achieved zero-emission for its industrial wastes since 1998. Table 9.22 shows the amounts for each of the wastes and their uses [46]. Moreover, the emission factor of wastes has itself also decreased from 0.205 kg/L in 1996 to 0.140 kg/L in 2001. Wastewater is treated by a UASB reactor and activated sludge method in 10 out of 12 factories in this company. In fact, $18,860 \times 10^3$ m³ of wastewater generate 4800 tons of methane gas from UASB reactors, corresponding to 5200 kL of oil. The biogas is used for the fuel of boiler and cogeneration systems. In another big beer company in Japan, we can see similar situations. They treat wastewater in eight out of nine factories by UASB or EGSB reactors and activated sludge. They produce 8315 ton of methane from $14,652 \times 10^3$ m³ of wastewater.

9.6.5 Recycling of Garbage

Composting

Composting has long been a traditional technology, but new composting technologies have also been developed. Generally there are mainly two types of technology, both dependent on using microorganisms. One type uses a thermophilic bacillus, which is effective in enhancing the initial decomposing phase. Many types of microbial additives have been developed by

Table 9.20 Characteristics of Vegetable Processing Wastewater after Removing Solids

Unit operation	Raw material	Total COD		Soluble COD			
		Average	Range	Average	Range		
Steam peeling	Carrot	19.4	17.4–23.6	17.8	15.1–22.6	Exp.1	1–73 days
	Potato	27.4	13.7–32.6	14.2	11.7–17.5	Exp.2	38–46 days
Blanching	Carrot	45	26.3–71.4	37.6	22.1–45.8	Exp.1	1–73 days
	Potato	39.6	17.0–79.1	31.3	10.9–60.6	Exp.2	38–46 days
	Swede	49.8	40.5–59.1	49.4	40.5–58.3	Exp.2	1–38 days

Source: Ref. 43.

Table 9.21　Quality of Influent and Effluent for Thermophilic UASB in Vegetable Food Processing

	Influent		Effluent	
	BOD	COD	BOD	COD
Exp. 1	6,900–19,900	9,500–27,600	660–1,400	1,200–2,100
Exp. 2	6,700–7,600	10,500–11,800	1,400–1,600	2,000–2,300

Source: Ref. 43.

companies. Some employ a heat-shock process at 75°C for 3 hours before adding the microbes. Figure 9.9 shows a flow of this type adopted in Komagane City [47].

Another type of composting uses mixed culture mainly constituted by lactic acid bacteria (EM) under anaerobic condition. Figure 9.10 shows an example of this process adopted in a small town in Japan. Usually the addition of a relatively large amount of rice husk may be needed for the production of valuable compost.

A combination of both types of composting has also been tested. In one case, relatively fresh fish refuse was preheated at 80°C for several hours with the addition of a thermophilic bacillus, and then EM was added. In this case EM was expected to work after packaging in plastic bags. This product may be used as chicken feed.

Usually, garbage includes salt and oily materials, which makes it difficult to use as a sole source of compost. Often after drying or primary fermentation through consuming electric

Table 9.22　Recycling of Wastes from Beer Breweries (Kirin Beer Co. Ltd. in 1988)

Wastes	Amount (tons/year)	Uses
Spent grain	359,919	Feedstuff, fertilizer, heat recovery
Refined dregs	2,618	Feedstuff, fertilizer
Excess yeast	7,168	Foodstuff, medicine
Used diatomaceous earth	17,556	Cement feedstock, soil conditioner
Wastewater sludge	35,319	Fertilizer
Paper waste		
Labels	3,051	Recycled paper
Danball	2,167	
Bags for raw materials	103	
Used paper	400	
Wood pallet	3,231	Fuel
Glass wastes	57,409	Raw material of glass
Caps	337	Raw material of steel
Aluminum cans	493	Raw material of aluminum
Steel cans	1,008	Raw material of steel
Plastic wastes		
Plastic box	6,795	
Plastics	553	Reductant for high furnace
Used oil	141	Fuel, recycled oil
Incineration ash	152	Cement feedstock, soil conditioner
Others	1,869	
	500,289	Recycling rate: 100%

Source: Ref. 46.

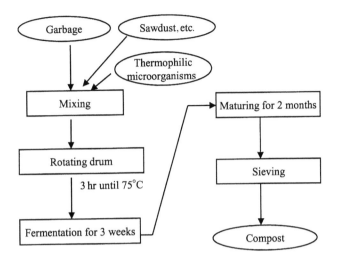

Figure 9.9 Composting process using thermotolerant microorganisms.

energy at home, restaurant, and supermarket, pretreated garbage is transported and mixed to livestock manure compost at a certain proportion, 20% for instance.

In Japan, many companies intend to produce compost using the garbage collected from commercial sectors. However, as mentioned previously, due to the difficult situation surrounding organic fertilizer, a high quality of compost is needed. Under these circumstances, a pilot study was conducted in Kitakyushu City to produce polylactic acid (PL) through glucosidation, lactic acid fermentation, butyl ester formation, hydrolysis, and polymerization [48]. The cost of

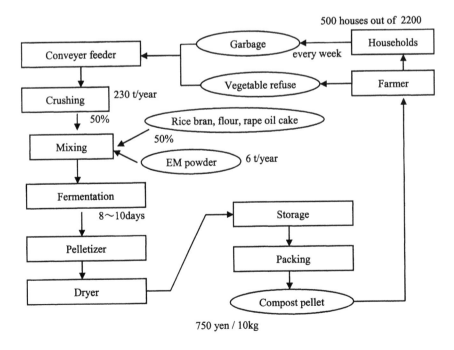

Figure 9.10 Composting process of garbage using EM in a small town in Japan.

3.5 \$/kg PL was estimated for the plant of 100 tons of garbage/day. The harvest rate of PL is assumed to be 5% vs. garbage. If the garbage treatment cost is 30 yen/kg, then it corresponds to $600/125 = 4.5$ \$/kg PL, and it is expected to be feasible. This project is challenging, however, and may encounter other difficulties.

Biogas Production

This system also seems feasible, especially since the capacities of night soil treatment plants have become excessive, accompanied by the spread of publicly owned sewerage systems. Therefore a part of these night soil plants have been reformed to treat night soil, sludge from septic tanks, and garbage together. Table 9.23 shows a list of such plants [49]. There are also several plants treating garbage solely by methane fermentation in Japan. It is said that the production of biogas corresponds to $100–200 \, \mathrm{N \, m^3}$/ton of garbage. Table 9.24 shows the results of comparing a biogas system to an incineration system with power generation. The energy efficiency of the biogas system is estimated to be better than that of an incineration system. If the heat value of garbage becomes higher, then incineration with power generation becomes advantageous [50].

9.7 CONCLUSION

This chapter discussed the structural point of view and the characterization of food processing relating to liquid, solid, and hazardous wastes. Focusing on case studies mainly in Japan the details of food waste generation and treatment have been presented, sometimes in comparison with cases in the United States. We can summarize the conclusions and recommendations on several points as follows.

As a main principle, the self-supply of food should be a goal for all countries. The local unbalance between food production and consumption makes food wastes hazardous and induces eutrophication of water bodies, nitrate pollution of groundwater, and ruins farmland soil because of limited recycling of minerals. Therefore, the free trade of primary products should be re-examined in light of environmental issues.

Table 9.23 Sludge Recycling Centers Using Anaerobic Treatment in Japan

Prefecture	Region	Construction period	Amount of wastes night soil and sludge (kL/day)	Garbage, etc. (t/day)	Energy use	Technology
Nara	Jouetsu	1999	240	8.0	a	Thermophilic
Nagano	Shimoina	1999	16	8.0	a	Thermophilic
Niigata	Higashi Kanbara	1999	22.3	3.5	b	Mesophilic
Nara	Ikoma	2000	80	1.3	a	Thermophilic
Miyazaki	Kusima	2000	35	0.9	b	Mesophilic
Nara	Nara	2001	90	3.0	b	Mesophilic
Hokkaido	Minami Souya	2002	15	16.0	a	Thermophilic
Hokkaido	Nisi Tenpoku	2002	20	8.0	b	Thermophilic
Miyagi	Mutsunokuni	2002	105	1.0	b	Thermophilic
Nagasaki	Kamigotou	2002	69	3.0	b	Thermophilic

a: power generation with cogeneration; b: heat use only.
Source: Ref. 49.

Table 9.24   Energy Efficiency of Biogas Production System and Incineration with Power Generation

	Unit	Incineration with power generation	Biogas production system
Biogas generated	Nm^3/t		100
Calorific value generated	Kcal/t	1,000,000[a]	5,000,000
Electric power generated	KWh/t	150	150
Efficiency assumed	%	13	26
Self-consumption of electricity	KWh/t	100	50[b]
Net gain of energy	KWh/t	50	100

[a]Assuming garbage of 1000 kcal/kg; [b]Not including wastewater treatment.
Source: Ref. 50.

Generally, food processing does not generate chemical hazards. However, attention should be given to chlorine used for cleaning and sanitation leading to chlorinated byproducts in wastewater. If contaminated by poisonous materials or pathogens, food may change to hazardous wastes. Treatment of BMP by cement kiln is a typical example of this.

As previously discussed, incineration of food wastes together with other miscellaneous wastes is not a suitable solution because of the generation of hazardous ash containing DXNs and heavy metals; doing so also threatens food recycling efforts. A recommended option would be composting followed by the combination of biogas production and composting of the sludge.

During the past 30 years, wastewater from food processing in Japan has been significantly improved due to better wastewater treatment systems and conversion of raw materials or processes. It is notable that primary processes generating much waste have been shifted to other countries where the raw materials are produced and sometimes to developing countries where cheap labor is supplied.

Anaerobic treatment systems lost their popularity in wastewater treatment fields for failing to meet strict environmental criteria. Recently, however, these systems have regained attention because of their ability to save energy and also reflecting the development of UASB or EGSB technology.

In relation to ISO 14000, some companies have targeted and attained zero emission in their industries. However, zero emission should also cover all products, including containers and wrappers.

ACKNOWLEDGMENTS

The authors thank Dr. H. Nakanishi, Emeritus Professor of Yamaguchi University, and Dr. I. Fukunaga, Professor of Osaka Human Science University, and other individuals for giving us valuable input.

REFERENCES

1. Ukita, M.; Prasertsan, P. Present state of food and feed cycle and accompanying issues around Japan. *Water Sci. Technol.*, **2002**, *45*(*12*), 13–21.
2. Ushikubo, A. Present state of food wastes and countermeasures for Food Recycling Act; http://mie.lin.go.jp/summary/recycle/recyle.htm (in Japanese), 2002.

3. Ministry of Agriculture, Forestry and Fisheries. *International Statistics on Agriculture, Forestry and Fisheries*; MAFF, 1998.

4. Fukunaga, I. Recent advances of the treatment and disposal of wastewater and solid waste in food industry. *Foods and Food Ingredients J. Japan*, **1995**, *165*, 21–30 (in Japanese).

5. Mavrov; Belieres. Reduction of water consumption and wastewater quantities in the food industry by water recycling using membrane processes. *Desalination* **2000**, *131*, 75–86.

6. Ridgway; Henthorn; Hull. Controlling of overfilling in food processing. *J. Mater. Process. Technol.*, **1999**, *93*, 360–367.

7. Norcross, K.L. Sequencing batch reactors – An overview. *Water Sci. Technol.* **1992**, *26* (*9–11*), 2523–2526.

8. Wentzel, M.C.; Ekama, G.A.; Marais, G.V.R. Kinetic of nitrification denitrification biological excess phosphorus removal systems: A review. *Water Sci. Technol.* **1990**, *17* (*11–12*), 57–71.

9. Zhou, H.; Smith, D.E. Advanced technologies in water and wastewater treatment. *J. Environ. Eng. Sci.* **2002**, *1*, 247–264.

10. Sumino, T. Immobilization of nitrifying bacteria by polyethylene glycol pre-polymer, *J. Ferment. Bioeng.* **1991**; *73*, 37–42.

11. Lettinga, G. *et al.* Use of upflow sludge blanket reactor concept for biological wastewater treatment, especially for anaerobic treatment. *Biotechnol. Bioeng.* **1980**, *22*, 699–734.

12. Kato, M.; Field, J.A.; Versteeg, P.; Lettinga, G. Feasibility of the expanded granular sludge bed (EGSB) reactors for the anaerobic treatment of low strength soluble wastewaters. *Biotechnol. Bioeng.* **1994**, *44*, 469–479.

13. US Environmental Protection Agency. *Wastewater Technology Fact Sheet Sequencing Batch Reactors*, EPA/832/F-99/073; US Environmental Protection Agency, Office of Water: Washington, DC, 1999.

14. Nippon Suido Consultant Co. Ltd. *Report of the Investigation on Instruction for Reducing Nitrogen and Phosphorus Load*(in Japanese); Nippon Suido Consultant Co. Ltd., 1982.

15. Ministry of Environment Japan. *Report on the Discharge of Water Pollutant in 2002* (in Japanese); Ministry of Environment Japan, 2003.

16. USEPA. *Electronic Code of Federal Regulations*, http://ecfr1.access.gpo.gov/otcgi/cfr/, 2003.

17. Central Council of Environment. *Draft on the Regulation of COD, Nitrogen and Phosphorus* (in Japanese), http://www.env.go.jp/press/press.php3?serial = 1317 (in Japanese), 2000.

18. Bureau of Environment Tokyo Metropolitan Government. *Criteria of wastewater* (in Japanese); Bureau of Environment: Tokyo, 2002. http//www.kankyo.metro.tokyo.jp/kaizen/kisei/mizu/kijun/np.htm.

19. The Research Group for Biological Organic Wastes. *Present State and Problems on Biological Organic Wastes* (in Japanese); Research Group for Biological Organic Works, 1999.

20. Japan Livestock Industry Association. *Report on Promoting Feed Use of Un-utilized Resources* (in Japanese); Japan Livestock Industry Association, 1996.

21. Teijin Ltd. *Technical Report on PET recycling*; Teijin Ltd., 2000.

22. USEPA. *EPCRA Section 313 Reporting Guidance for Food Processors*; USEPA: Washington, DC, 1998.

23. The final report of the investigation on the food poisoning accidents by Yukijirushi Milk Company; http://www.mhlw.go.jp/topics/001/tp1220-1.html (in Japanese).

24. Endo, K. The recode of the discussion in budget committee in Japanese Parliament, The 1st Mar. (in Japanese), 2002.

25. Ukita, M. *Fundamental Research on the Behavior of Nitrogen and Phosphorus and on the Mechanisms of Eutrophication in Japan*; Thesis of Kyoto University (in Japanese), 1987.

26. Titapiwatanakun, B. *Report of Strategy Agricultural Comodity Project*: Cassava; Department of Agricultural and Resource Economics, Faculty of Economics: Kasetsart University, 1997.

27. Klanorang, S.; Kuakoon, P.; Sittichoke, W.; Christopher, O. Cassava Starch Technology: The Thai Experience. *Trends Food Sci. Technol.*, **2000**, *52*, 439–449.

28. Balagopalan, C.; Padmaja, G. Cyanogen accumulation in environment during processing of casaba (*Manihot esculenta* Grantz) for starch and sago. *Water Air Solid Pollu.* **1998**, 102, 407–413.

29. Nagano, A. *Study on the Decolorization of Molasses Wastewater*; Thesis of Nagaoka University of Technology, Japan, 2000.

30. Kumar, M.N.V.R. *Chitin and Chitosan for Versatile Applications*, http://www.geocities.com/mnvrk/chitin.html, 2003.

31. Souza, M.E.; Fuzaro, G.; Polegato, A.R. Thermophilic anaerobic digestion of vinasse in pilot plant UASB reactor. *Wat. Sci. Technol.*, **1992**, *25* (*7*), 213.

32. Wiegant, W.M.; Claasseen, J.A.; Lettinga, G. Thermophilic anaerobic digestion of high strength wastewaters. *Biotechnol. Bioeng.*, **1986**, *27* (*9*), 1374.

33. Harada, H.; Uemura, S.; Chen, A.C.; Jayadevan, J. Anaerobic treatment of a recalcitrant distillery wastewater by a thermophilic UASB reactor. *Biores. Technol.*, **1996**, *55*, 215–221.

34. Rintala, J.; Martin, J.S.L.; Lettinga, G. Thermophilic anaerobic treatment of sulfate-rich pulp and paper integrate process water. *Wat. Sci. Technol.*, **1991**, *24* (*3/4*), 149.

35. Zuxuan, Z.; Zepeng, C.; Zeshu, Q. Status quo and prospects on the study of anaerobic digestion for industrial wastewater in China In *Proceedings of 4th International Symposium on Anaerobic Digestion*, 1985; 259.

36. Ohtsuki, T.; Tominaga, S.; Morita, T.; Yoda, M. Thermophilic UASB system start-up and management change in sludge characteristics in the start-up procedure using mesophilic granular sludge In *Proceedings of 7th International Symposium on Anaerobic Digestion*, 1994; 348.

37. Daoming, S.; Forster, C.F. An examination of the start-up of a thermophilic upflow sludge blanket reactor treating a synthetic coffee waste. *Environ. Technol.*, **1994**, *14*, 965.

38. Lepisto, S.S.; Rintala, J. The removal of chlorinated phenolic compounds from chlorine bleaching effluents using thermophilic anaerobic processes. *Water Sci. Technol.*, **1994**, *29* (*5/6*), 373.

39. Harada, H.; Syutsubo, K.; Ohashi, A.; Sekiguchi, Y.; Tagawa, T. Realization of super high-rate anaerobic wastewaters treatment by a novel multi-staged thermophilic UASB reactor. *Environ. Eng. Res.*, **1997**, *34*, 327–336.

40. Syutsubo, K.; Harada, H.; Ohashi, A.; Suzuki, H. An effective start-up of thermophilic UASB reactor by seeding mesophilically grown granular sludge. In *Proceedings 8th International Conference on Anaerobic Digestion*, 1997; Vol. 1, 388–396.

41. Rintala, J.; Lepisto, S.S. Anaerobic treatment of thermomechanical pulping wastewater at 35–70°C. *Water Res.*, **1992**, *26* (*10*), 1297.

42. Zhang J.; Shanghai Communication University. Personal communication, 2003.

43. Satu S. Lepisto. Start-up and operation of laboratory-scale thermophilic upflow anaerobic sludge blanket reactors treating vegetable processing wastewaters. *J. Chem. Technol. Biotechnol.*, **1997**, *68*, 331–339.

44. Yushina, Y.; Husegawa, J. Process performance comparison of membrane introduced anaerobic digestion using food industry wastewater. *Desalination*, **1998**, *27*, 413–421.

45. Inoue, Y.; Doi, K.; Kamiyama, K. Granule formation facilitation of up flow type sludge blanket reactor using UF membrane. In *Proceedings of 45th JSCE Annual Conference*, 1991; Vol. 45/II, 1084–1085 (in Japanese).

46. Hibana, K. Tackling with zero emission in beer brewery. In *Practices of the Introducing Zero-Emission in Industries*; NTS Publisher, 2001; 147–168 (in Japanese).

47. Sakai, S. Booklet "Let's make compost and fermented feed from garbage" (in Japanese), 2000.

48. Shirai, Y. *The introduction of research*, http://www.lsse.kyutech.ac.jp/lsse_j/kyokan_shirai.html, 2003.

49. Li, G. *The application of methane fermentation technology in sludge recycling center, New century of environmental engineering*; The Committee of Environmental Engineering JSCE, 2003; 73–74 (in Japanese).

50. Kawano, T. *Biogas production from garbage, new century of environmental engineering*; The Committee of Environmental Engineering JSCE, 2003; 73–74 (in Japanese).

Index